Methods of Graph Decompositions

Methods of Graph Decompositions

Edited by

Vadim Zverovich
and
Pavel Skums

OXFORD
UNIVERSITY PRESS

Great Clarendon Street, Oxford, OX2 6DP,
United Kingdom

Oxford University Press is a department of the University of Oxford.
It furthers the University's objective of excellence in research, scholarship,
and education by publishing worldwide. Oxford is a registered trade mark of
Oxford University Press in the UK and in certain other countries

Published in the United States of America by Oxford University Press
198 Madison Avenue, New York, NY 10016, United States of America

British Library Cataloguing in Publication Data
Data available

Library of Congress Control Number: 2024025041

ISBN 9780198882091

DOI: 10.1093/oso/9780198882091.001.0001

Printed and bound by
CPI Group (UK) Ltd, Croydon, CR0 4YY

The image on page v and cover image by Lucy SM Johnston

In memory of Professor Regina Tyshkevich (1929–2019) who made significant contribution to graph theory

Preface

In general terms, a graph decomposition is a partition of a graph into parts (e.g. subgraphs) satisfying some special conditions. This book discusses some state-of-the-art decomposition methods of graph theory, which are highly instrumental when dealing with a number of fundamental concepts such as unigraphs, isomorphism, reconstruction conjectures, k-dimensional graphs, degree sequences, line graphs and line hypergraphs. In the first part of the book, we explore the algebraic theory of graph decomposition, whose major idea is to define a binary operation that turns the set of graphs, or objects derived from graphs, into an algebraic semigroup. If an operation and a class of graphs are appropriately chosen, then, just as for integers, each graph has a unique factorization (or canonical decomposition) into a product of prime factors. The unique factorization property makes this type of decomposition especially efficient for problems associated with graph isomorphism, and several such examples are described in the book.

The second part is devoted to Krausz-type decompositions, which are special coverings of graphs by cliques directly associated with the representation of graphs as line graphs of hypergraphs. The book discusses various algorithmic and structural results associated with the existence, properties and applications of such decompositions. In particular, it demonstrates how Krausz-type decompositions are directly related to topological dimension, information complexity and self-similarity of graphs, thus allowing to establish links between combinatorics, general topology, information theory and studies of complex systems.

The book is structured as follows. In the introductory chapter, we give a brief overview of the selected themes of graph theory considered in the book. In particular, we discuss the algebraic theory of graph decomposition. Chapter 2 is devoted to an exhaustive description of the structure of unigraphs based on the decomposition theorem. In Chapter 3, we consider some important subclasses of unigraphs such as matrogenic, matroidal and threshold graphs and, in particular, discuss their decomposition, enumeration and recognition. Further applications of the operator decomposition are described in Chapter 4, for example the proof of the reconstruction conjecture for several substantial families of graph classes. Line graphs and line hypergraphs are discussed in Chapter 5. Finally, in Chapter 6 we will explore the notions of graph dimensions that arise from the theory of intersection graphs described in the previous chapters.

The book is ideal for researchers, engineers and specialists, who are interested in the fundamental problems of contemporary graph theory and proof techniques to tackle them. Advanced students in discrete mathematics in its applications may use the problems and research methods presented in this book to develop their final-year projects, masters' theses or doctoral dissertations; however, to use the information effectively, knowledge of graph theory would be required. It is the authors' hope that this publication will serve as a motivation for women who wish to do research in the field.

The presented topics are united by the role played in their development by Prof Dr Regina Tyshkevich, and this book is a tribute to her memory. Regina Tyshkevich was one of the leading graph theory experts in Eastern Europe, and the pioneering woman who made significant contributions to the aforementioned research areas. She was born in 1929 in the city of Minsk, Belarus. During the Second World War, she and her mother, Maria, became refugees and found asylum in Udmurtia in the interior of the Soviet Union. It was a hungry time, full of hardships, but it turned out that a number of the best teachers of mathematics from Leningrad were also evacuated to Udmurtia to work at schools. They showed Regina the beauty of mathematics and actually determined her future career.

In 1946, after the war, Regina returned to Belarus alone because her mother was not allowed to leave the factory where she worked. The following year, when she graduated from a secondary school, she had no hesitation in choosing her future career path as she could not imagine her future life without mathematics. She entered the Faculty of Physics and Mathematics of Belarus State University, graduated in 1952 with honours and in 1956 began teaching at the same faculty. At that time, Regina Tyshkevich met Dmitry Suprunenko—a prominent scientist and one of the founders of the algebraic school in Belarus, who became her PhD supervisor. He helped to shape Tyshkevich's scientific interests and, under his supervision, in 1959 she defended her PhD thesis devoted to nilpotent groups. Soon after that, Tyshkevich, in collaboration with Suprunenko, published her first book *Commutative Matrices*, which is still widely used and cited to this day.

In the 70s, the focus of Tyshkevich's research interests shifted from algebra to discrete mathematics and, in particular, to one of its most promising areas—graph theory. To some extent, this happened under the influence of the Annual Scientific Seminar on Discrete Mathematics in Odessa, Ukraine, which had been led by Professor Alexander Zykov. It was a world-class event organized by graph-theory enthusiasts, which attracted researchers from the entirety of Eastern Europe, including the Soviet Union, Czechoslovakia, Poland and Hungary. However, algebra continued to have a significant influence on Tyshkevich's research in graph theory. As a result, a distinctive feature of all her further studies was a synthesis of combinatorial and algebraic methods. Such an interdisciplinary approach helped her to obtain a number of world-class results, including the solution to the problem of a complete characterization of unigraphs (i.e. graph defined up to isomorphism by their degree sequences), which had remained open for more than 30 years. In 1984, at the Glushkov Institute of Cybernetics in the Ukrainian Academy of Sciences, Tyshkevich defended her dissertation for the 'Doctor of Sciences' degree. The dissertation was devoted to algebraic methods for graph theory, and it became the first Doctor of Sciences dissertation in the Soviet Union on graph theory. In 1986, Tyshkevich was awarded the academic title of professor.

Regina Tyshkevich was the founder of the Belarusian School of Graph Theory, which was one of the leading research groups in the former Soviet Union and had recognition all over the world. She has been a principal investigator of numerous research projects, both domestic and international. Her international partners and collaborators were researchers from the universities of Rostock and Kaiserslautern (Germany), Vienna and Graz (Austria), Amiens (France) and the Austrian Academy of Sciences. Tyshkevich published more than 150 scientific and educational works, including 13 textbooks and monographs, three of which have also been published in English. Her former PhD students work in academia and industry in countries of four different continents such as the USA, the UK, Belarus, Vietnam and Guinea. She was a recipient of many awards, including the State Prize of the Republic of Belarus (for the book *Lectures on Graph Theory* published in 1990) and the Order of Francysk Skaryna—the country's highest award for achievements in arts and sciences. Besides being a world-class mathematician, Tyshkevich was a person of high culture and erudition with keen interests in literature, history and theatrical art.

The Authors

Dr Vadim Zverovich (PhD, 1993; Docent, 2000) is an Associate Professor of Mathematics and the Head of the Mathematics and Statistics Research Group at the University of the West of England (UWE) in Bristol. An accomplished researcher, he is also a Fellow of the United Kingdom Operational Research Society, and was previously a Fellow of the prestigious Alexander von Humboldt Foundation in Germany. In 2016, Dr Zverovich was awarded Higher Education Academy fellowship status, and the Faculty of Environment and Technology of the UWE named him its researcher of the year in 2017. Dr Zverovich's research interests include graph theory and its applications, networks, probabilistic methods, combinatorial optimization and emergency responses. With 30 years of research experience, he has published many research articles and three books on the above subjects, and established an internationally recognized academic track record in the mathematical sciences covering both theoretical and applied aspects. The previous book, *Modern Applications of Graph Theory*, was published by Oxford University Press in 2021.

Email: vadim.zverovich@uwe.ac.uk

Dr Pavel Skums is an Associate Professor at the Computer Science and Engineering Department of the University of Connecticut. He is a recipient of the prestigious National Science Foundation CAREER award, GSU Dean of College of Arts and Sciences Early Career Award and US Centers for Disease Control and Prevention Charles C. Shepard Science Award. His research concentrates on graph and network theories and their applications in genomics, epidemiology and immunology. He has published more than 60 papers and book chapters, and has been a guest editor of more than 20 special issues of leading journals.

Email: pavel.skums@uconn.edu

Contents

1
Introduction

V. Zverovich & P. Skums

In this chapter, we will give a brief overview of selected themes of graph theory discussed in the book. A range of such topics extends from decomposition and reconstruction of graphs to isomorphism and line hypergraphs. More precisely, the first two sections focus on the algebraic theory of graph decomposition and its practical use in tackling a number of challenging graph-theoretic problems such as the reconstruction conjecture and unigraph characterization. Then, we will discuss line graphs, line hypergraphs and their relations to so-called Krausz decompositions and the Whitney theorem on edge isomorphisms. We will also explore the dimensionality of graphs and its connection with line hypergraphs and the equivalence covering number.

1.1 Algebraic Theory of Graph Decomposition

In general terms, a *graph decomposition* involves partitioning a graph into subgraphs satisfying certain conditions. The conditions may be quite different; they are determined by problems being investigated.

Decomposition-based graph methods have a long history, and their role in both theory and applications keeps increasing. In particular, the importance of elaborating the theory of graph decomposition is emphasized in several monographs (e.g. see [15, 41, 43]). Nowadays, theories of various graph decompositions represent some of the most intensively developing branches of discrete mathematics and combinatorics. Indeed, for solving a considerable number of theoretical and applied problems, it is necessary to construct and investigate mathematical models using discrete structures (e.g. graphs), which are characterized by a large number of elements and enormous collections of relations between them. As a result, one cannot expect to obtain an adequate characterization of the entire structure in acceptable terms, whereas the corresponding algorithmic problems usually are NP-hard. In such situations, it is reasonable to decompose the structure into parts with prescribed properties in order to solve the problem for the individual parts and then build the general solution from the partial solutions. In applications, a vivid

V. Zverovich & P. Skums *Introduction.* In: *Methods of Graph Decompositions.* Edited by: Vadim Zverovich and Pavel Skums, Oxford University Press. DOI: 10.1093/oso/9780198882091.003.0001

example of such approach is the design and training of artificial neural networks using so-called convolutional methods [25]. In theoretical studies, to name just a few examples, various types of tree decompositions (including treewidth, cliquewidth, branchwidth) [20], modular decompositions [15], skew partitions (including clique cutsets and star cutsets) [17] have been instrumental in both structural and algorithmic graph theory. The particularly fascinating instance is the history of the famous conjecture posed by Berge in the early 1960s:

Strong Perfect Graph Conjecture *A graph G is perfect if and only if neither G nor its complementary graph $\overline{G}$ contains induced odd cycles C_n, where $n \geq 5$.*

The conjecture had remained open for 40 years, and its proof was based on decomposition methods [16]. These and many other examples, including those presented later in this book, confirm the relevance and necessity of development of graph decomposition methods.

In this and some following chapters, we will consider a particular type of decompositions that will be referred to as an 'algebraic' or 'operator' decomposition. It should be noted that other well-developed algebraic graph theories have appeared long before the theory discussed here [12, 19, 27, 40]. In contrast to existing monographs, our focus is the decomposition methodology. Roughly speaking, the major idea of the algebraic decomposition is to define a binary operation on the set of graphs or objects derived from graphs that turns the set of graphs into an *algebraic system with a semigroup of operators.* The multiplication in the semigroup is category theoretic: the operators are multiplied as morphisms. Furthermore, the graphs themselves serve as operators. Under this approach, an arbitrary graph is decomposed (or, speaking more properly, factorized) into a product of 'prime' multipliers with respect to the selected operation. Such decomposition in some sense is similar to the canonical decomposition of a natural number. Hence, the decomposition of a graph can also be called canonical. This approach is well known in general algebra but at the time of its development it was quite new for graph theory.

The most important feature of canonical decompositions arising in various mathematical branches is the so-called unique factorization property that consists in the existence of the unique factorization up to exchange of commuting factors. This property does not hold for all graph operations, but it does hold for many of them in quite general settings. This fact makes operator decompositions highly instrumental for problems associated with graph isomorphism such as characterization of unigraphs (i.e. graphs defined up to isomorphism by their degree sequences), reconstruction conjectures or construction of automorphism groups. The applications are by no means restricted to these types of problems. Basically, many properties of graphs hold for a graph if and only if they or their close variants hold for all prime factors in its decomposition. This allows for description of structures and enumeration of a number of graph classes, as well as deep penetration into their properties [41]. The examples of such results include many structural and algorithmic results such as characterization and recognition of threshold graphs, difference graphs, matroidal graphs, matrogenic graphs, box-threshold graphs, domishold graphs and pseudo-split graphs. Some of these results will be described in the following chapters.

1.1.1 Notation and Terminology

All graphs considered here are *simple*; that is, finite, undirected, without loops and multiple edges. The vertex set and the edge set of a graph G are denoted by $V(G)$ and $E(G)$, respectively. Let us denote by $G(X)$ the subgraph of G induced by a subset $X \subseteq V(G)$. We write $u \sim v$ ($u \not\sim v$) if the vertices u and v are adjacent (non-adjacent). If $U, W \subseteq V(G)$, then $U \sim W$ denotes that $u \sim w$ for every pair of vertices $u \in U$ and $w \in W$; $U \not\sim W$ means that there are no adjacent vertices $u \in U$ and $w \in W$. To simplify the notation, we write $u \sim W$ ($u \not\sim W$) instead of $\{u\} \sim W$ ($\{u\} \not\sim W$). The graph complementary to G is denoted by $\overline{G}$. The neighbourhood of a vertex v in a graph G is denoted by $N_G(v)$, or simply $N(v)$.

Let us recall two well-known graph-theoretic operations. Let $G_i = (V_i, E_i)$ be two graphs ($i = 1, 2$) and $V_1 \cap V_2 = \emptyset$. The graph $G_1 \cup G_2$ is defined as follows:

$$V(G_1 \cup G_2) = V_1 \cup V_2 \quad \text{and} \quad E(G_1 \cup G_2) = E_1 \cup E_2,$$

and it is called the *disjoint union* of G_1 and G_2. The graph

$$G_1 \nabla G_2 = G_1 \cup G_2 + \{ab : a \in V_1, b \in V_2\}$$

is called the *join* of G_1 and G_2. In other words, to construct $G_1 \nabla G_2$, we take the disjoint union $G_1 \cup G_2$ and add the set of all possible edges between the sets V_1 and V_2.

Let G be an arbitrary graph. The subset $M \subseteq V(G)$ is a *module* or a *homogeneous set* of G if either $v \sim M$ or $v \not\sim M$ for each vertex $v \in V(G) - M$. Thus, if M is a module of G, we have a natural partition

$$V(G) = A \cup B \cup M, \quad A \sim M, \quad B \not\sim M. \tag{1.1}$$

In this partition, it is possible that $A = \emptyset$ or $B = \emptyset$. The partition (1.1) is *associated with the module M.*

For every graph G, the set $V(G)$, the empty set and singleton subsets of $V(G)$ are modules, which are called *trivial.* Other modules are *non-trivial.* The subgraph $G(M)$ of the graph G induced by a module M is called a *module* as well, and sometimes it is denoted by the same letter M.

1.1.2 Semigroup of Operators

We shall deal with triads $\tau = (G, A, B)$, where G is an arbitrary graph and (A, B) is an ordered partition of the set $V(G)$ into two disjoint subsets (*parts*). The graph G is called the *base graph* of the triad τ; (A, B) is the *bipartition* of the triad τ and the graph G; the sets A and B are called the *upper* and the *lower* parts of τ, respectively. One of the parts may be empty. If $A = \emptyset$ then τ has *B-type*, if $B = \emptyset$ then τ is of *A-type*.

For $i = 1, 2$, let

$$\tau_i = (G_i, A_i, B_i)$$

be triads, and

$$\varphi : V(G_1) \to V(G_2)$$

be an isomorphism of their base graphs. We call φ an *isomorphism of triads* if it preserves the bipartition; that is, $\varphi(A_1) = A_2$ and, consequently, $\varphi(B_1) = B_2$. Notice that the base graphs of non-isomorphic triads can be isomorphic.

Denote the set of all triads distinguished up to isomorphism of triads by Σ. Also, let us denote the set of all graphs distinguished up to isomorphism of graphs by $\mathbb{G}$. Define the *action of a triad*

$$\sigma = (H, A, B) \in \Sigma$$

on a graph $G \in \mathbb{G}$ by the following formula:

$$\sigma \circ G = (H \cup G) + \{av : a \in A, v \in V(G)\}. \tag{1.2}$$

In this formula, the edges of the complete bipartite graph with the parts A and $V(G)$ are added to the disjoint union $H \cup G$, as can be seen in Figure 1.1. The edges of G and H are not shown in the figure. The bold line stands for the set of all edges of the complete bipartite graph with the relevant parts, and the dash line means that there are no edges between the corresponding parts. A similar convention will be used in Figure 1.2.

The action (1.2) induces on Σ a binary algebraic operation $\circ$, which is called the *multiplication of triads*:

$$(H_1, A_1, B_1) \circ (H_2, A_2, B_2) = ((H_1, A_1, B_1) \circ H_2,\ A_1 \cup A_2,\ B_1 \cup B_2). \tag{1.3}$$

This operation is illustrated in Figure 1.2.

It can easily be checked that the operation $\circ$ is associative:

Lemma 1.1 *For arbitrary triads* $\alpha, \beta, \gamma \in \Sigma$ *and* $G \in \mathbb{G}$, *the following holds:*

$$(\alpha \circ \beta) \circ \gamma = \alpha \circ (\beta \circ \gamma), \quad (\alpha \circ \beta) \circ G = \alpha \circ (\beta \circ G).$$

In other words, the set Σ *with respect to the triad multiplication (1.3) is a semigroup acting on the set* $\mathbb{G}$ *according to formula (1.2).*

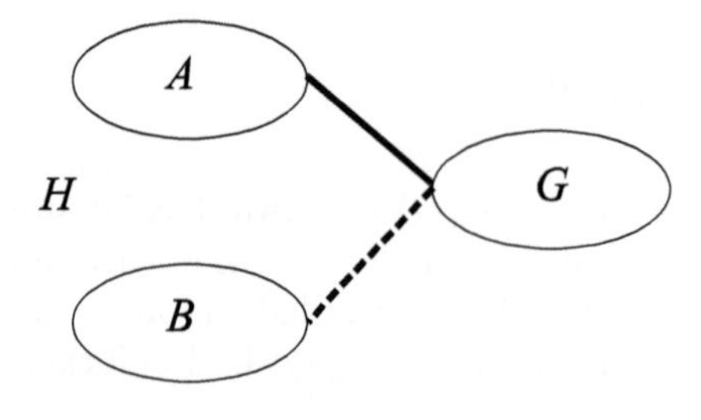

Figure 1.1 The action of the operator σ on the graph G.

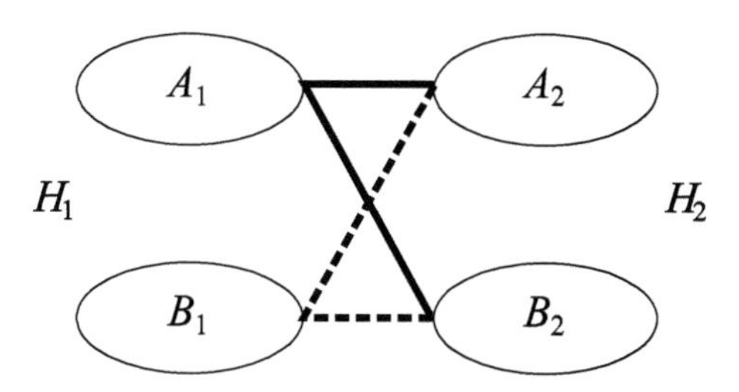

Figure 1.2 The multiplication scheme for the triads: $A_1 \sim A_2$, $A_1 \sim B_2$, $B_1 \not\sim A_2$, $B_1 \not\sim B_2$.

Now, we can consider the triads from Σ as *operators* on the set $\mathbb{G}$. An operator $\alpha \in \Sigma$ is called *decomposable* if there exist $\beta, \gamma \in \Sigma$ such that $\alpha = \beta \circ \gamma$; a graph $G \in \mathbb{G}$ is called *decomposable* if there exist $\alpha \in \Sigma$ and $H \in \mathbb{G}$ such that $G = \alpha \circ H$. Otherwise, the triad α (the graph G) is *indecomposable* (or *prime*).

The definition (1.2) of the action of $\sigma = (H, A, B)$ on a graph G is inspired by the definition of a module. By the action of σ on G, we embed G as a module in a larger graph $F = \sigma \circ G$, for which (G, A, B) is the *associated partition* of type (1.1) (see Figure 1.1). Thus, we have the following property:

Lemma 1.2 *A graph is decomposable if and only if it contains a non-trivial module.*

Consider the situation when the base graph of a certain triad has a non-trivial module. So, let $\tau = (H, A, B) \in \Sigma$, let M be a module of H and $V(H) = M \cup A' \cup B'$ be the corresponding associated partition. The module M is called a τ*-module* if $A' \subseteq A$ and $B' \subseteq B$. In general, a single-vertex module of H is not always a τ-module, so here it is worth to exclude single-vertex modules from the list of trivial modules. The following straightforward facts link τ-modules with the decomposability:

Lemma 1.3 *The triad τ is decomposable if its base graph has a non-trivial τ-module.*

Lemma 1.4 *Let τ_1 and τ_2 be two triads, $\tau = \tau_1 \circ \tau_2$ be their product and M be the module associated with this product. Then*

(1) τ_1 is indecomposable if and only if M is a maximal τ-module with respect to inclusion;

(2) τ_2 is indecomposable if and only if M is a minimal τ-module with respect to inclusion.

1.1.3 Operator Decomposition and Unique Factorization Property

It is obvious that every triad σ can be represented as a product of indecomposable (or prime) triads σ_i:

$$\sigma = \sigma_1 \circ \sigma_2 \circ \ldots \circ \sigma_k, \quad k \geq 1. \tag{1.4}$$

Such a representation is called a *decomposition of* σ.

Our major goal now is to establish an analogue of the unique factorization property for the operator decomposition. With this goal in mind, we will introduce several definitions. A prime factor σ_i of the product (1.4) with empty lower (upper) part has *A-type* (*B-type*). Factors σ_i and σ_j of A-type with $i < j$ are called *undivided* if every prime factor σ_k with $i < k < j$ is also of A-type. The same terminology is used for B-type. The following fact is easy to check:

Lemma 1.5 *Prime triads σ_i and σ_j commute if and only if σ_i and σ_j have the same A-type or B-type and they are undivided.*

Having multiplied all neighbouring undivided factors of A-type in the decomposition (1.4) as well as all neighbouring undivided factors of B-type, we obtain the *canonical decomposition* of σ:

$$\sigma = \alpha_1 \circ \alpha_2 \circ \ldots \circ \alpha_l, \quad l \le k. \tag{1.5}$$

The factors α_i from (1.5) are called *canonical components* of σ. A canonical component α_i is either a product of factors of A-type, a product of factors of B-type or an indecomposable operator in which both parts are not empty.

Theorem 1.1 [47] *The decomposition of a triad into prime factors is determined uniquely up to permutation of undivided factors of A-type or undivided factors of B-type. In other words, the canonical decomposition of a triad is unique.*

Proof: The statement is obviously true for indecomposable triads. Further, we will apply induction on the number of vertices.

Let

$$\sigma = \sigma_1 \circ \sigma_2 \circ \ldots \circ \sigma_k, \quad k > 1,$$

and

$$\sigma = \sigma_1' \circ \sigma_2' \circ \ldots \circ \sigma_l', \quad l > 1,$$

be two decompositions of a triad σ. Assume that

$$(G_1, A_1, B_1) = \sigma_1 \neq \sigma_1' = (G_1', A_1', B_1').$$

Setting

$$\sigma_2 \circ \sigma_3 \circ \ldots \circ \sigma_k = \mu = (H, C, D)$$

and

$$\sigma_2' \circ \sigma_3' \circ \ldots \circ \sigma_l' = \mu' = (H', C', D'),$$

we have

$$\sigma = \sigma_1 \circ \mu = \sigma_1' \circ \mu'.$$

By Lemma 1.4, the sets $M = C \cup D$ and $M' = C' \cup D'$ are maximal σ-modules. It is easy to see that the intersection $M' \cap (A_1 \cup B_1) = I$ is a σ_1-module of $G(A_1 \cup B_1)$.

Now, Lemma 1.3 implies that the module I is trivial. On the other hand, $I \neq \emptyset$ since M' is maximal. So, $I = A_1 \cup B_1$ and, by symmetry,

$$A_1 \cup B_1 \subseteq M', \quad A_1' \cup B_1' \subseteq M. \tag{1.6}$$

By (1.6), the triad σ_1 is the first indecomposable component in some decomposition of μ'. Without loss of generality, we may suppose by the induction assumption that $\sigma_1 = \sigma_2'$ and, therefore,

$$\sigma = \sigma_1' \circ \sigma_1 \circ \sigma_3' \circ \sigma_4' \circ \ldots \circ \sigma_l'.$$

Further, we have $A_1 \sim M$, $B_1 \not\sim M$, $A_1' \sim M'$, $B_1' \not\sim M'$. This together with (1.6) implies that $A_1 = A_1' = \emptyset$ or $B_1 = B_1' = \emptyset$. In both situations, we have

$$\sigma_1' \circ \sigma_1 = \sigma_1 \circ \sigma_1', \qquad \sigma = \sigma_1 \circ \sigma_1' \circ \sigma_3' \circ \ldots \circ \sigma_l'$$

and

$$\sigma_2 \circ \sigma_3 \circ \ldots \circ \sigma_k = \sigma_1' \circ \sigma_3' \circ \ldots \circ \sigma_l'.$$

By the induction assumption, one can conclude that $k = l$, and under the appropriate ordering, we have $\sigma_1' = \sigma_2$ and $\sigma_i = \sigma_i'$ for $i = 3, 4, \ldots, l$. □

Let us now turn to graphs. It is obvious that a decomposable graph G can be represented as

$$G = \sigma \circ G_0,$$

where σ is an operator and G_0 is an indecomposable graph. We call G_0 the *indecomposable part of G* and σ the *operator part*. If (1.5) is the canonical decomposition of σ, then the representation

$$G = (\alpha_1 \circ \alpha_2 \circ \ldots \circ \alpha_l) \circ G_0 \tag{1.7}$$

is called the *canonical decomposition of G*. However, unlike triads, the canonical decomposition of graphs may not be unique. Indeed, for the indecomposable part G_0, we have $G_0 = G(M)$, where M is a minimal non-trivial module of G, and for every such module there exists a relevant operator part α. Thus, we have the following property:

Corollary 1.1 *Every minimal non-trivial module of a graph determines a unique canonical decomposition of this graph.*

1.1.4 (P, Q)-decomposition

As we saw in the previous section, a graph in general can have several canonical operator decompositions. Furthermore, in many situations we need only the decompositions whose operator parts satisfy some prescribed properties determined by the nature of a particular problem. Thus, the goal of this section is twofold: to formalize the definition of 'prescribed properties' for operator decompositions and to study when such restricted operator decompositions lead to the unique factorization property.

The major object considered here is the (P,Q)-*decomposition*. It is constructed in the way similar to the method described above, but now it is not the case that arbitrary modules are allowed. The 'prescribed properties' of decomposition are 'encoded' in the choice of modules. Some additional terminology is needed to explain this idea.

A *class of graphs*, which may be considered as a graph-theoretic property, is an arbitrary non-empty set of graphs considered up to isomorphism. A class of graphs P is called *hereditary* (with respect to induced subgraphs) if the following implication holds:

$$(G \in P,\ U \subseteq V(G)) \ \Rightarrow \ G(U) \in P.$$

Let (P,Q) be an ordered pair of hereditary classes of graphs P and Q. A pair (P,Q) is called a *closed hereditary pair* if the following conditions hold:

(i) the class P is closed with respect to the join of graphs;
(ii) the class Q is closed with respect to the disjoint union of graphs.

For any closed hereditary pair (P,Q), we can formulate analogues of all basic definitions from the previous sections. Specifically, if the vertex set of a graph G can be split into two parts A and B such that $G(A) \in P$ and $G(B) \in Q$, then the graph G is called a (P,Q)-*split graph* and the triad $\alpha = (G,A,B)$ is called a (P,Q)-*split triad*. Note that one of the parts could be empty. We will denote the set of all (P,Q)-split graphs by $(P,Q)Split$, and the set of all (P,Q)-split triads by $(P,Q)\sum$.

Similarly, let M be a module of a graph G and (M,A,B) be the associated partition. If $G(A) \in P$ and $G(B) \in Q$, then M is called a (P,Q)-*module*. Similar to τ-modules, single-vertex modules should be considered as *non-trivial*, since not every single-vertex module is a (P,Q)-module.

By definition, if $\alpha, \alpha_1, \alpha_2$ are triads and $\alpha = \alpha_1 \circ \alpha_2$, then $\alpha \in (P,Q)\sum$ if and only if $\alpha_1, \alpha_2 \in (P,Q)\sum$. Thus, the following fact holds:

Proposition 1.1 *The set $(P,Q)\sum$ is a subsemigroup in $\sum$ if (P,Q) is a closed hereditary pair.*

In light of Proposition 1.1, we can naturally define the (P,Q)-decomposition of a graph—a restricted version of the general operator decomposition. Namely, a graph G is (P,Q)-*decomposable* or *decomposable at the level* (P,Q) if G can be represented in the form

$$G = \sigma \circ H, \quad \sigma \in (P,Q)\sum, \quad H \in \mathbb{G}. \tag{1.8}$$

Otherwise, G is (P,Q)-*indecomposable*.

If the graph H in (1.8) is (P,Q)-indecomposable and (1.4) is a decomposition of the triad σ into indecomposable factors, then

$$G = \sigma_1 \circ \sigma_2 \circ \ldots \circ \sigma_k \circ H, \qquad \sigma_i \in (P,Q)\sum. \tag{1.9}$$

The decomposition (1.9) is called a (P,Q)-*decomposition* of G. The *canonical* (P,Q)-*decomposition* is defined analogously to the canonical decomposition of a triad.

There are many examples of closed pairs that have already been studied within different domains of graph theory. The list below is by no means exhaustive.

- Suppose that K is the class of all complete graphs and O is the class of all empty graphs. In this case, (K, O)-split graphs are exactly *split* graphs [23]; that is, the graphs whose vertex set can be partitioned into a clique and an independent set. Split graphs form a well-known and important graph class that has been used and explored in numerous studies. It will also play a key role here. The class of split graphs will be denoted simply by *Split*. It is obvious that $K \subseteq P$ and $O \subseteq Q$ for any closed hereditary pair (P,Q). In other words, for such a pair (P,Q),

  $$Split \subseteq (P,Q)Split.$$

  Thus, split graphs form the simplest class of (P,Q)-split graphs. In what follows, the (K, O)-decomposition will play a prominent role and will be referred to as 1-*decomposition*.

- If k and l are some constants, then P is the class of graphs whose largest independent set is of size at most k, and Q is the class of graphs whose largest clique is of size at most l.

- Let $P^{(\alpha)}$ be the class of complete multipartite graphs G with at most α vertices in each part, and $Q^{(\beta)}$ be the class of disjoint unions of complete graphs with at most β vertices in each. In this case, $(P^{(\alpha)}, Q^{(\beta)})Split$ is the class of so-called (α, β)-*polar graphs* that have been studied in many different papers [41]. The corresponding $(P^{(\alpha)}, Q^{(\beta)})$-decomposition is sometimes simply called (α, β)-*decomposition* [41].

- If P is the class of P_4-free graphs (*cographs*) and $Q = P$, then the class $(P,Q)Split$ is the class of P_4-*bipartite* graphs introduced in [29]. Interestingly, the problem of recognition of P_4-*bipartite* graphs is NP-complete [29], even though the cograph recognition problem can be solved in linear time [18].

- Let Q be the class of disconnected graphs and P be the class of graphs with disconnected complement. Then $(P,Q)Split$ is the class of graphs with *skew partition*. It was introduced in [17] in the context of structural properties of perfect graphs and later became fundamental for the proof of the Strong Perfect Graph Conjecture [16].

Now, we are ready to discuss the unique factorization property. In general, it still may not hold—in fact, every non-trivial (P,Q)-module of G determines a

(P,Q)-decomposition of G and vise versa. However, the property holds under a quite general condition described in the following theorem:

Theorem 1.2 [47] *Let (P,Q) be an arbitrary closed hereditary pair and $G \in \mathbb{G}$. Let G have a minimal non-trivial (P,Q)-module M_1 such that $G(M_1) \notin P \cup Q$. Then, G has a unique (P,Q)-decomposition up to permutation of undivided factors of A-type or undivided factors of B-type.*

Proof: It is clear that every (P,Q)-decomposition (1.9) is associated with a minimal non-trivial (P,Q)-module M and vise versa. By Corollary 1.1, it remains to show that M_1 is a unique minimal non-trivial (P,Q)-module in G. Let there exist another minimal non-trivial (P,Q)-module M_2. Then,

$$G = \sigma_1 \circ H_1, \quad \sigma_1 \in (P,Q)\sum, \quad V(H_1) = M_1, \quad \sigma_1 = (G_1, A_1, B_1),$$

$$G = \sigma_2 \circ H_2, \quad \sigma_2 \in (P,Q)\sum, \quad V(H_2) = M_2, \quad \sigma_2 = (G_2, A_2, B_2).$$

Assume that $M_1 \cap M_2 \neq \emptyset$ and set

$$M_0 = M_1 \cap M_2, \quad M_1' = M_1 - M_2, \quad M_2' = M_2 - M_1.$$

Since M_2 is a (P,Q)-module, the triad

$$\sigma_2' = (G(M_1'), A_2 \cap M_1', B_2 \cap M_1')$$

is a (P,Q)-triad and we have $G = \sigma_1 \circ \sigma_2' \circ H_1(M_0)$, where $\sigma_1 \circ \sigma_2' \in (P,Q)\sum$. It contradicts the minimality of M_1. Hence, we conclude that $M_1 \cap M_2 = \emptyset$. In this case, since $M_1 \sim M_2$ or $M_1 \not\sim M_2$, we obtain $M_1 \subseteq A_2$ or $M_1 \subseteq B_2$. Therefore, $G(M_1) \in P$ or $G(M_1) \in Q$, which gives us a contradiction to the conditions of the theorem. □

Graphs satisfying the conditions of Theorem 1.2 can be called (P,Q)-*heterogeneous.* In particular, such are the graphs $G \notin (P,Q)Split$.

Corollary 1.2 *Let*

$$G = \sigma_1 \circ \sigma_2 \circ \ldots \circ \sigma_k \circ H \quad \textit{and} \quad G' = \sigma_1' \circ \sigma_2' \circ \ldots \circ \sigma_l' \circ H'$$

be canonical (P,Q)-decompositions of graphs G and G', respectively. Suppose that there exists the unique minimal non-trivial (P,Q)-module in G (in particular, G is (P,Q)-heterogeneous). Then, $G \simeq G'$ if and only if the following conditions hold:

(i) $k{=}l$;
(ii) $\sigma_i \simeq \sigma_i', \ i = 1, 2, \ldots, k$;
(iii) $H \simeq H'$.

Note that the set of closed hereditary classes has the cardinality of the continuum, whereas for a fixed closed hereditary pair (P,Q) the class of operators

$(P,Q)\sum$ is large enough too. For example, one may consider a disjoint union of graphs $G \in P$ and $H \in Q$ and add an arbitrary subset of edges between $V(G)$ and $V(H)$ to obtain an arbitrary operator from $(P,Q)\sum$. Thus, the construction of operator decomposition yields a variety of essentially different decompositions of a graph. Furthermore, different (P,Q)-decompositions form a natural hierarchy that allows for a large degree of flexibility in selection of a decomposition level suitable for a particular problem. Indeed, on the set of all closed hereditary pairs (P,Q) we can define an order relation $\subseteq$ as follows: we set

$$(P_i, Q_i) \subseteq (P_j, Q_j)$$

if

$$P_i \subseteq P_j \quad \text{and} \quad Q_i \subseteq Q_j.$$

This relation transforms the above set into a lattice $\mathbf{Z}$, where

$$\sup\Big((P_i, Q_i), (P_j, Q_j)\Big) = (P_i \cup P_j, Q_i \cup Q_j),$$

$$\inf\Big((P_i, Q_i), (P_j, Q_j)\Big) = (P_i \cap P_j, Q_i \cap Q_j).$$

The pair (K, O) is the *minimum element* in $\mathbf{Z}$. In particular, (α, β)-decompositions form a sublattice of $\mathbf{Z}$, where $(P^{(\alpha)}, Q^{(\beta)}) \subseteq (P^{(\alpha')}, Q^{(\beta')})$ whenever $\alpha \leq \alpha'$ and $\beta \leq \beta'$.

Now, let us choose an arbitrary element (P,Q) in the lattice. This determines the subsemigroup $(P,Q)\sum$ of (P,Q)-acceptable operators. The semigroup can be used to decompose a given triad or graph into (P,Q)-indecomposable factors. If this decomposition is not satisfactory enough (e.g. indecomposable factors are too large), then in the lattice $\mathbf{Z}$ we may turn to a higher element $(P', Q') \supseteq (P,Q)$ that can make some factors decomposable, thus producing a finer partition of a graph. If we are lucky and the pair (P', Q') is coherent enough with the particular class of graphs in question, then either our problem will be solved simultaneously for all graphs in this class or it will be partitioned into similar problems for smaller graphs. In the latter situation, the problems can be solved independently.

1.1.5 Graph Decomposition and Reconstruction Conjecture

One of the traditional problems in different branches of mathematics is to investigate relationships between the structure of an object and its substructures. The principal questions are how far the structure of an object is determined by the structure of its parts and whether it is possible to reconstruct an object from its parts.

To understand all substructures of an object, often it suffices to know only the maximal substructures of this object. If the structure under consideration is a graph G, then its maximal substructures may be proper maximal induced subgraphs of G. The question about reconstruction of the graph on the basis of such subgraphs yields the content of the well-known Kelly–Ulam Conjecture (also known as the reconstruction conjecture). Formulated in 1941, the conjecture has not been proved or disproved yet, even though it has a simple statement. This is one of the most famous open problems in graph theory.

Let $G = (V(G), E(G))$ be a simple graph and G_v be a subgraph obtained from G by deleting a vertex v. The collection

$$D(G) = (G_v)_{v \in V(G)}$$

is called the *deck* of G, and the elements of the collection $D(G)$ are called *cards* of the deck.

A graph H with the deck

$$D(H) = (H_u)_{u \in V(H)}$$

is called a *reconstruction* of G if there exists a bijection

$$f : V(G) \to V(H)$$

such that

$$G_v \simeq H_{f(v)}$$

for all $v \in V(G)$. In this case, the decks $D(G)$ and $D(H)$ are called *equal.* The graph G is called *reconstructible* if it is isomorphic to any of its reconstructions.

Kelly–Ulam Conjecture [32, 64] *Every graph with at least three vertices is reconstructible.*

Many publications devoted to the reconstruction conjecture are available, for example [12–14]. In [47], operator decomposition of graphs was applied for analysing the conjecture. The main idea is as follows. Let G be a graph whose reconstructibility should be proved, and assume that G has a decomposition with the following properties:

1. The graph is determined by the decomposition components up to isomorphism.
2. The components of the graph are determined by its deck up to isomorphism.

Then G is reconstructible. Often, the graph decomposition has the first property for large enough classes of graphs. Hence, when using operator decomposition for proving the reconstructibility of a graph, it suffices to show that the second property holds. Note that a similar approach was first applied by Tyurin for proving the reconstructibility of 1-decomposable graphs [63], which are discussed in the next section.

The following theorem was proved with the help of graph decomposition:

Theorem 1.3 [47] *(P,Q)-decomposable non-(P,Q)-split graphs are reconstructible for every closed hereditary pair (P,Q).*

Using Theorem 1.3 and graph decomposition, the following sufficient condition for reconstructibility is obtained:

Corollary 1.3 *A graph G is reconstructible if it contains a non-trivial non-(P,Q)-split (P,Q)-module for a closed hereditary pair (P,Q).*

Here we reach a high level of decomposition but we do not need to know this level exactly—we only have to be sure that a decomposition at such level exists.

1.2 1-Decomposition and Unigraphs

The 1-decomposition is the simplest example of algebraic graph decomposition. In this section, we discuss 1-decomposition and how it can be used for a characterization of unigraphs. Also, a recognition algorithm for unigraphs is described, as well as the algorithm for constructing 1-decomposition.

The list d of vertex degrees of a graph G is called its *degree sequence*, and G is called a *realization of the sequence d*. An integer sequence is called *graphical* if it has at least one realization. If all realizations of a graphical sequence are isomorphic, then the sequence is called *unigraphical* and its realization is called a *unigraph*.

1.2.1 1-Decomposition

The algebraic theory of graph decomposition started in [52], where the notion of canonical decomposition based on split graphs was introduced (we call it 1-decomposition). The definition of the canonical decomposition can be obtained from our general definition of the algebraic graph decomposition if the semigroup of all triads $\sum$ is replaced by the class S of triads whose base graphs are split. It is obvious that the class S coincides with the closed hereditary pair (K, O), where K and O are the classes of complete and empty graphs, respectively. Hence, the canonical decomposition coincides with the operator (K, O)-decomposition.

Furthermore, the definition of a closed hereditary pair implies the following result:

Proposition 1.2 *The pair (K, O) is the unique minimal element in the lattice of all closed hereditary pairs.*

As discussed above, the original canonical decomposition from [52] is called 1-*decomposition*; other studies use the term *Tyshkevich decomposition.* There are two features of 1-decomposition:

1. The semigroup of operators S is a free semigroup whose alphabet consists of indecomposable operators. Every S-decomposable graph contains only one S-module.
2. The 1-decomposition of an arbitrary graph is uniquely determined: decomposable graphs can be represented as

$$\sigma_1 \circ \sigma_2 \circ \cdots \circ \sigma_k \circ G_0,$$

where σ_i is an indecomposable operator and G_0 is an indecomposable graph. Different graphs yield different decompositions.

The uniqueness theorem for 1-decomposition was announced in [52] with a general scheme of the proof. The complete proof can be found in [53], where 1-decomposition was investigated thoroughly. In particular, it was proved in [53] that the structure of a graph does not influence its 1-decomposition, which is determined by the degree sequence of the graph. The following result is true.

Proposition 1.3 [53] *For any graphical sequence d and its realization G, every prime factor in the 1-decomposition is the subgraph induced in G by the set of its vertices with certain degrees. These degrees are determined by the sequence d and do not depend on the choice of realization.*

Thus, the operation $\circ$ and the 1-decomposition of a graph are transferred to graphical sequences. An algorithm for constructing a 1-decomposition of a graphical sequence of length n in time $O(n)$ was developed in [60], see Section 1.2.4. The 1-decomposition allows to reduce solutions of many algorithmic graph problems to the indecomposable case. This can be achieved if the considered properties of graphs are hereditary with respect to the composition $\circ$. The aforementioned algorithmic problems include the graph isomorphism problem, the Hamiltonicity problem, calculating the chromatic number and constructing the automorphism group. If a graph is 1-decomposable, then the problem in question can be decomposed into several independent problems of smaller size.

The two features mentioned above make 1-decomposition convenient for handling classes of graphs that are determined by vertex degrees. On this basis, it is possible to tackle some characterization and classification problems in those complex situations where the number of objects under consideration is very large. The most imposing example of this sort is discussed below.

1.2.2 Unigraph Characterization Based on 1-Decomposition

In the set of 1-indecomposable triads Σ^*, let us consider the following two operations: complement and inverting (part shifting). For a triad (G, A, B), the *inverted triad* is defined as follows:

$$(G, A, B)^{\mathrm{I}} = (G^{\mathrm{I}}, B, A),$$

where

$$G^{\mathrm{I}} = G - \{xy : x, y \in A\} + \{uv : u, v \in B\}. \tag{1.10}$$

The *complementary triad* $\overline{(G, A, B)}$ is equal to $(\overline{G}, B, A)$, where $\overline{G}$ is the graph complementary to G. Notice that $\overline{G^{\mathrm{I}}} = \overline{G}^{\mathrm{I}}$.

Theorem 1.4 [53] *Let*

$$G = (G_1, A_1, B_1) \circ (G_2, A_2, B_2) \circ \ldots \circ (G_k, A_k, B_k) \circ G_0 \tag{1.11}$$

be a 1-decomposition of a graph G. Then, G is a unigraph if and only if its indecomposable part G_0 and all graphs G_i $(1 \leq i \leq k)$ are unigraphs.

In [53], the catalogue L of all 1-indecomposable unigraphs was obtained. Based on this catalogue, one can find out for an arbitrary graph G whether G is a unigraph in linear time. The aforementioned algorithm for constructing a 1-decomposition [60] separates components of the decomposition in turn starting from the first one. It only remains to compare every separated component with the graphs from the catalogue L. The graphs from L are described below.

Indecomposable Split Unigraphs

A *multistar* is an indecomposable split graph M with the bipartition (A, B) such that

$$|A| = p \geq 2, \quad B \neq \emptyset \quad \text{and} \quad \deg b = 1$$

for every vertex $b \in B$, and the part $A = \{a_1, a_2, \ldots, a_p\}$ contains all possible edges. Split the vertex set of the lower part B into groups B_i such that $B_i \sim a_i$. Denote $q_i = |B_i|$ and assign a row vector $Q = (q_1, q_2, \ldots, q_p)$ to M. It is clear that the multistar M is determined by Q up to isomorphism. We will denote this multistar by $M(q_1, q_2, \ldots, q_p)$ or $M(Q)$, see Figure 1.3.

If $M(Q)$ is a multistar, then the *inverted multistar* $I(Q) = M(Q)^{\mathrm{I}}$ is obtained from $M(Q)$ by part shifting:

$$I(Q) = M(Q) - \{xy : x, y \in A\} + \{uv : u, v \in B\}, \quad B_i = N_{I(Q)}(a_i).$$

Similar to the star $M(Q)$, the graph $I(Q)$ is determined by Q up to isomorphism. Both introduced graphs are indecomposable.

Denote the complementary graph to $M(Q)$ by $R(Q)$:

$$R(Q) = \overline{M(Q)}.$$

It is obvious that $R(Q)$ is a split graph with the same bipartition (B, A) as the inverted multistar $I(Q)$. The graphs $I(Q)$ and $R(Q)$ differ from each other only in

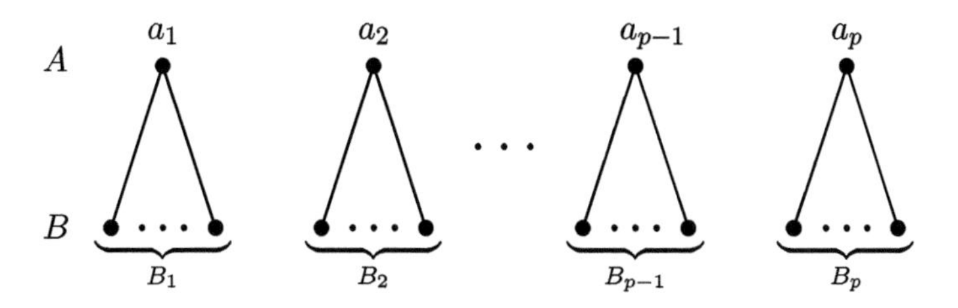

Figure 1.3 The multistar $M(q_1, q_2, \ldots, q_p)$, $p \geq 2$, $|B_i| = q_i$, $q_i \in \mathbb{N}$.

edges whose end points belong to different parts. Now, let us denote the inverted complementary graph to $M(Q)$ by $IR(Q)$:

$$IR(Q) = R^{\mathrm{I}}(Q) = \overline{I(Q)}.$$

The difference between graphs $R(Q)$ and $IR(Q)$ is that their upper and lower parts exchange their roles.

For some values of Q, the multistar $M(Q)$ is a subgraph of another indecomposable unigraph. Namely, let there exist an index p' that satisfies the following assumptions:

(a) $2 \leq p' < p$;

(b) the elements of Q are as follows:

$$q_i = \begin{cases} r & \text{if } \ i \leq p'; \\ r+1 & \text{if } \ i > p'. \end{cases}$$

In this situation, the vector Q is denoted by Q', and the graphs determined by it are called *exclusive*, see Figures 1.4 and 1.5. Let us denote

$$A' = \{a_1, a_2, \dots, a_{p'}\} \quad \text{and} \quad A'' = \{a_{p'+1}, a_{p'+2}, \dots, a_p\}.$$

Add a new vertex b to the lower part B of the multistar $M(Q')$ and introduce all edges from b to A'. Denote the obtained graph by $M_3(r, p', p)$. This graph is shown in Figure 1.4.

Further, let us add a new vertex a to the upper part of the graph $M_3(r, 2, p)$, so that this part induces a complete graph, and introduce all edges from a to $B - b$. The resulting graph is shown in Figure 1.5 and denoted by $M_4(r, p'')$. The graphs I_3, I_4, R_3, R_4 and IR_3, IR_4 are constructed in a similar way.

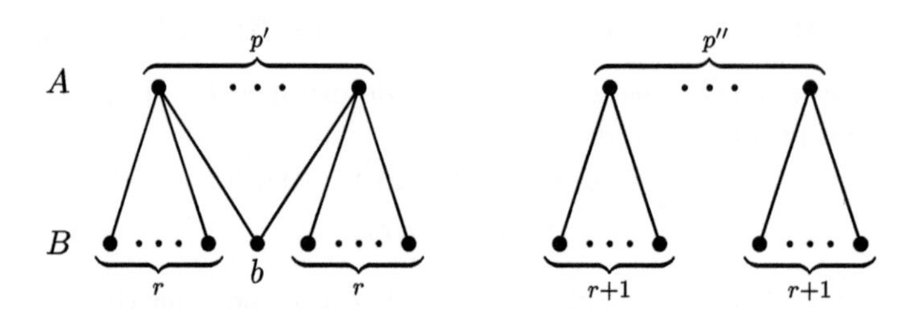

Figure 1.4 The exclusive graph $M_3(r, p', p)$, $r \geq 1$, $p > p' \geq 2$, $p'' = p - p'$.

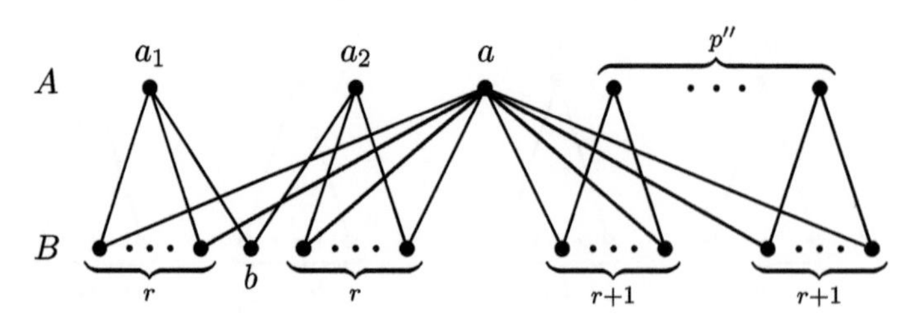

Figure 1.5 The exclusive graph $M_4(r, p'')$, $r \geq 1$, $p'' \geq 1$.

Lemma 1.6 *An indecomposable split graph G is a unigraph if and only if at least one of the graphs G, $\overline{G}$, G^{I}, $\overline{G}^{\mathrm{I}}$ belongs to the following set:*

$$M = \{K_1, M(Q), M_3(r, p', p), M_4(r, p'')\}.$$

Indecomposable Non-split Unigraphs

Let us consider the following two classes of graphs. For arbitrary positive integers $p \geq 1$ and $q \geq 2$, let $U_2(p, q)$ denote the disjoint union of the perfect matching pK_2 and the star $K_{1,q}$ (see Figure 1.6):

$$U_2(p, q) = pK_2 \cup K_{1,q}.$$

Now, let $U_3(p)$ denote the disjoint union of the cycle C_4 and the perfect matching pK_2, $p \geq 1$, where all edges connecting a fixed vertex of C_4 with vertices of pK_2 are added. The resulting graph is shown in Figure 1.6.

Lemma 1.7 *An indecomposable non-split graph G is a unigraph if and only if either G or $\overline{G}$ is contained in the following set:*

$$U = \{C_5, rK_2, U_2(p, q), U_3(p) : r \geq 2, p \geq 1, q \geq 2\}. \tag{1.12}$$

The following theorem describes the list L of all 1-indecomposable unigraphs.

Theorem 1.5 [55–58] *Let $\widetilde{M}$ be the closure of the set M with respect to the triad complement and the inversion operations and $\widetilde{U}$ be the closure of the set U with respect to taking complementary graphs. Then*

$$L = \widetilde{M} \cup \widetilde{U}.$$

We can construct an arbitrary non-split unigraph when we independently replace G_0 in the decomposition (1.11) with a graph from the list $\widetilde{U}$, and G_i $(i \neq 0)$ with

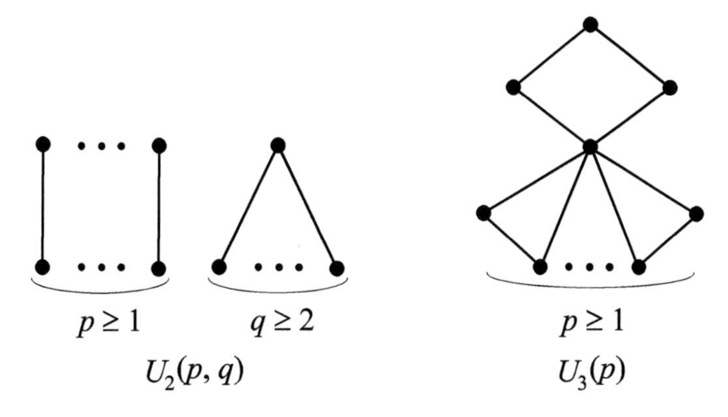

Figure 1.6 The classes of graphs $U_2(p, q)$ and $U_3(p)$.

graphs from the list $\widetilde{M}$. After deleting the last indecomposable part G_0, we obtain a decomposable unigraph. The unigraphs constructed in this way are pairwise non-isomorphic.

In fact, 1-decomposition and the comprehensive unigraph characterization given above have appeared earlier than the papers [52, 53]. An interest to unigraphs had been aroused in scientific literature in the 70th of the XX century. In 1975–1976, Kleitman, Li and Koren developed an algorithm that answers the question whether a given graphical sequence is unigraphical in linear time with respect to its length [33–35, 38]. Johnson's survey [31] appeared in 1980. In 1978–1979, in the series of papers by Tyshkevich and Chernyak [56–58] and also in [55], 1-decomposition was defined and, on its basis, the structure of unigraphs was described; that is, the characterization given above was obtained. Moreover, the following lower and upper bounds for the number u_n of n-vertex unigraphs were found:

$$2.3^{n-2} \le u_n \le 2.6^{n}.$$

However, the main characteristic feature of 1-decomposition was then still unknown: the 1-decomposition is determined not by the graph structure but by its vertex degrees only. More precisely, to obtain a 1-decomposition of a graph, it is necessary to decompose not the graph itself but its degree sequence. The unigraph characterization in [55–58] required significant efforts and an enumeration of a great number of variants. In fact, for every variant, a separate decomposition was constructed.

Proposition 1.3 was proved in [52], hence this paper is the starting point of operator decomposition theory. The unigraph characterization was revised in [53], where the decomposition is described in the language of degree sequences and 1-indecomposable graphs are characterized.

1.2.3 Recognition Algorithm for Unigraphs

The recognition algorithm for unigraphs is based on Theorem 1.4 and the list L of indecomposable unigraphs. Thus, the geometry of an arbitrary unigraph is known: it is decomposed into indecomposable components, and the relations between the components as well as the components themselves are clearly described. In [53], the existence of an algorithm that recognizes unigraphs in linear time was announced, whereas in [60] such a linear time algorithm was elaborated. It starts from the degree sequence of an arbitrary graph and constructs its unique 1-decomposition. The degree sequence is divided by the algorithm into 'segments' that correspond to the components of its 1-decomposition. Every factor of the 1-decomposition is the subgraph induced by the vertices whose degrees compose a relevant segment.

Here, we will discuss the linear algorithm for recognizing the property 'being a unigraph', which was developed on the basis of the unigraph characterization and the algorithm in [60]. Let us recall that a *chain* of a graph is a set of its vertices with equal degrees.

Proposition 1.4 [60] *Every chain of a graph is completely contained in a component of the 1-decomposition of the graph. If this component is a triad, then a chain is completely contained either in the upper part or in the lower part of the triad.*

Let us write the degree sequence of a graph G in the *box form*:

$$C = (c_1^{k_1}, c_2^{k_2}, \dots, c_N^{k_N}), \quad c_1 > c_2 > \dots > c_N, \tag{1.13}$$

where k_i is the number of vertices whose degrees are equal to c_i. The corresponding chain sequence is as follows:

$$D = (D_1, D_2, \dots, D_N), \quad D_i = \{v \in V(G) : \deg v = c_i\}. \tag{1.14}$$

In what follows, the term 'sequence' will mean exactly the box form of a degree sequence.

Let G be an arbitrary graph with the degree sequence (1.13) and the corresponding chain sequence (1.14). Proposition 1.5 follows directly from the definition of decomposition. Let us remind ourselves what is the 1-decomposition of the graph G:

$$G = (G_1, A_1, B_1) \circ (G_2, A_2, B_2) \circ \dots \circ (G_r, A_r, B_r) \circ H.$$

Proposition 1.5 *The following statements are true:*

(i) *The degree of an arbitrary vertex in the upper part of the graph G_l, $1 \le l \le r$, is equal to the degree of this vertex in the graph G minus the sum of the orders of all upper parts of the graphs*

$$G_1, G_2, \dots, G_{l-1}, G_{l+1}, G_{l+2}, \dots, G_r$$

minus the sum of the orders of all lower parts of the graphs

$$G_{l+1}, G_{l+2}, \dots, G_r$$

minus the order of the module H.

(ii) *The degree of an arbitrary vertex in the lower part of G_l, $1 \le l \le r$, is equal to the degree of this vertex in G minus the sum of the orders of the upper parts of $G_1, G_2, \dots, G_{l-1}$.*

(iii) *The degree of an arbitrary vertex of the induced subgraph H is equal to the degree of this vertex in G minus the sum of the orders of the upper parts of $G_1, G_2, \dots, G_r$.*

The procedure that is based on Proposition 1.5 constructs the list of the degree sequences

$$S_{G_1}, S_{G_2}, \dots, S_{G_r}, S_H$$

for the 1-decomposition of G. This procedure is called the *C-procedure*.

Lemma 1.8 *The following statements are true:*

(i) *The degree sequence of any graph from* $U_2(p,q)$ *has the form*

$$d(U_2(p,q)) = (q^1, 1^{2p+q}),$$

where $p \geq 1$, $q \geq 2$.

(ii) *The degree sequence of any graph from* $U_3(p)$ *has the form*

$$d(U_3(p)) = \left((2p+2)^1, 2^{2p+3}\right),$$

where $p \geq 1$.

(iii) *The degree sequence of any graph from* $M(Q)$ *has the form*

$$d(M(Q)) = \left((p+q_1-1)^{k_1}, \ldots, (p+q_t-1)^{k_t}, 1^{q_1+q_2+\ldots+q_p}\right).$$

Here, we assume that t *is the number of distinct elements of the vector*

$$Q = (q_1, q_2, \ldots, q_p),$$

and the first t *elements* $q_1, q_2, \ldots, q_t$ *are different. Also,*

$$p \geq 2, \ \sum_{i=1}^{t} k_i = p, \ \ q_i \geq 1, \ \ k_i \geq 1 \ \ \text{for} \ 1 \leq i \leq t.$$

(iv) *The degree sequence of any graph from* $M_3(r,p',p)$ *has the form*

$$d(M_3(r,p',p)) = \left((p+r)^p, p'^1, 1^{p(r+1)-p'}\right),$$

where $p \geq 3$, $p > p' \geq 2$ *and* $r \geq 1$.

(v) *The degree sequence of any graph from* $M_4(r,p'')$ *has the form*

$$d(M_4(r,p'')) = \left((p(r+2)-2)^1, (p+r+1)^p, 2^{p(r+1)-1}\right),$$

where $p \geq 3$ *and* $r \geq 1$.

Now, the following result is obtained:

Theorem 1.6 [60] *Using the degree sequence of a graph, the unigraph recognition problem can be solved in linear time. If the degree sequence is given in the box form, then this problem can be solved in linear time with respect to the number of different elements in the sequence.*

Finally, the recognition algorithm [60] for unigraphs is given below. Its algorithmic complexity is $O(N)$, where N is the number of elements in the sequence (1.13).

Algorithm 1.1 Recognition algorithm for unigraphs.

Input: The degree sequence of a graph G and the corresponding sequence of chains.

Output: The answer to the question whether G is a unigraph.

(1) Construct the 1-decomposition $T_1 \circ T_2 \circ \ldots \circ T_r \circ H$ of G, where $T_i = (G_i, A_i, B_i)$ (see Algorithm 1.2).

(2) Execute the procedure that creates the list of the degree sequences

$$S_{G_1}, S_{G_2}, \ldots, S_{G_r}, S_H$$

of the graphs $G_1, G_2, \ldots, G_r$ and the induced subgraph H.

(3) Using the degree sequence S_H of H, construct the sequence $\overline{S}_H$.

(4) For the sequence S_H, execute the procedure that verifies whether this sequence belongs to the list of indecomposable non-split unigraphs. Then, run the same procedure for $\overline{S}_H$. If any of these executions returns 'Yes', then go to Step 5; otherwise, go to Step 8.

(5) Set $i = 1$.

(6) Using the degree sequence S_{G_i} of G_i, construct the degree sequences $\overline{S}_{G_i}$, $S^{\mathrm{I}}_{G_i}$, $\overline{S}^{\mathrm{I}}_{G_i}$.

(7) For the sequences $\overline{S}_{G_i}$, $S^{\mathrm{I}}_{G_i}$, $\overline{S}^{\mathrm{I}}_{G_i}$, execute the procedure that verifies whether they belong to the list of indecomposable split unigraphs. If none of those sequences belong to the list, then go to Step 9. Otherwise, if $i = r$ then go to Step 8, if $i < r$ then the next triad T_i should be verified: set $i = i + 1$ and go to Step 6.

(8) Report 'Yes, G is a unigraph'. Algorithm stops.

(9) Report 'No, G is not a unigraph'. Algorithm stops.

1.2.4 Algorithm for Constructing 1-Decomposition

The fact that 1-decomposition is determined by a graph degree sequence established in Proposition 1.3 can be used to develop a linear algorithm to decide whether a graph is 1-decomposable and construct a decomposition if it exists. Here, we will describe an implementation of the algorithm proposed in [60]. It starts with the degree sequence of a graph and constructs the corresponding 1-decomposition by dividing the degree sequence into 'segments' that correspond to the factors of the decomposition. Every factor is then the subgraph induced by the vertices whose degrees compose a relevant segment.

Suppose that

$$G = (G_1, A_1, B_1) \circ (G_2, A_2, B_2) \circ \ldots \circ (G_r, A_r, B_r) \circ G_0$$

is the 1-decomposition of the graph G, where $G_0 = G(M)$ and M is a module of G. Let us recall that a *chain* of a graph is a set of its vertices with equal degrees. If the degree sequence of G is represented in the box form (1.13), then the corresponding chain sequence is given by (1.14).

The following fact follows directly from the definition of 1-decomposition:

Proposition 1.6 [60] *Let D_i be a chain of a graph G. Then, one of the following alternatives holds:*

(i) D_i is completely contained in a single factor of the 1-decomposition of G. If this factor is a triad, then D_i is contained either in the upper part or in the lower part of this triad.

(ii) D_i is a clique with $B_r = \emptyset$, $M = \{v\}$ and $D_i = A_r \cup \{v\}$.

(iii) D_i is an independent set with $A_r = \emptyset$, $M = \{v\}$ and $D_i = B_r \cup \{v\}$.

In addition, we will need the following corollary of Theorem 2.2. Notice that the definition of a good pair is given in the statement of this theorem.

Corollary 1.4 *Suppose that $G = \sigma \circ H$ is the 1-decomposition of a graph G, and (p, q) is the corresponding good pair of integers. The multiplier H is 1-indecomposable if and only if the numbers p and q are maximal values of the first and second coordinates of all good pairs.*

Proposition 1.6 and Corollary 1.4 form the basis for the algorithm that constructs 1-decomposition. Proposition 1.6 implies that each set A_i, B_i or M is associated with some subset of chains. Hence, the algorithm will consider these sets as subsets of chain indices. We will also assume that the vertices of G are ordered by their degrees.

Note that the sequences S and K at Step 1 can be constructed in time $O(N)$, and the running time of Step 5 is $O(q_D - \texttt{Bound})$. The condition (c1) and the way the variable Bound is updated at Step 6.5 guarantee that each decomposition factor T_i can be constructed in time $O(q_D + p_D) = O(|A_i| + |B_i|)$. Furthermore, after T_i is found, Step 6.2 decreases the number of chains that were not allocated to triads by $q_D + p_D$. Therefore, the total running time of the algorithm is $O(N)$.

Algorithm 1.2 Construction of 1-decomposition of a graph G.

Input: The box form (1.13) of the degree sequence of a graph G and the corresponding chain sequence (1.14).

Output: Subsets $A_1, B_1, \ldots, A_r, B_r, M \subseteq V(G)$ such that

$$G = (G(A_1 \cup B_1), A_1, B_1) \circ \cdots \circ (G(A_r \cup B_r), A_r, B_r \circ G(M).$$

(1) Construct two sequences $S = (S_0, S_1, \ldots, S_N)$ and $K = (K_0, K_1, \ldots, K_N)$ as follows:

$$S_0 = 0, \;\; K_0 = 0, \;\; S_i = S_{i-1} + C_i k_i, \;\; K_i = K_{i-1} + k_i, \;\; i = 1, 2, \ldots, N.$$

For any $n_0 \leq n_1 \leq r$, we have

$$\sum_{\substack{v \in \bigcup D_i \\ n_0 \leq i \leq n_1}} \deg v = S_{n_1} - S_{n_0-1}, \qquad \sum_{j=n_0}^{n_1} |D_j| = K_{n_1} - K_{n_0-1}.$$

(2) Suppose that the first $i-1$ factors of the 1-decomposition

$$T_1 = (G(A_1 \cup B_1), A_1, B_1), \ldots, T_{i-1} = (G(A_{i-1} \cup B_{i-1}), A_{i-1}, B_{i-1})$$

were already constructed. Denote by f_D (by l_D) the minimal (maximal) index of chains that do not belong to $A_1 \cup B_1 \cup \ldots \cup A_{i-1} \cup B_{i-1}$. Furthermore, let f (l) be the minimal (maximal) index of vertices from D_{f_D} (D_{l_D}). At the beginning, for $i=1$, these variables are set as $f_D = 1$, $f=1$; $l_D = N$, $l=n$.

(3) If $C_{f_D} = l - 1$, then

 (3.1) Set $A_i = \{f_D\}, B_i = \{0\}$.
 (3.2) Set $f_D = f_D + 1$, $f = f + k_{f_D}$.
 (3.3) Proceed to Step 2.

(4) If $C_{l_D} = f - 1$, then

 (4.1) Set $A_i = \{0\}, B_i = \{l_D\}$.
 (4.2) Set $l_D = l_D - 1$, $l = l - k_{l_D}$.
 (4.3) Proceed to Step 2.

(5) Denote by p_D and q_D the sizes of the subsets A_i and B_i that are being constructed, and by p and q the numbers of vertices in the upper and lower parts of the corresponding triad T_i, i.e.

$$p = \Big| \bigcup_{j \in A_i} D_j \Big| \quad \text{and} \quad q = \Big| \bigcup_{j \in B_i} D_j \Big|.$$

(5.1) Set $p_D = 1$, $p = k_{f_D}$, $\texttt{Bound} = 0$.

(5.2) Find an integer q_D corresponding to p_D. It is obvious that $l_D - f_D + 1$ is the number of chains D_i that do not belong to the previously constructed triads $T_1, T_2, \ldots, T_{i-1}$. Thus, q_D should satisfy the following conditions:

(c1) $\texttt{Bound} < p_D + q_D < l_D - f_D + 1$;

(c2) $C_{l_D - q_D + 1} < p + (f - 1)$;

(c3) $C_{l_D - q_D} \geq p + (f - 1)$.

(5.3) If such q_D does not exist, then the number of triads in the 1-decomposition of G is $i-1$. In this case go to Step 7.

(5.4) $q = K_{l_D} - K_{l_D - q_D}$.

(6) Check whether the pair (p_D, q_D) is good. If the equality

$$(S_{f_D+p_D-1} - S_{f_D-1}) - p(f-1) = p(l-f-q) + (S_{l_D} - S_{l_D-q_D}) - q(f-1)$$

holds, then (p_D, q_D) is good, and we can identify the triad T_i and update the variables as follows:

(6.1) $A_i = \{f_D, f_D + 1, \ldots, f_D + p_D - 1\}$,
$B_i = \{l_D - q_D + 1, l_D - q_D + 2, \ldots, l_D\}$.

(6.2) $f_D = f_D + p_D$, $f = f + p$, $l_D = l_D - q_D$, $l = l - q$.

(6.3) Proceed to Step 2.

If (p_D, q_D) is not good, then

(6.4) $p = p + k_{f_D + p_D}$, $p_D = p_D + 1$.

(6.5) $\texttt{Bound} = q_D$.

(6.6) Proceed to Step 5.2.

(7) At this step, all triads $T_1, T_2, \ldots, T_r$ are constructed. Set $M = \{f_D, f_D + 1, \ldots, l_D\}$.

1.2.5 A_4-Structure, Residue and Unigraphs

The concept of an A_4-structure plays an important role in 1-decomposition theory. A 4-uniform hypergraph H is called the *A_4-structure* of a graph G if $V(H) = V(G)$ and the edges of H are the vertex subsets of G that induce one of the graphs $2K_2$, C_4 or P_4. This definition was introduced by Barrus and West [5] and it was inspired by the known definition of the P_4-structure introduced by Chvátal in 1984.

The problem of internal characterization of the property 'being 1-indecomposable' was analysed in terms of A_4-structures:

Theorem 1.7 [5] *A graph G is 1-indecomposable if and only if its A_4-structure is a connected hypergraph. Consequently, if*

$$G = (G_1, A_1, B_1) \circ (G_2, A_2, B_2) \circ \ldots \circ (G_k, A_k, B_k) \circ G_0 \tag{1.15}$$

is the canonical decomposition of G, then its A_4-structure H contains only $k+1$ connected components $A_i \cup B_i$, $i = 1, 2, \ldots, k$ and $V(G_0)$.

The proof of this result is based on the uniqueness theorem for canonical decomposition and is quite involved.

Now, let us turn to the concept of the residue of a graph and its relation to unigraphs. The *residue* $r(G)$ of a graph G is determined by the degree sequence $d = (d_1, d_2, \ldots, d_n)$ of G sorted in the ascending order and by the Gavel–Hakimi algorithm, which constructs a realization of the sequence. If s is the number of steps in the algorithm transforming d to the sequence of n zeros, then

$$r(G) = r(d) = n - s.$$

Notice that $r(G) = 1$ if and only if $G = K_n$.

It is known that $r(G) \leq \alpha(G)$, where $\alpha(G)$ is the independence number of G. This estimate for the independence number can be both accurate and rough. Hence, the following question is of interest: how large can be the difference between

$$\min\{\alpha(G) : G \text{ is a realization of a graphical sequence}\}$$

and $r(G)$? The following theorem yields an exhaustive answer for unigraphs:

Theorem 1.8 [4] *Let $G = \alpha_1 \circ \alpha_2 \circ \ldots \circ \alpha_k \circ G_0$, $k \geq 1$, be the canonical decomposition of a unigraph G. Then*

(i) $r(G) \leq \alpha(G) \leq r(G) + 1$,
(ii) $\alpha(G) = r(G) + 1$ *if and only if* $G_0 = \overline{U_3(p)}$ *for some* $p \geq 1$.

This implies that for many unigraphs the independence number coincides with the Gavel–Hakimi residue.

1.3 Line Hypergraphs

This section is devoted to the multivalued function $\mathcal{L}$ called the line hypergraph. This function generalizes two classical concepts at once, namely the line graph and the dual hypergraph. In a certain sense, line graphs and dual hypergraphs are extreme values of the function $\mathcal{L}$. There are many publications about line graphs, but our focus is on the Krausz global characterization of line graphs and the Whitney theorem on edge isomorphisms. The notion of the line hypergraph is quite natural and it enables one to unify the classical theorems on line graphs and to obtain their more general versions in a simpler way.

1.3.1 Terminology

With minor adaptations, we adopt the terminology of Berge [9]. A *hypergraph* is a pair $(V, \mathcal{E})$, where V (the *vertex set*) is a finite non-empty set and $\mathcal{E}$ (the *edge family*) is a finite family of non-empty subsets of V. Thus, $\mathcal{E} = (e_i : i \in I)$, where $\emptyset \neq e_i \subseteq V$ and I is a finite set of *indices*. The set of vertices, the family of edges and the set of indices of a hypergraph H are denoted by $V(H)$, $\mathcal{E}(H)$ and $I(H)$, respectively. This definition differs from that of Berge [9], where isolated vertices are not permitted, and of Zykov [67], where empty edges are permitted.

The family

$$\mathcal{E}(v) = \mathcal{E}_H(v) \subseteq \mathcal{E}(H)$$

of edges incident to a vertex $v \in V(H)$ is called the *star of the vertex* v. Note that $\mathcal{E}(v)$ is a family of edges but not a partial hypergraph induced by $\mathcal{E}(v)$ as in [9]. The family

$$\mathcal{S}(H) = (\mathcal{E}(v) : v \in V(H))$$

is called the *family of stars* of H. Vertices u and v are called *similar* if $\mathcal{E}(u) = \mathcal{E}(v)$. A hypergraph having no multiple edges is called *simple*. It is convenient to ignore index sets when handling simple hypergraphs. Thus, a simple hypergraph is a pair $H = (V, E)$, where V is as above and E (the *edge set*) is a set of some non-empty subsets of V. The edge set of H is denoted by $E(H)$. For a finite family $(H_\lambda : \lambda \in \Lambda)$ of simple hypergraphs, the *set-theoretic union* $\bigcup H_\lambda$ is defined as the simple hypergraph H such that

$$V(H) = \bigcup_{\lambda \in \Lambda} V(H_\lambda) \quad \text{and} \quad E(H) = \bigcup_{\lambda \in \Lambda} E(H_\lambda).$$

Let H_1 and H_2 be two hypergraphs:

$$H_k = (V_k, \mathcal{E}_k), \quad \mathcal{E}_k = (e_i^k : i \in I_k), \quad k = 1, 2.$$

An *isomorphism*

$$(\alpha, \beta) : H_1 \to H_2$$

is a pair of bijections $\alpha : V_1 \to V_2$ and $\beta : I_1 \to I_2$ such that for any edge $e_i^1 = \{v_1, v_2, \ldots, v_d\}$, the following equality holds:

$$\alpha(e_i^1) = \{\alpha(v_1), \alpha(v_2), \ldots, \alpha(v_d)\} = e_{\beta(i)}^2.$$

If hypergraphs H and G are isomorphic, then we write $H \simeq G$. For simple hypergraphs H_1 and H_2, their isomorphism is defined more easily. This is a bijection $\alpha : V(H_1) \to V(H_2)$ such that $\alpha(e) \in E(H_2)$ if and only if $e \in E(H_1)$ for any subset $e \subseteq V(H_1)$.

A hypergraph H is called an *r-uniform* hypergraph if all its edges have the same degree r. The *complete r-uniform hypergraph* K_n^r is the simple hypergraph of order n such that its edge set coincides with the set of all r-subsets of the vertex set. For $r \geq 2$, the *clique of rank* r is K_n^r; the *clique of rank* 1 is K_1^1; and the *clique of rank* 0 is the one-vertex graph without loops. (This definition differs from that of Berge [8].) The vertex set of a clique will be referred to as a clique too.

Let H be a hypergraph. A finite family

$$Q = (Q_\lambda : \lambda \in \Lambda) \tag{1.16}$$

of cliques Q_λ is called a *clique covering* of H if $H = \bigcup_{\lambda \in \Lambda} Q_\lambda$. The cliques Q_λ are called the *clusters* of Q. A cluster is *trivial* if its rank is equal to 0. The minimum number of clusters taken over all clique coverings of H is called the *clique covering number* $\mathrm{cc}(H)$. It is evident that only simple hypergraphs have clique coverings.

If (1.16) is a clique covering of H and $P_\lambda = V(Q_\lambda)$, then the family of cliques

$$P = (P_\lambda : \lambda \in \Lambda) \tag{1.17}$$

is a *clique covering of the vertex set* $V(H)$. Now, suppose that (1.16) is a clique covering of a hypergraph H, (1.17) is the corresponding clique covering of the set $V(H)$ and ϕ is a permutation on the set $V(H)$. We set

$$\phi(P) = (\phi(P_\lambda) : \lambda \in \Lambda).$$

Obviously, $\phi(P)$ is a covering of the set $V(H)$, but $\phi(P_\lambda)$ may not be a clique. However, if $\phi(P_\lambda)$ is a clique and rank $\phi(P_\lambda) =$ rank P_λ, then we write

$$\phi(Q) = (\phi(Q_\lambda) : \lambda \in \Lambda).$$

A vertex subset of a hypergraph H having no pair of adjacent vertices is called *independent*. A partition

$$V(H) = V_1 \cup V_2 \cup \cdots \cup V_s, \quad s \leq r$$

of the vertex set $V(H)$ into independent subsets is called a *(strong) r-colouring*. A hypergraph for which there exists an r-colouring is *r-colourable*. A hypergraph is called *r-chromatic* if it is r-colourable but not $(r-1)$-colourable.

Finally, a hypergraph possesses the *Helly property* if any edge subfamily F such that every two edges in F have non-empty intersection is contained in a star.

1.3.2 Line Graphs and the Function 'Line Hypergraph'

Let H be a hypergraph without isolated vertices. The *line graph* $L(H)$ is the graph whose vertex set $V(L(H))$ is the edge set of H, and two vertices are adjacent in $L(H)$ if and only if the corresponding edges are adjacent in H. The concept of the line graph is so natural that it has been introduced under different names by many authors. Whitney [65] was the first to consider line graphs. The term 'line graph' was later introduced by Hoffmann [30]. In the book of Berge [8], these graphs are called *representative graphs*.

The class of line graphs of simple graphs has an attractive peculiarity. It is isomorphically complete by the Whitney theorem on edge isomorphisms [65]; that is, the recognition problem for graph isomorphisms is reduced in polynomial time to the analogous problem for line graphs. On the other hand, using Alexeev's technique [1, 2], it is not difficult to see that the class of line graphs is one of the 'minimal classes' among all infinite hereditary classes of graphs. To put it more precisely, let P be an infinite hereditary class of labelled graphs and P_n be the set of graphs from P of order n. The class P is called *non-trivial* if it does not coincide with the class of all graphs. Alexeev proved in [1] that for any infinite hereditary class P there exists an entropy, denoted by $h(P)$:

$$h(P) = \lim_{n \to \infty} (\log_2 |P_n|) \binom{n}{2}^{-1}.$$

If the class P is non-trivial, then

$$h(P) = 1 - \frac{1}{k}$$

for some natural number $k \in \mathbb{N}$. Conversely, for any $k \in \mathbb{N}$, there exists an infinite hereditary class with $h(P) = 1 - 1/k$ [2]. This means that

$$|P_n| = 2^{\binom{n}{2}(1-1/k)+o(n^2)} \quad \text{or} \quad |P_n| = o(2^{\binom{n}{2}}).$$

Thus, a non-trivial infinite hereditary class of graphs 'almost contains no graphs'. However, a class with entropy 0 contains 'fewer graphs':

$$|P_n| = 2^{o(n^2)}.$$

Proposition 1.7 *The entropy of the class of line graphs is equal to 0.*

Proof: Let $\mathcal{B}_{i,j}$ be the class of graphs whose vertex set can be divided into i cliques and j independent sets. It is proved in [1] that $h(P) = 0$ if and only if

$$\mathcal{B}_{0,2}, \mathcal{B}_{1,1}, \mathcal{B}_{2,0} \not\subseteq P. \tag{1.18}$$

The graphs depicted in Figure 1.7 are not line graphs for any simple graph [7]. Also, $K_{1,3} \in \mathcal{B}_{0,2} \cap \mathcal{B}_{1,1}$ and $W_4^- \in \mathcal{B}_{2,0}$. Consequently, (1.18) holds for the class P of line graphs of simple graphs. □

Obviously, the line graph $L(H)$ of a hypergraph H can be represented in the following form:

$$L(H) = \bigcup_{v \in V(H)} F_v,$$

where F_v is the clique whose vertex set is the star $\mathcal{E}(v)$ of H. Hence, $L(H)$ coincides with the 2-section of the dual hypergraph H^*: $L(H) = [H^*]_2$.

If G is a graph without isolated vertices, then $L(G^*) \simeq G$. Thus, any graph is the line graph of some hypergraph. In this context, the problem of representing graphs as line graphs of hypergraphs with prescribed properties arises. A comparison of the concepts of line graph $L(H)$ and dual hypergraph H^* has prompted the following

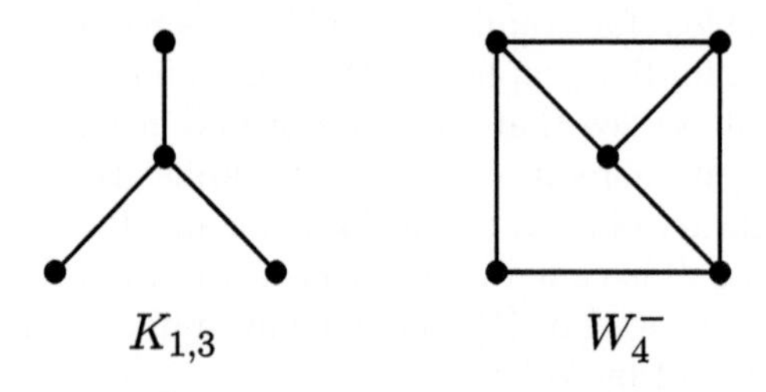

Figure 1.7 Graphs $K_{1,3}$ and W_4^-.

definition of the line hypergraph as a multivalued function on the set of hypergraphs [37, 54, 59].

Let H be a hypergraph without isolated vertices, $V(H) = \{v_1, v_2, \ldots, v_n\}$ be its vertex set and

$$\mathbf{1}_H = (\deg\, v_i : 1 \leq i \leq n)$$

be the degree sequence of H. We set

$$\mathbf{0}_H = (0_{v_i} : 1 \leq i \leq n), \quad \text{where} \quad 0_{v_i} = \begin{cases} 0 & \text{if } \deg v_i = 1; \\ 2 & \text{if } \deg\, v_i > 1. \end{cases}$$

Furthermore, let $\mathbb{Z}_+^n$ be the lattice of integer-valued vectors

$$\boldsymbol{x} = (x_1, x_2, \ldots, x_n), \quad x_i \geq 0, \quad 1 \leq i \leq n,$$

with the following order:

$$\boldsymbol{x} \leq \boldsymbol{y} \ \Leftrightarrow \ x_i \leq y_i, \ 1 \leq i \leq n.$$

Now, let $\mathcal{D}_H = [\mathbf{0}_H, \mathbf{1}_H]$ be an interval in $\mathbb{Z}_+^n$, and

$$\mathcal{D} = (d_1, d_2, \ldots, d_n) \in \mathcal{D}_H.$$

For $v_i \in V(H)$, let F_{v_i} denote the clique of rank d_i with the vertex set $\mathcal{E}(v_i)$, and let us consider the following hypergraph:

$$\mathcal{L}_{\mathcal{D}}(H) = \bigcup_{i=1}^{n} F_{v_i}.$$

The hypergraph $\mathcal{L}_{\mathcal{D}}(H)$ is called the *line hypergraph of H with respect to the vector $\mathcal{D}$*. If we write $\mathcal{D}$ in the form $\mathcal{D} = (d_v : v \in V(H))$, where $d_v = d_i$ for $v = v_i$, then the previous definition takes the form

$$\mathcal{L}_{\mathcal{D}}(H) = \bigcup_{v \in V(H)} F_v. \tag{1.19}$$

Let us denote by $\mathcal{H}$ the set of hypergraphs without isolated vertices and define the multivalued function $\mathcal{L}$ on $\mathcal{H}$ as follows:

$$\mathcal{L}(H) = \{\mathcal{L}_{\mathcal{D}}(H) : \mathcal{D} \in \mathcal{D}_H\}, \quad \text{where} \quad H \in \mathcal{H}.$$

The function $\mathcal{L}$ is called the *line hypergraph*. Any element in the image $\mathcal{L}(H)$ is called a *line hypergraph of H*. Notice that the hypergraphs in $\mathcal{L}(H)$ are simple.

It is evident that

$$L(H) = \mathcal{L}_{\mathbf{0}_H}(H) = [\mathcal{L}_{\mathcal{D}}(H)]_2$$

for any $\mathcal{D} \in \mathcal{D}_H$, and $\mathcal{L}_{\mathbf{1}_H}(H)$ is obtained from the dual hypergraph H^* if multiple edges are replaced by a single edge. We have

$$\mathcal{L}_{\mathbf{1}_H}(H) = H^*$$

if H does not contain similar vertices, that is, $\mathcal{E}(v_i) \neq \mathcal{E}(v_j)$ for $i \neq j$. It is also evident that $|\mathcal{L}(H)| = 1$ if and only if $\mathbf{1}_H = (2, 2, \ldots, 2)$.

Example 1 For $H = K_2$, we have $\mathbf{1}_H = (1,1)$, $\mathbf{0}_H = (0,0)$ and

$$\mathcal{D}_H = \{(d_1, d_2) : 0 \le d_i \le 1\} = \{\mathcal{D}_i : 1 \le i \le 4\},$$

where

$$\mathcal{D}_1 = \mathbf{1}_H, \quad \mathcal{D}_2 = (1,0), \quad \mathcal{D}_3 = (0,1), \quad \mathcal{D}_4 = \mathbf{0}_H.$$

Hence, $\mathcal{L}(H)$ takes two values:

$$\mathcal{L}_{\mathcal{D}_1}(H) = \mathcal{L}_{\mathcal{D}_2}(H) = \mathcal{L}_{\mathcal{D}_3}(H) = K_1^1 \quad \text{and} \quad \mathcal{L}_{\mathcal{D}_4}(H) = L(H) = K_1.$$

Example 2 For $H = K_3$, we have

$$\mathbf{1}_H = \mathbf{0}_H = (2,2,2) \quad \text{and} \quad \mathcal{L}_{(2,2,2)}(K_3) = L(K_3) = (K_3)^* = K_3.$$

Example 3 Let $H = K_{1,3}$. Since $\mathbf{1}_H = (3,1,1,1)$, we have $\mathbf{0}_H = (2,0,0,0)$ and

$$\mathcal{D}_H = \{(d_1, d_2, d_3, d_4) : \ 2 \le d_1 \le 3, \ 0 \le d_i \le 1 \text{ for } i = 2,3,4\}, \quad |\mathcal{D}_H| = 16.$$

However, some pairs of vectors from $\mathcal{D}_H$ yield isomorphic line hypergraphs. So, the function $\mathcal{L}$ takes eight values shown in Figure 1.8. These values correspond to the following vectors:

$$\mathcal{D}_1 = (3,1,1,1), \quad \mathcal{D}_2 = (3,1,1,0), \quad \mathcal{D}_3 = (3,1,0,0), \quad \mathcal{D}_4 = (3,0,0,0),$$

$$\mathcal{D}_5 = (2,1,1,1), \quad \mathcal{D}_6 = (2,1,1,0), \quad \mathcal{D}_7 = (2,1,0,0), \quad \mathcal{D}_8 = (2,0,0,0).$$

Notice that $\mathcal{D}_1 = \mathbf{1}_H$ and $\mathcal{D}_8 = \mathbf{0}_H$.

It follows from (1.19) that the family of cliques $(F_v : v \in V(H))$ is a covering of the line hypergraph $\mathcal{L}_{\mathcal{D}}(H)$, which is a simple hypergraph. Let us denote this covering by $Q(H, \mathcal{D})$.

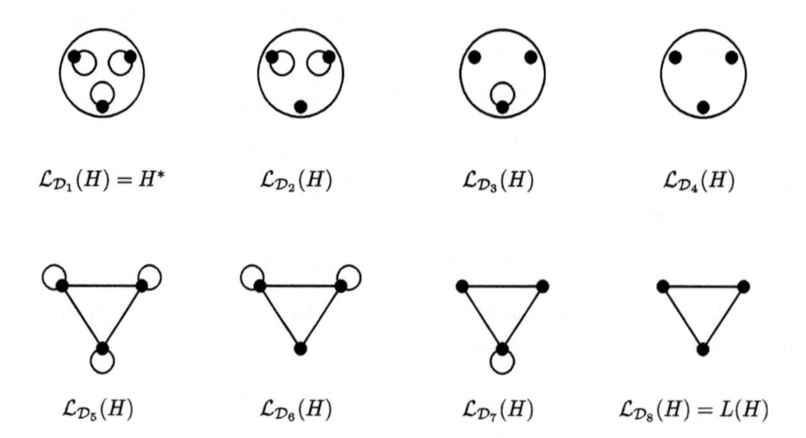

Figure 1.8 Line hypergraphs of star $K_{1,3}$.

Theorem 1.9 [62] *Let H_i, $i = 1, 2$, be hypergraphs without isolated vertices whose line hypergraphs $G_i = \mathcal{L}_{\mathcal{D}_i}(H_i)$ are isomorphic. Then*

(1) *$H_1 \simeq H_2$ if and only if there exists a bijection*

$$\beta : V(G_1) \to V(G_2)$$

such that $\beta(\mathcal{S}(H_1)) = \mathcal{S}(H_2)$, where $\mathcal{S}(H_i)$ is the family of stars of the hypergraph H_i.

(2) *An isomorphism $\gamma : G_1 \to G_2$ is induced by an isomorphism $H_1 \to H_2$ if and only if*

$$\gamma(Q(H_1, \mathcal{D}_1)) = Q(H_2, \mathcal{D}_2).$$

Let us consider a simple hypergraph G. The *inverse image* $\mathcal{L}^{-1}(G)$ is defined as the set of hypergraphs H such that $G \simeq \mathcal{L}_{\mathcal{D}}(H)$ for some $\mathcal{D} \in \mathcal{D}_H$. The aim of further considerations is to describe the set $\mathcal{L}^{-1}(G)$ for a given G. The concept of a canonical hypergraph introduced by Berge [8] plays an important role here.

Let G be a simple hypergraph, $\mathcal{A}_G$ be the set of its clique coverings and

$$Q = (Q_i : 1 \leq i \leq t) \in \mathcal{A}_G.$$

Define the hypergraph $F = F(Q)$ as follows:

$$V(F) = V(G), \quad \mathcal{E}(F) = (e_i : 1 \leq i \leq t), \quad e_i = V(Q_i). \tag{1.20}$$

The hypergraph F does not contain isolated vertices, hence the dual hypergraph F^* exists. The hypergraph F^* is called *canonical* (*with respect to* Q) and is denoted by $C(Q)$. If G is a hypergraph without isolated vertices and Q is an edge covering of G, then $C(Q) = G^*$. Thus, canonical hypergraphs are a generalization of dual hypergraphs.

Theorem 1.10 [62] *The following statements are true:*

(1) $\operatorname{rank} Q \in \mathcal{D}_{C(Q)}$ *for any* $Q \in \mathcal{A}_G$.

(2) $\mathcal{L}_{\operatorname{rank} Q}(C(Q)) \simeq G$.

(3) *If $H \in \mathcal{L}^{-1}(G)$ and $\mathcal{L}_{\mathcal{D}}(H) \simeq G$, then there exists $Q \in \mathcal{A}_G$ such that*

$$(H, \mathcal{D}) \simeq (C(Q), \operatorname{rank} Q).$$

This theorem is an evolution of Berge's idea. It was proved in [8] for the case when G is a graph.

Corollary 1.5 *Statements (1) and (2) are true:*

(1) *For any simple hypergraph G,*

$$\mathcal{L}^{-1}(G) = \{C(Q) : Q \in \mathcal{A}_G\} \quad \text{and} \quad G \simeq \mathcal{L}_{\operatorname{rank} Q}(C(Q)).$$

(2) *For $Q_1, Q_2 \in \mathcal{A}_G$, $C(Q_1) \simeq C(Q_2)$ if and only if there exists a permutation ψ of $V(G)$ such that $\psi(P_1) = P_2$, where P_k is a clique covering of $V(G)$ corresponding to the covering Q_k, $k = 1, 2$.*

Corollary 1.6 *Statements (1) and (2) are true:*

(1) $\mathcal{L}^{-1}(G) \neq \emptyset$ *if and only if* G *is a simple hypergraph.*

(2) The minimal order of hypergraphs in $\mathcal{L}^{-1}(G)$ *is equal to* $\mathrm{cc}(G)$. *In particular, if* G *is a simple triangle-free graph, then the minimal order is equal to the number of edges in* G.

1.3.3 Theorems of Krausz and Whitney

The following two classical theorems concerning line graphs are well known. The first is the Krausz theorem giving a global characterization of line graphs; notice that some clusters in a clique covering may be single vertices.

Krausz Theorem [36] *A graph* G *is the line graph of some simple graph if and only if there exists a clique covering* Q *of* G *satisfying the next two conditions:*

(1) Each vertex of G *belongs to exactly two clusters of the covering* Q.

(2) Every edge of G *belongs to exactly one cluster of* Q.

Analogous characterizations in terms of clique coverings were obtained for line graphs of hypergraphs with prescribed properties; for example, for simple hypergraphs, r-uniform hypergraphs, linear hypergraphs, linear r-uniform hypergraphs and 2-colourable hypergraphs [8, 9, 61].

The second classical result is the Whitney theorem on edge isomorphisms.

Whitney Theorem [65] *If* G *and* H *are connected graphs and* $L(G) \simeq L(H)$, *then either* $G \simeq H$ *or* $\{G, H\} = \{K_3, K_{1,3}\}$. *Further, if the orders of* G *and* H *are greater than 4, then for any isomorphism* $\alpha : L(G) \to L(H)$ *there exists a unique isomorphism* $G \to H$ *inducing* α.

Berge and Rado [10] and Gardner [24] obtained results for hypergraphs analogous in some sense to this theorem. They used a relation for hypergraphs which is stronger than isomorphism. See Section 5.1 for further details.

The technique of coverings, namely Corollary 1.5, enables one to unify the theorems of Krausz and Whitney and state a more general version as a single theorem. Let P be a graph-theoretic property. We can envision P as a class of hypergraphs closed with respect to isomorphism. Put

$$P^* = \{H^* : H \in P,\ H \text{ has no isolated vertices}\}.$$

Further, let G be a simple hypergraph and let $F(Q)$ be the hypergraph determined by (1.20). If $F(Q) \in P$, then we say that Q *has property* P or $Q \in P$. Obviously, $C(Q) \in P$ if and only if $Q \in P^*$. Thus, Corollary 1.5 implies the following existence and uniqueness theorem.

Theorem 1.11 *The following statements are true:*

(1) *For any simple hypergraph G and property P, $\mathcal{L}^{-1}(G) \cap P \neq \emptyset$ if and only if $\mathcal{A}_G \cap P^* \neq \emptyset$.*
(2) *For a simple uniform hypergraph G, $|\mathcal{L}^{-1}(G) \cap P| = 1$ if and only if the group of automorphisms $\mathrm{Aut}(G)$ acts transitively on the set of coverings $\mathcal{A}_G \cap P^*$.*

The requirement that a hypergraph be uniform in part (2) of Theorem 1.11 is not essential and is assumed for simplicity. Also, if in part (1) we take P to be the class of simple graphs, then the Krausz theorem is obtained. Therefore, all characterizations of line hypergraphs of graphs with prescribed properties that follow directly from Theorem 1.11 (1) are called *Krausz-type characterizations* or *Krausz P-characterizations* if P is given. The corresponding clique coverings are called *Krausz P-coverings.*

Theorem 1.11 (2) enables one to obtain analogues of the Whitney theorem for different classes of hypergraphs.

Corollary 1.7 *Let (P, R) be a pair of properties such that $\mathrm{Aut}(G)$ acts transitively on the set of coverings $\mathcal{A}_G \cap P^*$ for any $G \in R$. Further, let*

$$H_i \in P, \quad G_i \in \mathcal{L}(H_i) \cap R, \quad i = 1, 2 \quad \textit{and} \quad G_1 \simeq G_2.$$

Then $H_1 \simeq H_2$, and for any isomorphism $\alpha : G_1 \to G_2$ there exists an isomorphism $H_1 \to H_2$ inducing α.

Under the conditions of Corollary 1.7, we say that the *Whitney-type theorem* holds for the pair (P, R). Put

$$\mathcal{L}(P) = \{G : G \in \mathcal{L}(H) \text{ for } H \in P\}.$$

Let P^ℓ denote the class of linear hypergraphs without isolated vertices and P_r^ℓ denote the class of linear r-uniform hypergraphs without isolated vertices. In particular, P_2^ℓ is the class of simple graphs. Suppose that P is one of these properties, $G \in \mathcal{L}(P)$ and C is a maximal clique in G. The clique C is called *P-large* if

(1) rank $C > 2$ for $P = P^\ell$,
(2) rank $C > 2$ or $|C| > r^2 - r + 1$ for $P = P_r^\ell$.

Lemma 1.9 [62] *Any P-large clique of a graph G is a cluster of every Krausz P-covering of G.*

Lemma 1.9 and Corollary 1.7 yield the following statement.

Corollary 1.8 *The Whitney-type theorem holds for the pair (P, R), where (P, R) is one of the following:*

(1) $P = P^\ell$; *R is the class of hypergraphs whose edge degrees are greater than 2.*
(2) $P = P_r^\ell$; *R is the class of hypergraphs whose maximal cliques of rank at most 2 have orders greater than* $r^2 - r + 1$.
(3) P is the class of hypergraphs possessing the Helly property such that the vertex stars are pairwise non-comparable with respect to inclusion; R is the class of all hypergraphs.

The classical Whitney theorem can also be obtained by this scheme. The covering of a hypergraph by cliques with the properties pointed out in the Krausz theorem is called a *strict linear 2-covering.*

Lemma 1.10 [59] *Any non-trivial clique of a connected hypergraph* $G \in \mathcal{L}(P_2^\ell)$ *is a cluster of at most one strict linear 2-covering of G.*

This lemma and Corollary 1.7 together imply the uniqueness of such a covering for all graphs G with the exception of the case when G is a simple K_4-free graph with maximum degree $\Delta(G) \leq 4$ such that each edge of G belongs to a triangle.

The next Theorem 1.12 follows from Lemmas 1.9 and 1.10.

Theorem 1.12 [62] *Let* H_1 *and* H_2 *be connected graphs whose line hypergraphs* $G_i = \mathcal{L}_{\mathcal{D}_i}(H_i)$ *are isomorphic. Then*

(1) Either $H_1 \simeq H_2$ *or* $G_i \simeq K_3$.
(2) Every isomorphism $G_1 \to G_2$ *is induced by an isomorphism* $H_1 \to H_2$ *if and only if* G_i *is not isomorphic to any of* K_3, $K_4 - e$, $\overline{3K_2}$ *or the wheel* W_4 *shown in Figure 1.9.*

1.3.4 Characterizations

Characterizations of some classes of line graphs in terms of forbidden induced subgraphs can be obtained on the basis of Krausz-type characterizations. There are characterizations for line graphs of the following classes of graphs: simple graphs [7], multigraphs [11], multigraphs with restricted multiplicities of edges [51], bipartite graphs [26] and bipartite multigraphs [61]. A procedure for characterizing line

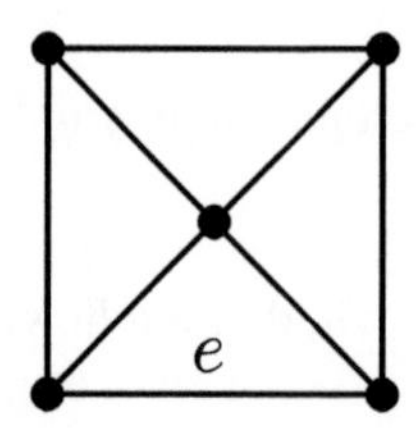

Figure 1.9 Wheel W_4.

graphs of a strict hereditary class in terms of forbidden induced subgraphs is given in [66]. This procedure is based on Beineke's characterization of line graphs of simple graphs [7] and the aforementioned Whitney theorem on edge isomorphisms.

The problem becomes more complicated when hypergraphs are considered instead of simple graphs. Denote by L_r and L_r^ℓ the classes of line graphs of r-uniform hypergraphs and line graphs of linear r-uniform hypergraphs, respectively. Lovász [39] posed the problem of characterizing the class L_3 and noted that this class cannot be characterized by a finite list of forbidden induced subgraphs.

Corollary 1.5 yields the following Krausz-type characterization.

Corollary 1.9 *We have $G \in L_r$ if and only if there exists a clique covering of G such that each vertex of G belongs to at most r clusters.*

This corollary implies the result about triangle-free graphs:

Corollary 1.10 *For a simple triangle-free graph G, the following statements are equivalent:*

(1) $G \in L_r$.
(2) Vertex degrees of G are at most r.
(3) The star $K_{1,r+1}$ is not a subgraph of G.

It was proved in [44] that the class L_3^ℓ cannot be characterized by a finite list of forbidden induced subgraphs. However, such characterizations are possible under some restrictions on vertex degrees. This problem will be explored in Chapter 5.

The representation of a graph as an intersection graph is one of the popular questions in graph theory. Let $F = (S_1, S_2, \ldots, S_k)$ be a family of pairwise distinct non-empty subsets of a set S. If S is finite, then the *intersection graph $\Omega(F)$ of the family F* is the line graph $L(H)$, where H is a simple hypergraph with

$$V(H) = \bigcup_{i=1}^{k} S_i \quad \text{and} \quad E(H) = F.$$

Hence, the next result follows from Corollary 1.6.

Corollary 1.11 *The following statements are true:*

(1) Any simple finite graph is isomorphic to some intersection graph [50].
(2) If $\omega(G)$ is the minimum number of elements in the set S such that $G \simeq \Omega(F)$ for some family F of subsets of S, then

$$\omega(G) \leq m(G) + \kappa(G), \tag{1.21}$$

where $m(G)$ is the number of edges in G and $\kappa(G)$ is the number of connected components of order at most 2.
(3) Inequality (1.21) is an equality if and only if G is triangle-free.

An important class of intersection graphs is the class of *clique graphs*; that is, intersection graphs of the families of all maximal cliques of simple graphs. Roberts and Spencer gave a Krausz-type characterization for this class of graphs:

Theorem 1.13 [45] *A simple graph is a clique graph if and only if it has a clique covering Q such that the hypergraph $F(Q)$ determined by (1.20) possesses the Helly property.*

An analogous characterization of intersection graphs of k-cliques can be found in [8]. Notice that the class of clique graphs is not hereditary. Moreover, any graph can be an induced subgraph of some clique graph.

Theorem 1.14 [22] *Any simple graph is an induced subgraph of some clique graph. Every triangle-free graph is a clique graph.*

To show this, let H be a simple graph. Put

$$V(G) = V(H) \cup E(H)$$

and

$$G = \bigcup_{v \in V(H)} G_v,$$

where G_v is a complete graph with vertex set $\{v \cup \mathcal{E}_H(v)\}$. Then, H is an induced subgraph of the clique graph of G.

1.3.5 k-Dimensional Graphs

A graph G is called a *k-dimensional cell graph* if $V(G)$ is a subset of the Cartesian product

$$A_1 \times A_2 \times \cdots \times A_k$$

of finite non-empty sets A_i, and two vertices

$$x = (x_1, x_2, \ldots, x_k) \quad \text{and} \quad y = (y_1, y_2, \ldots, y_k)$$

are adjacent if and only if $x_i = y_i$ for some index i. Any graph isomorphic to a k-dimensional cell graph is called *k-dimensional.* Evidently, every graph is k-dimensional for some k and every k-dimensional graph is $(k+1)$-dimensional. The minimal k for which a graph G is k-dimensional is called the *co-product dimension* but for simplicity we will call it *dimension* of G and denote by $\dim(G)$.

A characterization of k-dimensional graphs is given in the next theorem.

Theorem 1.15 [3] *For a simple graph G, the following statements are equivalent:*

(1) $\dim(G) \leq k$.
(2) $G \simeq \mathcal{L}(H)$, *where H is a simple k-colourable hypergraph.*
(3) $G \simeq \mathcal{L}(H)$, *where H is a simple k-uniform k-chromatic hypergraph.*

(4) *There exists a clique covering Q of G satisfying the following two conditions:*
 (i) *Q can be divided into $l \leq k$ parts such that each vertex of G belongs to at most one cluster of each part;*
 (ii) *The intersection of all clusters having a vertex v is $\{v\}$.*

(5) *There exists a clique covering Q of G satisfying the following two conditions:*
 (iii) *Q can be divided into k parts such that each vertex of G belongs to exactly one cluster of each part;*
 (iv) *Any k clusters have at most one vertex in common.*

Theorem 1.15 implies that the recognition problem '$\dim(G) \leq k$?' for a fixed $k > 2$ is NP-complete. Also, the recognition problem for graph isomorphism can be reduced in polynomial time to the analogous problem for 2-dimensional graphs. The last statement follows directly from the Whitney theorem if we make use of the König representation of graphs.

Corollary 1.12 [3] *For a simple graph G, the following statements are equivalent:*

(1) $\dim(G) \leq 2$.
(2) $G \simeq \mathcal{L}(H)$, *where H is a simple bipartite graph.*
(3) *G does not contain an induced subgraph isomorphic to $K_{1,3}$, $K_4 - e$ or C_{2n+1}, $n \geq 2$.*

The dimension of a graph is related to the following parameter. It is easy to see that any graph G can be represented as the union

$$G = G_1 \cup G_2 \cup \ldots \cup G_k$$

of graphs G_i whose connected components are complete graphs. The minimal number k taken over all such representations of G is called the *equivalence covering number* $\mathrm{eq}(G)$. This number was introduced by Duchet [21] and Behrendt [6]. It can be shown [3] that for any simple graph G,

$$\dim(G) \leq \mathrm{eq}(G) \leq \dim(G) + 1.$$

Let $\chi(H)$ and $\chi'(H)$ be the chromatic number and the chromatic index of H, respectively, and let $\mathcal{A}_G$ be the set of clique coverings of G. It was proved in [59] that for any graph G,

$$\mathrm{eq}(G) = \min_{Q \in \mathcal{A}_G} \{\chi(\Omega(Q))\} \leq \chi'(G).$$

Moreover, for a triangle-free graph G,

$$\mathrm{eq}(G) = \chi'(G).$$

This result implies that, similar to the dimension of G, the recognition problem '$\mathrm{eq}(G) \leq k$?' for any fixed $k > 2$ is NP-complete.

The *multiplication* of a vertex v in a graph G by a natural number $l \in \mathbb{N}$ means that we add the graph K_{l-1} and all edges between K_{l-1} and the set $N(v) \cup \{v\}$, where $N(v)$ denotes the neighbourhood of v. Denote by $\langle X \rangle$ the closure of a set of graphs X with respect to the multiplication of vertices, and let D_k be the set of k-dimensional graphs. Let $I(G)$ denote the set of all independent subsets of the vertex set of G ($\emptyset \in I(G)$). The pair

$$(V(G), I(G))$$

is called an *independence system.*

Theorem 1.16 [3] *For a simple graph G, the following statements are equivalent:*

(1) $\mathrm{eq}(G) \leq k$.
(2) *The independence system $(V(G), I(G))$ can be represented in the form of an intersection of k matroids.*
(3) $G \in \langle D_k \rangle$.
(4) $G = L(H)$*, where H is a k-uniform k-chromatic graph.*
(5) $G = L(H)$*, where H is a k-colourable graph.*
(6) *There exists a clique covering of G satisfying part (i) of Theorem 1.15.*
(7) *There exists a clique covering of G satisfying part (iii) of Theorem 1.15.*

The following corollary provides a characterization of graphs with the equivalence covering number at most 2.

Corollary 1.13 [59] *For a simple graph G, statements (1)–(4) are equivalent:*

(1) $\mathrm{eq}(G) \leq 2$.
(2) *G is the line graph of a bipartite multigraph.*
(3) $G \in \langle D_2 \rangle$.
(4) *G does not contain an induced subgraph isomorphic to $K_{1,3}$, W_4, $W_4 - e$ (see Figure 1.9) or C_{2n+1}, $n \geq 2$.*

The dimensionality of graphs will be further explored in Chapter 6.

Acknowledgements

Some material for the first two sections was supplied by Irina Suprunenko. Section 1.3 is based on the following publication: Acta Applicandae Mathematicae, 52, R. I. Tyshkevich and V. E. Zverovich, Line hypergraphs: a survey, 209–222, ©1998, reproduced with permission from Springer Nature.

1.4 References

[1] V. Alexeev, Hereditary classes and coding of graphs, *Problemy Kibernetiki*, **39** (1982), 151–164.
[2] V. Alexeev, Range of values for entropy of hereditary classes of graphs, *Diskretnaya Matematika*, **4** (2)(1992), 148–157.

[3] A. Babaitsev and R. I. Tyshkevich, k-Dimensional graphs, *Izvestiya Akademii Nauk Belarusi*, Ser. fiz.-mat. nauk, **3** (1996), 75–82.

[4] M. D. Barrus, Havel–Hakimi residues of unigraphs, *Information Processing Letters*, **112** (1–2)(2012), 44–48.

[5] M. D. Barrus and D. B. West, The A_4-structure of a graph, *Journal of Graph Theory*, **71** (2)(2012), 159–175.

[6] G. Behrendt, Semi-complete factorizations of graphs, *Bulletin of the London Mathematical Society*, **20** (1)(1988), 11–15.

[7] L. W. Beineke, Derived graphs and digraphs, in H. Sachs, H. Voss and H. Walter (eds), *Beiträge zur Graphentheorie*, Teubner, Leipzig, 1968, 17–33.

[8] C. Berge, *Graphs and Hypergraphs*, Amsterdam, The Netherlands: North-Holland Publishing Company, 1973.

[9] C. Berge, *Hypergraphs: Combinatorics of Finite Sets*, Amsterdam, The Netherlands: Elsevier Science Publishers B.V., 1989.

[10] C. Berge and R. Rado, Note on isomorphic hypergraphs and some extensions of Whitney's theorem to families of sets, *Journal of Combinatorial Theory, Ser. B*, **13** (1972), 226–241.

[11] J. Bermond and J. Meyer, Graphe représentaif des arêtes d'un multigraphe, *Journal de Mathématiques Pures et Appliquées*, **52** (9)(1973), 299–308.

[12] N. Biggs, *Algebraic Graph Theory*, Cambridge: Cambridge University Press, 1974.

[13] A. Bondy, A graph reconstructor's manual, *Surveys in Combinatorics*, **166** (1991), 221–252.

[14] A. Bondy and R. L. Hemminger, Graph reconstruction—a survey, *Journal of Graph Theory*, (1)(1977), 227–268.

[15] A. Brandstädt, V. B. Le and J. Spinrad, *Graph Classes: a Survey*, Philadelphia, PA: Society for Industrial and Applied Mathematics, 1999.

[16] M.Chudnovsky, N.Robertson, P. Seymour and R. Thomas, The Strong Perfect Graph Theorem, *Annals of Mathematics*, **164** (2006), 51–229.

[17] V. Chvatal, Star-cutsets and perfect graphs, *Journal of Combinatorial Theory, Ser. B*, **39** (3)(1985), 189–199.

[18] D. G. Corneil, Y. Perl and L. K. Stewart, A linear recognition algorithm for cographs, *SIAM Journal on Computing*, **14** (4)(1985), 926–934.

[19] D. M. Cvetkovich, M. Doob and H. Sachs, *Spectra of Graphs: Theory and Application*, Berlin: Deutscher Verlag der Wissenschaften, 1980.

[20] R. Diestel, *Graph Theory*, 5th edition, Heidelberg: Springer–Verlag, 2016.

[21] P. Duchet, Représentations, noyaux en théorie des graphes et hypergraphes, *Doctoral dissertation*, Éditeur inconnu, Université de Paris, VI, 1979.

[22] V. A. Emelichev, O. I. Melnikov, V. I. Sarvanov and R. I. Tyshkevich, *Lectures on Graph Theory*, Mannheim–Leipzig–Wein–Zurich: B.-I.-Wissenschaftsverlag, 1994 (translated from Russian, Moscow: 'Nauka' Publishing Company, 1990).

[23] S. Földes and P. L. Hammer, Split graphs, *Congressus Numerantium*, **19** (1977), 311–315.

[24] M. L. Gardner, Hypergraphs and Whitney theorem on edge isomorphisms of graphs, *Discrete Mathematics*, **51** (1984), 1–9.

[25] I. Goodfellow, Y. Bengio and A. Courville, *Deep Learning*, Massachusetts, USA: MIT Press, 2016.

[26] F. Harary and C. Holzmann, Line graphs of bipartite graphs, *Rev. Soc. Mat. Chile*, **1** (1)(1974), 19–22.

[27] F. Harary and E. L. Palmer, *Graphical Enumeration*, New York, London: Academic Press, Inc., 1973.

[28] N. Hartsfield and G. Ringel, *Pearls in Graph Theory: a Comprehensive Introduction*, Mineola, NY: Dover Publications, Inc., 1990.

[29] C. T. Hoàng and V. B. Le, P_4-free colorings and P_4-bipartite graphs, *Discrete Mathematics and Theoretical Computer Science*, **4** (2001), 109–122.

[30] A. J. Hoffmann, On the line graph of the complete bipartite graph, *The Annals of Mathematical Statistics*, **35** (2)(1964), 883–885.

[31] R. H. Johnson, Properties of unique realizations—a survey, *Discrete Mathematics*, **31** (2)(1980), 185–192.

[32] P. J. Kelly, *On Isometric Transformations*, PhD thesis, University of Wisconsin, 1942.

[33] D. J. Kleitman and S.-Y. Li, A note on unigraphic sequences, *Studies in Applied Mathematics*, **54** (4)(1975), 283–287.

[34] M. Koren, Pairs of sequences with a unique realization by bipartite graphs, *Journal of Combinatorial Theory, Ser. B*, **21** (3)(1976), 224–234.

[35] M. Koren, Sequences with unique realization, *Journal of Combinatorial Theory, Ser. B*, **21** (3)(1976), 235–244.

[36] J. Krausz, Démonstration nouvelle d'une theoreme de Whitney sur les réseaux, *Matematikai és Fizikai Lapok*, **50** (1943), 75–85.

[37] A. G. Levin and R. I. Tyshkevich, Line hypergraphs, *Discrete Mathematics and Applications*, **3** (4)(1993), 407–428.

[38] S.-Y. Li, Graphic sequences with unique realization, *Journal of Combinatorial Theory, Ser. B*, **19** (1)(1975), 42–68.

[39] L. Lovász, Problem 9, in *Beiträge zur Graphentheorie und deren Anwendungen.* Vorgetragen auf dem International Kolloquium in Oberhof (DDR), 1977, 313.

[40] L. Lovász and M. D. Plummer, *Matching Theory*, Budapest: Akadémiai Kiadó, 1986.

[41] N. V. R. Mahadev and U. N. Peled, *Threshold Graphs and Related Topics*, Annals of Discrete Mathematics, Vol. 56, Amsterdam: Elsevier, 1995.

[42] O. V. Maksimovich and R. I. Tyshkevich, A normal form of a dominant-threshold graph, *Doklady Akademii Nauk Belarusi*, **53** (2)(2009), 16–18.

[43] T. A. McKee and F. R. McMorris, *Topics in Intersection Graph Theory*, SIAM Monographs on Discrete Mathematics and Applications, Philadelphia, PA: Society for Industrial and Applied Mathematics, 1999.

[44] R. N. Naik, S. B. Rao, S. S. Shrikhande and N. M. Singhi, Intersection graphs of k-uniform linear hypergraphs, *European Journal of Combinatorics*, **3** (2)(1982), 159–172.

[45] F. S. Roberts and J. H. Spencer, A characterization of clique graphs, *Journal of Combinatorial Theory, Ser. B*, **10** (2)(1971), 102–108.

[46] P. V. Skums, S. V. Suzdal and R. I. Tyshkevich, Algebraic graph decomposition theory, *Trudy Instituta Matematiki*, **18** (1)(2010), 99–115.

[47] P. V. Skums, S. V. Suzdal and R. I. Tyshkevich, Operator decomposition of graphs and the reconstruction conjecture, *Discrete Mathematics*, **310** (3)(2010), 423–429.

[48] P. V. Skums and R. I. Tyshkevich, Reconstruction conjecture for graphs with restrictions on 4-vertex paths, *Diskretnyi Analiz i Issledovanie Operatsiy*, **16** (4)(2009), 87–96.

[49] S. V. Suzdal and R. I. Tyshkevich, (P, Q)-decomposition of graphs, Rostock University, 1999, 1–16. (Preprint aus dem Fachbereich Informatik, No 5-1999).

[50] E. Szpilrajn-Marczewski, Sur deux propriétés des classes d'ensembles, *Fundamenta Mathematicae*, **33** (1945), 303–307.

[51] V. Tashkinov, A characterization of line graphs of P-graphs, *All-Union Conference on Problems of Theoretical Cybernetics*, 1980, 135–137.

[52] R. I. Tyshkevich, Canonical decomposition of a graph, *Doklady Akademii Nauk BSSR*, **24** (8)(1980), 677–679.

[53] R. I. Tyshkevich, Decomposition of graphical sequences and unigraphs, *Discrete Mathematics*, **220** (2000), 201–238.

[54] R. I. Tyshkevich, Line hypergraph and Whitney's theorem, *Doklady Akademii Nauk Belarusi*, **38** (4)(1994), 8–10.

[55] R. I. Tyshkevich and A. A. Chernyak, Decomposition of graphs, *Kibernetika*, **21** (2)(1985), 231–242.

[56] R. I. Tyshkevich and A. A. Chernyak, Unigraphs I, *Izvestiya Akademii Nauk BSSR*, Ser. fiz.-mat. nauk, (5)(1978), 5–11.

[57] R. I. Tyshkevich and A. A. Chernyak, Unigraphs II, *Izvestiya Akademii Nauk BSSR*, Ser. fiz.-mat. nauk, (1)(1979), 5–12.

[58] R. I. Tyshkevich and A. A. Chernyak, Unigraphs III, *Izvestiya Akademii Nauk BSSR*, Ser. fiz.-mat. nauk, (2)(1979), 5–11.

[59] R. I. Tyshkevich and A. Levin, Line hypergraphs and matroid intersection, *Preprint of the Institute of Engineering Cybernetics of the Belarusian Academy of Sciences*, **23** (1990), 1–36.

[60] R. I. Tyshkevich and S. V. Suzdal, Decomposition of graphs, *Izbrannye Trudy Belorusskogo Gosudarstvennogo Universiteta (Selected Scientific Proceedings of Belarus State University)*, **6** (2001), 482–504.

[61] R. I. Tyshkevich, O. Urbanovich and I. E. Zverovich, Matroidal decomposition of a graph, *Combinatorics and Graph Theory, Banach Center Publications*, **25** (1)(1989), 195–205.

[62] R. I. Tyshkevich and V. E. Zverovich, Line hypergraphs, *Discrete Mathematics*, **161** (1996), 265–283.

[63] V. Tyurin, Reconstructibility of decomposable graphs, *Izvestiya Akademii Nauk BSSR*, Ser. fiz.-mat. nauk, (3)(1987), 16–20.

[64] S. M. Ulam, *A Collection of Mathematical Problems*, Vol. 29, New York: Wiley (Interscience), 1960.

[65] H. Whitney, Congruent graphs and the connectivity of graphs, *American Journal of Mathematics*, **54** (1)(1932), 150–168.

[66] I. E. Zverovich, An extension of hereditary classes and line graphs, *Izvestiya Akademii Nauk Belarusi*, Ser. fiz.-mat. nauk, (3)(1994), 108–113.

[67] A. Zykov, Hypergraphs, *Russian Mathematical Surveys, (Uspekhi Matematicheskih Nauk)*, **29** (6)(1974), 89–154.

2

Decomposition of Graphical Sequences and Unigraphs

Based on R. Tyshkevich's article (edited by V. Zverovich)

A binary operation $\circ$ is introduced on the set of graphs and is transferred to graphical sequences. Then, the decomposition theorem is proved, which states that any graph and any graphical sequence can be uniquely decomposed into indecomposable components with respect to the operation $\circ$. An exhaustive description of the structure of unigraphs based on this theorem is given. This chapter is a reproduction of the original article by Regina Tyshkevich published in *Discrete Mathematics*, **220** (2000), 201–238 (reprinted by permission from Elsevier, ©2000).

2.1 Preliminaries

All graphs are finite, undirected, without loops and multiple edges. This chapter is divided into three parts. First, we discuss some known facts and preliminary results. The second section deals with a proof of the decomposition theorem for graphical sequences announced in [15]. The proof has never been published. Roughly speaking, a binary operation $\circ$ is introduced on the set of graphs and is transferred to graphical sequences. The mentioned theorem states that every graphical sequence and every graph can be uniquely decomposed into indecomposable components with respect to the operation $\circ$. An indecomposability criterion is given. On this basis, one can obtain the list of indecomposable components for a graph and the corresponding graphical sequence in linear time.

For graphs, the operation $\circ$ was defined in [17], although the relevant construction had been used in [11]. The operation is based on the concept of a split graph introduced by Földes and Hammer [4]. Later, a far-reaching generalization of splitness was introduced, and a more general decomposition theorem for graphs was obtained on this basis [22]. However, that theorem cannot be transferred to graphical sequences. The theorem and its applications were explored by Mahadev and Peled in [14].

Based on R. Tyshkevich's article (edited by V. Zverovich), *Decomposition of Graphical Sequences and Unigraphs.* In: *Methods of Graph Decompositions.* Edited by: Vadim Zverovich and Pavel Skums, Oxford University Press. © Vadim Zverovich & Pavel Skums (2024).
DOI: 10.1093/oso/9780198882091.003.0002

It may be pointed out that the decomposition theorem is a convenient tool for studying graph classes closed under the operation $\circ$. Using this theorem, the following classes of graphs were characterized and enumerated: box-threshold graphs [21], dominant-threshold graphs [22] and matrogenic graphs [16] (see also Section 3.1.3).

The third section of the chapter is devoted to a description of the structure of unigraphs based on the decomposition theorem. This is the most convincing example of the successful application of the theorem.

Properties of unigraphical sequences were studied by Kleitman, Koren and Li [9–12]. An algorithm for recognizing unigraphicity in linear time was obtained in [9] with the use of arguments from [10–12]. Note that a description of the structure of unigraphs is a geometric aspect of the problem. This research was started by Johnson [7]. A complete description of the structure of unigraphs was given in the series of papers [17–20], even though at that time the decomposition theorem was unknown. On this basis, the following bounds for the number u_n of unlabelled unigraphs of order n were obtained in [20]:

$$2.3^{n-2} \leq u_n \leq 2.6^{n}.$$

Notice that the articles [17–20] were published in Russian, and hence they were not known to many graph theorists—the papers [1, 2, 8, 13] indirectly indicate this.

The above bounds show that unigraphs are not too scarce. This class of graphs is rather interesting in some aspects. For instance, as the degree sequence of a graph is known to be reconstructible, the well-known Kelly–Ulam Reconstruction Conjecture holds in the class of unigraphs.

In this chapter, the accents are placed in such a way that the decomposition theorem plays a decisive role, thus simplifying the proofs of main results and making them conceptually transparent. Roughly speaking, by means of the decomposition theorem, the set of unigraphs is turned into something close to a free semigroup. The alphabet of this 'semigroup' is completely described. In Section 2.1.1, relevant definitions and notation are given, whereas in Section 2.1.2 an important notion of a split graph is considered. The decomposition theorem is given in Section 2.2.1, and the main theorem on the structure of unigraphs is stated in Section 2.2.2. Sections 2.2.3–2.3.2 are devoted to the proof of the latter theorem.

2.1.1 Notation, Terminology and Known Facts

The vertex set and the edge set of a graph G are denoted by $V(G)$ and $E(G)$, respectively. The number $|V(G)|$ is the *order* of a graph G. A graph of order 1 is called *trivial*. We write $a \sim b$ ($a \not\sim b$) if the vertices a and b are adjacent (non-adjacent). If $A \subseteq V(G)$ and $B \subseteq V(G)$, then $A \sim B$ ($A \not\sim B$) means that $a \sim b$ ($a \not\sim b$) for every $a \in A$ and every $b \in B$. For $c \in V(G)$, we write $c \sim A$ ($c \not\sim A$) if $\{c\} \sim A$ ($\{c\} \not\sim A$). Also, $\overline{G}$ denotes the complement of a graph G.

Let $G(X)$ be the subgraph induced by a vertex set X. If $G(X)$ is a complete (an empty) graph, then X is a *clique* (an *independent set*). Also, $K(X)$ ($O(X)$) is the complete (the empty) graph with the vertex set X. The graph $G + ab$ is obtained from G by adding a new edge ab, where $a, b \in V(G)$. The graph $G - ab$ is obtained from G by deleting the edge ab (without deleting the vertices a and b).

The set of vertices adjacent to a vertex c in a graph G is the *neighbourhood* of c; it is denoted by $N_G(c)$ or briefly $N(c)$. The *degree* of a vertex c is $\deg c = |N(c)|$. For $X \subset V(G)$, let us denote $\deg_X c = |X \cap N(c)|$. The set of all vertices of the same degree in a graph is called a *regular subset of rank* 0 (or simply a *regular subset*). A graph is *regular* (*biregular*) if it has a unique regular subset (at most two regular subsets). If G and H are graphs, then $G \cup H$ denotes the *disjoint union* (i.e. $V(G) \cap V(H) = \emptyset$); mK_2 is the disjoint union of m copies of K_2; P_n is the simple n-vertex path and C_n is the chordless n-vertex cycle.

The *degree sequence* of a graph is the list of its vertex degrees. An integer sequence is called *graphical* if there exists a graph (a *realization* of the sequence) such that the sequence is its degree sequence. In what follows, a sequence of length n is called an *n-sequence*. The i-th member of a sequence d is denoted by d_i. An n-sequence d is called *proper* if

$$n - 1 \geq d_1 \geq d_2 \geq \ldots \geq d_n \geq 0.$$

Obviously, a graphical sequence can be assumed to be proper.

Fact 1 [5] *For any term d_i of a proper graphical n-sequence d, there is a realization of d in which some vertex of degree d_i is adjacent to d_i vertices of the maximal possible degrees, that is, of the following degrees*

$$d_1, d_2, \ldots, d_{d_i} \quad \textit{if} \quad d_i < i,$$

$$d_1, d_2, \ldots, d_{i-1}, d_{i+1}, \ldots, d_{d_i+1} \quad \textit{if} \quad d_i \geq i,$$

where $i = 1, 2, \ldots, n$.

A graphical sequence is called *unigraphical* if all its realizations are isomorphic. A *unigraph* is a realization of a unigraphical sequence.

Fact 2 [7, 11] *All regular unigraphs are the following graphs:*

$$K_n, \; O_n, \; mK_2, \; \overline{mK_2}, \; C_5.$$

A bipartite graph G together with a fixed ordered partition (*a bipartition*) $V(G) = A \cup B$ is called 2-*coloured*. In what follows, such a graph is denoted by $G(A, B)$. The 2-coloured graphs $G(A, B)$ and $H(C, D)$ are called *isomorphic* if there exists a graph isomorphism $f : G \to H$ which *preserves colouring* (i.e. $f(A) = C$ and $f(B) = D$).

A pair (c, d) of integer sequences is called *graphical* if it has a *realization*; that is, if there exists a 2-coloured graph $G(A, B)$ having c and d as the lists of vertex degrees in the parts A and B, respectively. A graphical pair is *unigraphical* provided that all its realizations are isomorphic. A *coloured unigraph* is a realization of a unigraphical pair.

A graphical pair

$$(c,d), \quad c=(c_1,c_2,\ldots,c_p), \quad d=(d_1,d_2,\ldots,d_q) \tag{2.1}$$

is called *biregular* if

$$c_1 = c_2 = \ldots = c_p, \quad d_1 = d_2 = \ldots = d_q.$$

Fact 3 [10] *A biregular graphical pair (2.1) is unigraphical if and only if*

$$c_1 \in \{0,1,q-1,q\} \quad \text{or} \quad d_1 \in \{0,1,p-1,p\}.$$

The following well-known concept of 2-switch plays an important role in the theory of degree sequences. Let a, b, c and d be four pairwise distinct vertices of a graph G. If $a \sim c$, $a \not\sim d$, $b \not\sim c$ and $b \sim d$, then G is said to *admit the 2-switch* $t = abcd$. The graph tG, the image of G under t, is obtained from G after replacing the edge pair ac, bd by the edge pair ad, bc; that is,

$$tG = G - ac - bd + ad + bc.$$

The following *symmetry* property of 2-switch is obvious:

$$abcd = badc = cdab = dcba.$$

It is easy to see that if a graph G admits a 2-switch $abcd$, then its complement admits $abdc$.

It is known that every two realizations of a graphical sequence can be obtained from each other by a sequence of 2-switches [5]. Hence, the following statement is true.

Fact 4 [7] *A graph G is a unigraph if and only if $tG \simeq G$ for every 2-switch t.*

Threshold graphs are known to be the simplest class of unigraphs. Threshold graphs have been introduced for different reasons, and an exhaustive description of them was given in [14]. There are several equivalent definitions of a threshold graph. The following definition is convenient for our purposes. A graph is called *threshold* if it can be obtained from K_1 by adding isolated or dominating vertices [3].

Fact 5 [14] *For any pair of regular subsets A and B in a threshold graph, either $A \sim B$ or $A \not\sim B$ holds.*

Fact 6 [14] *A graph is threshold if and only if it admits no 2-switch.*

Facts 1–6 are essential for proving Theorem 2.4 which characterizes unigraphs.

2.1.2 Split Graphs

A graph G is called *split* [4] if there exists a partition

$$V(G) = A \cup B \tag{2.2}$$

of its vertex set into a clique A and an independent set B. The partition, the clique and the independent set will be called a *bipartition*, an *upper part* and a *lower part*, respectively. One of the parts may be empty, but not both.

A graphical sequence is called *split* if it has a split realization. A splitness criterion for a graph is formulated below in terms of vertex degrees. It implies that all realizations of a split sequence are split. Two rather different variants of the criterion were simultaneously and independently obtained by Hammer and Simeone [6] and Tyshkevich et al. [23]. A joint version is given below.

For a proper graphical n-sequence d, let us set

$$m(d) = \max\{i : d_i \geq i - 1\} \ [6].$$

This parameter plays an important role in what follows.

Theorem 2.1 [6, 23] *Statements (i)–(iii) are true.*

(i) A proper graphical n-sequence d is split if and only if for some p, $0 \leq p \leq n$, the following equality holds:

$$\sum_{i=1}^{p} d_i = p(p-1) + \sum_{i=p+1}^{n} d_i. \tag{2.3}$$

For $p=0$ or n, equality (2.3) has, respectively, the forms

$$\sum_{i=1}^{n} d_i = 0 \quad \text{or} \quad \sum_{i=1}^{n} d_i = n(n-1).$$

(ii) If (2.3) holds, then an arbitrary realization of d is a split graph with a bipartition for which

$$(d_i : i = 1, 2, \ldots, p) \ \text{ and } \ (d_i : i = p+1, p+2, \ldots, n)$$

are the lists of vertex degrees in the upper part and the lower part, respectively.

(iii) If d is split, then the maximal p satisfying (2.3) is equal to $m(d)$.

Proof: First, let us prove (i) and (ii). Let G be an arbitrary realization of the sequence d and

$$V(G) = \{v_i : i = 1, 2, \ldots, n\}, \quad \deg v_i = d_i.$$

For a fixed p, let us denote

$$A = \{v_i : i \leq p\}, \quad B = V(G) - A$$

and partition the sum of degrees of vertices in A into two parts α and β, where α is the contribution of all edges that are not incident to vertices in B, whereas β is the contribution of all edges having an endpoint in B. Obviously,

$$\alpha \leq p(p-1) \quad \text{and} \quad \beta \leq \sum_{i=p+1}^{n} d_i.$$

Equality (2.3) holds if and only if both inequalities above hold as equalities. The latter is equivalent to saying that A is a clique and B is an independent set; that is, G is a split graph with the bipartition $V(G) = A \cup B$. Thus, both (i) and (ii) are proved.

(iii) Let G be a split realization of d, and let us fix bipartition (2.2) having the maximal number p of vertices in the upper part A. Obviously, $\deg a \geq p-1$ and $\deg b \leq p$ for $a \in A$ and $b \in B$. Further, the case $\deg b = p$ would contradict the choice of A. Hence, $p = m(d)$. □

It is convenient to consider split graphs together with fixed bipartitions. For a split graph G with bipartition (2.2), we shall call the triple (G, A, B) a *splitting* of G or a *splitted graph.*

The theorem above implies that a proper split sequence d can be divided into two parts d^A and d_B, which are the lists of vertex degrees for the upper and lower parts of its realizations, respectively. Note that one of the parts may be empty. The sequence d written in the form

$$d = (d^A; d_B)$$

is called a *splitting of d* or a *splitted sequence.* A splitted graph having d^A and d_B as the lists of vertex degrees for its upper and lower parts, respectively, is called a *realization* of the splitted sequence d.

The following statement is obvious.

Corollary 2.1 *A proper split n-sequence d with $d_{m(d)} > m(d) - 1$ has the unique splitting*

$$(d_1, d_2, \ldots, d_{m(d)}; d_{m(d)+1}, d_{m(d)+2}, \ldots, d_n). \tag{2.4}$$

If $d_{m(d)} = m(d) - 1$, the sequence d has exactly two splittings, namely, (2.4) and

$$(d_1, d_2, \ldots, d_{m(d)-1}; d_{m(d)}, d_{m(d)+1}, \ldots, d_n).$$

The concept of isomorphism of splitted graphs appears naturally. Let (G, A, B) and (H, C, D) be two splitted graphs and $f : G \to H$ be a graph isomorphism. If f *preserves the parts* (i.e. $f(A) = C$ and $f(B) = D$), then f is called an *isomorphism of splitted graphs* (G, A, B) and (H, C, D). In this case, we write $f : (G, A, B) \to (H, C, D)$ and $(G, A, B) \simeq (H, C, D)$. It may happen that $(G, A, B) \not\simeq (H, C, D)$, although $G \simeq H$; for example, two splitted graphs resulting from $K_4 - e$.

Corollary 2.2 *If* (G, A, B) *and* (H, C, D) *are splitted graphs and* $G \simeq H$, *then*

$$||A| - |C|| \leq 1. \tag{2.5}$$

Moreover, $(G, A, B) \simeq (H, C, D)$ *provided that* $|A| = |C|$.

Proof: The previous corollary immediately implies (2.5). Let $f : G \to H$ be a graph isomorphism such that $f(a) = d$ for $a \in A$, $d \in D$ and $|A| = |C| = p$. Then, $f(b) = c \in C$ for some $b \in B$. For $A' = A - \{a\}$, $B' = B - \{b\}$, $C' = C - \{c\}$ and $D' = D - \{d\}$, we have $f(A') = C'$ and $f(B') = D'$. Moreover,

$$p - 1 \leq \deg a = \deg d \leq p,$$

and a similar observation is true for c and b.

If $\deg a = p - 1$, then

$$N(a) = A', \quad b \not\sim a, \quad \deg b = p - 1, \quad N(b) = A'.$$

The transposition (a, b) is an automorphism of G. The composition $f(a, b) = g$ is a graph isomorphism $G \to H$ such that $g(A) = C$ and $g(B) = D$. In a similar way, when $\deg a = p$, we have $N(d) = C$, $N(c) = C' \cup \{d\}$ and the transposition $s = (c, d)$ is an automorphism of H such that $sf(A) = C$ and $sf(B) = D$. □

Corollary 2.3 *Every regular subset of a split graph is either a clique or an independent set.*

Proof: For a split graph G, let us fix a splitting (G, A, B) with the maximal number m of vertices in the upper part. Let U and r be a regular subset and the degree of vertices in U, respectively. Then, $r < m - 1$ implies $U \subseteq B$, and $r \geq m$ implies $U \subseteq A$. Consider the case $r = m - 1$. If $|U \cap A| \geq 2$, then $U \subseteq A$. If $U \cap A = \{a\}$, then one can relocate a into the lower part. □

Given a splitted sequence

$$d = (d^A; d_B), \quad d^A = (c_1, c_2, \ldots, c_p), \quad d_B = (d_1, d_2, \ldots, d_q), \tag{2.6}$$

we define the pair of sequences

$$(\varphi, \psi), \quad \varphi = (c_1 - p + 1, c_2 - p + 1, \ldots, c_p - p + 1), \quad \psi = d_B. \tag{2.7}$$

This pair is graphical. The correspondence

$$(d^A; d_B) \to (\varphi, \psi)$$

defines a bijection γ of the set of splitted sequences onto the set of graphical pairs. A realization $\gamma(G, A, B) = H(A, B)$ of pair (2.7) is associated with a realization (G, A, B) of sequence (2.6). Here, the graph H is obtained from G by deleting the edges in $K(A)$, i.e. $H = G - E(G(A))$.

A splitted sequence is called *unigraphical* if all its realizations are isomorphic as splitted graphs. We shall call a realization of a splitted unigraphical sequence a *splitted unigraph.* A splitted sequence (2.6) is called *biregular* if

$$c_1 = c_2 = \ldots = c_p, \qquad d_1 = d_2 = \ldots = d_q.$$

It is obvious that a splitted sequence (2.6) is unigraphical if and only if pair (2.7) is unigraphical. Therefore, Fact 3 implies the following:

Corollary 2.4 *A biregular splitted sequence (2.6) is unigraphical if and only if*

$$c_1 \in \{p-1, p, p+q-2, p+q-1\} \quad \text{or} \quad d_1 \in \{0, 1, p-1, p\}.$$

2.2 Basic Results

2.2.1 Decomposition Theorem

In what follows, graphs are considered up to isomorphism and splitted ones are considered up to isomorphism of splitted graphs. The sets of splitted graphs and simple graphs are denoted by Σ and Γ, respectively. Let us define the *composition* $\circ : \Sigma \times \Gamma \to \Gamma$ as follows: if $\sigma \in \Sigma$, $\sigma = (G, A, B)$ and $H \in \Gamma$, then

$$\sigma \circ H = (G \cup H) + \{av : a \in A, v \in V(H)\}. \tag{2.8}$$

In other words, the edge set of the complete bipartite graph with the parts A and $V(H)$ is added to the disjoint union $G \cup H$ (see Figure 2.1). If, in addition, H is

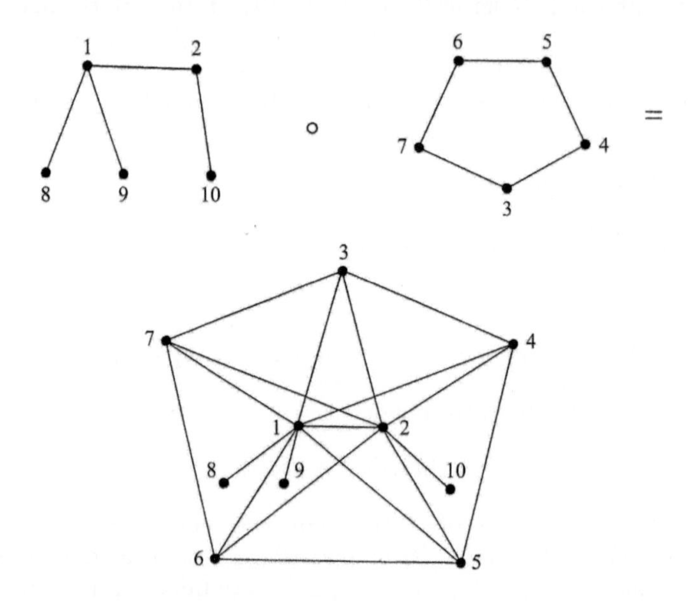

Figure 2.1 Illustration of the composition $\circ$, where $A = \{1, 2\}$ and $B = \{8, 9, 10\}$.

a split graph with a bipartition $V(H) = C \cup D$, then the composition $\sigma \circ H = F$ is split as well with the bipartition $V(F) = (A \cup C) \cup (B \cup D)$. In this case, we suppose that

$$(G, A, B) \circ (H, C, D) = (F, A \cup C, B \cup D). \tag{2.9}$$

Formula (2.9) defines a binary algebraic operation on the set Σ of triples, which is called the *multiplication* of triples. It is clear that this operation is associative. In what follows, Σ is regarded as a semigroup with multiplication (2.9).

Formula (2.8) defines an *action* of the semigroup Σ on the set of graphs; that is,

$$(\sigma \circ \rho) \circ G = \sigma \circ (\rho \circ G) \quad \text{for} \quad \sigma, \rho \in \Sigma,\ G \in \Gamma.$$

An element $\sigma \in \Sigma$ is called *decomposable* if there are $\alpha, \beta \in \Sigma$ such that $\sigma = \alpha \circ \beta$. Otherwise, σ is *indecomposable.* Analogously, a graph G is called *decomposable* if $G = \sigma \circ H$, $\sigma \in \Sigma$ and $H \in \Gamma$. Otherwise, G is *indecomposable.* Let us denote by Σ^* and Γ^* the set of indecomposable elements in the semigroup Σ and the set of indecomposable graphs, respectively.

Theorem 2.2 *Statements (i)–(iii) are true.*

(i) An n-vertex graph F with a proper degree sequence d is decomposable if and only if there exist non-negative integers p and q such that

$$0 < p + q < n, \qquad \sum_{i=1}^{p} d_i = p(n - q - 1) + \sum_{i=n-q+1}^{n} d_i. \tag{2.10}$$

(ii) Let us call a pair (p, q) satisfying the conditions (2.10) good. *With every good pair (p, q), one can associate the decomposition*

$$F = (G, A, B) \circ H, \tag{2.11}$$

where

$$(d_1, d_2, \ldots, d_p),\ (d_{p+1}, d_{p+2}, \ldots, d_{n-q}) \text{ and } (d_{n-q+1}, d_{n-q+2}, \ldots, d_n)$$

are the lists of vertex degrees for A, $V(H)$ and B, respectively. Moreover, every decomposition of form (2.11) is associated with some good pair.

(iii) Let us denote by p_0 the minimum of the first components in good pairs and set $q_0 = |\{i : d_i < p_0\}|$ if $p_0 \neq 0$ and $q_0 = 1$ if $p_0 = 0$. Then, the splitted component (G, A, B) in (2.11) is indecomposable if and only if the relevant good pair (p, q) coincides with (p_0, q_0).

Proof: Observe that for each graph of form (2.11) with $|A| = p$ and $|B| = q$, the following inequalities hold:

$$\deg a \geq n - q - 1 \geq \deg v \geq p \geq \deg b, \tag{2.12}$$

where $a \in A$, $b \in B$ and $v \in V(H)$.

For $p = 0$, condition (2.10) means that

$$0 < q < n \quad \text{and} \quad \sum_{i=n-q+1}^{n} d_i = 0.$$

So, a pair $(0, q)$ is good if and only if $d_i = 0$ for $i \geq n - q + 1$. In this case, F has q isolated vertices. By (2.12), for each decomposition of form (2.11), the part B contains an isolated vertex. Hence, the component (G, A, B) in (2.11) is indecomposable if and only if $A = \emptyset$ and $|B| = 1$; that is, the relevant good pair is $(0, 1)$.

In a similar way, a pair $(p, 0)$ is good if and only if $d_i = n - 1$ for $i \leq p$. The graph F can be represented in form (2.11) with $|A| = p$ and $B = \emptyset$. For each decomposition of form (2.11), the part A contains a dominating vertex. The component (G, A, B) is indecomposable if and only if the relevant good pair is $(1, 0)$.

Let $d_1 < n - 1$ and $d_n > 0$; that is, $p \neq 0$ and $q \neq 0$ for each pair (p, q). We set

$$V(F) = \{v_1, v_2, \ldots, v_n\}, \quad \deg\, v_i = d_i.$$

For a pair (p, q) with $p + q < n$, let us consider the sum of the degrees of the first p vertices:

$$S = \sum_{i=1}^{p} d_i.$$

Let us divide S into two parts: $S = R + T$, where R is the contribution of the edges $v_i v_j$ with $i \leq p$ and $j \leq n - q$, and T is the contribution of the edges with $i \leq p$ and $j > n - q$. It is clear that

$$R \leq p(n - q - 1) \quad \text{and} \quad T \leq \sum_{i=n-q+1}^{n} d_i.$$

The equality in (2.10) holds if and only if both inequalities above hold as equalities. The latter is true if and only if F has form (2.11) with

$$A = \{v_1, v_2, \ldots, v_p\}, \quad V(H) = \{v_{p+1}, v_{p+2}, \ldots, v_{n-q}\}$$

and

$$B = \{v_{n-q+1}, v_{n-q+2}, \ldots, v_n\}.$$

Now, let F have form (2.11) with $|A| = p \neq 0$ and $|B| = q \neq 0$, and let us consider inequalities (2.12). If $\deg\, a = n - q - 1$, then $a \not\sim B$ and

$$(G, A, B) = (G - a, A - \{a\}, B) \circ (K_1, \{a\}, \emptyset).$$

If $\deg\, b = p$, then we obtain analogously

$$(G - b, A, B - \{b\}) \circ (K_1, \emptyset, \{b\}).$$

Hence, for an indecomposable triple (G, A, B), inequalities (2.12) can be refined:

$$\deg a > n - q - 1 \geq \deg v \geq p > \deg b, \tag{2.13}$$

where $a \in A$, $b \in B$ and $v \in V(H)$. Consequently, each of the sets A, B and $V(H)$ is a union of regular subsets in F.

Further, if

$$F = (G, A, B) \circ H \quad \text{and} \quad F = (G', A', B') \circ H' \tag{2.14}$$

are two decompositions of form (2.11) with indecomposable triples (G, A, B) and (G', A', B'), then either $A \subseteq A'$ or $A' \subseteq A$. The same holds for B and B'. Let $A' \subseteq A$, and let (p, q) and (p', q') be good pairs associated with decompositions (2.14). Then, $p' \leq p$ and, by (2.13), $B' \subseteq B$. If at least one of the subsets $A - A'$ or $B - B'$ is non-empty, then

$$(G, A, B) = (G', A', B') \circ (G - A' - B').$$

This contradicts the indecomposability of (G, A, B). Hence, $A = A'$, $B = B'$ and $(p, q) = (p', q')$. Now, it is obvious that only one of the good pairs (p, q), namely, (p_0, q_0), corresponds to an indecomposable component. □

Note that operations (2.8) and (2.9) can naturally be transferred to graphical sequences. For a splitted sequence α and a graphical sequence e with realizations σ and H, respectively, let us set

$$\alpha \circ e = d, \tag{2.15}$$

where d is the degree sequence of $\sigma \circ H$. Analogously, for splitted sequences α_1 and α_2 with realizations σ_1 and σ_2, respectively, we set

$$\alpha_1 \circ \alpha_2 = \alpha, \tag{2.16}$$

where α is the splitted sequence realized by $\sigma_1 \circ \sigma_2$. A graphical sequence d is called *decomposable* if it can be represented in form (2.15). A splitted sequence α is called *decomposable* if it can be represented in form (2.16). Otherwise, the sequences are *indecomposable.*

By Theorem 2.2, if a graphical sequence d has form (2.15), then each of its realizations can be represented in form (2.11), where (G, A, B) and H are realizations of α and e.

Let us recall the following well-known graphicity criterion:

Graphicity Criterion [5] *A proper n-sequence d is graphical if and only if*

$$\sum_{i=1}^{p} d_i \leq p(n - q - 1) + \sum_{i=n-q+1}^{n} d_i \tag{2.17}$$

for each pair (p, q), where $p, q \in \{0, 1, \ldots, n\}$ and $0 < p + q \leq n$.

Inequalities (2.17) will be called the *FHM-inequalities.* Formally, the statement of this criterion slightly differs from the criterion in [5]. We added the FHM-inequalities for p or q (but not both) equal to 0 to the statement of the criterion, but it is clear that the added inequalities hold for every proper sequence. So, in fact, the above criterion is equivalent to that in [5].

Theorems 2.1 and 2.2 imply:

Corollary 2.5 *Statements (i) and (ii) are true.*

(i) A proper graphical n-sequence d is split if and only if some FHM-inequality with $q = n - p$ holds as an equality. In this case, d has a splitting

$$d = (d_1, d_2, \ldots, d_p; d_{p+1}, d_{p+2}, \ldots, d_n).$$

(ii) A proper graphical n-sequence d is decomposable if and only if some FHM-inequality with $q < n - p$ holds as an equality. In this situation,

$$\begin{aligned} d &= (d^A; d_B) \circ c, \\ d_A &= (d_1 - n + p + q, d_2 - n + p + q, \ldots, d_p - n + p + q), \\ d_B &= (d_{n-q+1}, d_{n-q+2}, \ldots, d_n), \\ c &= (d_{p+1} - p, d_{p+2} - p, \ldots, d_{n-q} - p). \end{aligned}$$

Suppose that a graph F can be represented as a composition of indecomposable components:

$$F = (G_1, A_1, B_1) \circ (G_2, A_2, B_2) \circ \ldots \circ (G_k, A_k, B_k) \circ F_0, \tag{2.18}$$

where (G_i, A_i, B_i) are indecomposable splitted graphs and F_0 is an indecomposable graph. Notice that if F is indecomposable, then there are no splitted components in (2.18). Decomposition (2.18) is called the *canonical decomposition*[1] of F.

Corollary 2.6 (Graph Decomposition Theorem)

(i) Every graph F *has canonical decomposition (2.18).*

(ii) Let graphs F and F' have canonical decompositions (2.18) and

$$F' = (G'_1, A'_1, B'_1) \circ (G'_2, A'_2, B'_2) \circ \cdots \circ (G'_l, A'_l, B'_l) \circ F'_0.$$

Then, F and F' are isomorphic if and only if the following conditions hold:

(1) $F_0 \simeq F'_0$;

(2) $k = l$;

(3) $(G_i, A_i, B_i) \simeq (G'_i, A'_i, B'_i)$ for $i = 1, 2, \ldots, k$.

Proof: Part (i) is obvious.

(ii) The sufficiency of conditions (1)–(3) for the isomorphism of F and F' is evident. Let us prove the necessity. Fix an isomorphism $f : F \to F'$. By Theorem

[1]In Chapter 1, the canonical decomposition (2.18) is called 1-decomposition.

2.2, the degrees of vertices in F that constitute the part A_1 coincide with the degrees of vertices in F' constituting A'_1. The same holds for B_1 and B'_1. Hence,

$$f(A_1) = A'_1, \quad f(B_1) = B'_1, \quad f\big(V(F) - (A_1 \cup B_1)\big) = V(F') - (A'_1 \cup B'_1).$$

Therefore, f induces the isomorphisms

$$(G_1, A_1, B_1) \to (G'_1, A'_1, B'_1)$$

and

$$F - A_1 - B_1 \to F' - A'_1 - B'_1.$$

The rest of the proof follows by induction on the number of vertices. □

Corollary 2.7 *The component H in decomposition (2.11) is indecomposable if and only if, for the associated good pair (p, q), the parameters p and q are the maxima of the first and the second coordinates in good pairs, respectively.*

By the decomposition theorem, each element σ in the semigroup Σ of splitted graphs can be uniquely decomposed into the product

$$\sigma = \sigma_1 \circ \sigma_2 \circ \ldots \circ \sigma_k, \quad k \geq 1, \quad \sigma_i \in \Sigma^*,$$

and every decomposable graph F can be uniquely represented as the decomposition

$$F = \sigma \circ F_0, \quad \sigma \in \Sigma, \quad F_0 \in \Gamma^*$$

of the *splitted part* σ and the *indecomposable part* F_0. In other words, the following corollary holds.

Corollary 2.8 *The set Σ of splitted graphs is a free semigroup over the alphabet Σ^* with respect to multiplication (2.9). A free action of this semigroup is defined by (2.8).*

Now, let us turn to canonical decomposition (2.18). The sequence

$$F_1, F_2, \ldots, F_k, F_0 \tag{2.19}$$

with

$$F_i = \begin{cases} G_i & \text{if} \quad |G_i| > 1 \\ (G_i, A_i, B_i) & \text{if} \quad |G_i| = 1 \end{cases}$$

for $i = 1, 2, \ldots, k$ is called the *canonical sequence* for F.

Theorem 2.3 *The canonical sequence determines a graph up to isomorphism.*

The theorem follows immediately from Corollary 2.6 and Lemma 2.1.

Lemma 2.1 *The following statements are true:*

(i) An indecomposable split sequence with at least two members has exactly one splitting.

(ii) Any non-trivial indecomposable split graph has exactly one splitting.

Proof: (i) Let an n-sequence d have two distinct splittings. By Corollary 2.1, one of them has the form

$$(d_1, d_2, \ldots, d_{m-1}; d_m, d_{m+1}, \ldots, d_n),$$

where $m = m(d)$ and $d_m = m - 1$. If (G, A, B) is a realization of the splitting and $v \in B$ with $\deg v = d_m$, then

$$G = (G - v, A, B - \{v\}) \circ K_1$$

is a decomposable graph. Thus, (i) is proved.

Statement (ii) follows from (i) and Corollary 2.2. □

Corollary 2.6 implies the following result:

Corollary 2.9 (Decomposition Theorem for Sequences)

(i) Each proper graphical sequence d can be uniquely represented as a composition of indecomposable components:

$$d = \alpha_1 \circ \alpha_2 \circ \cdots \circ \alpha_k \circ d_0,$$

where α_i are indecomposable splitted sequences and d_0 is an indecomposable graphical sequence.

(ii) An arbitrary realization F of the sequence d can be represented in form (2.18), where (G_i, A_i, B_i) and F_0 are realizations of α_i and d_0, respectively. All combinations of realizations (G_i, A_i, B_i) and F_0 yield realizations of d.

Next, let us define two unary algebraic operations in the semigroup Σ, *complement* and *inverting*, as follows:

$$\overline{(G, A, B)} = (\overline{G}, B, A), \qquad (G, A, B)^{\mathrm{I}} = (G^{\mathrm{I}}, B, A).$$

Here, $\overline{G}$ is the complementary graph of G, and G^{I} is obtained from G by deleting the set of edges $\{a_1a_2 : a_1, a_2 \in A\}$ and adding the set of edges $\{b_1b_2 : b_1, b_2 \in B\}$.

The graph G^{I} is not uniquely determined by G, it depends upon a bipartition choice as well. For instance, $G = K_4 - e$ yields two graphs: $G^{\mathrm{I}} = P_3 \cup K_1$ and $G^{\mathrm{I}} \simeq G$. The former graph is obtained from G whose bipartition has a three-vertex upper part, and the latter is obtained from G having a two-vertex upper part. Nevertheless, by Lemma 2.1 (ii), any non-trivial indecomposable split graph has exactly one splitting. Hence, G^{I} is uniquely determined for a non-trivial indecomposable split graph G.

Obviously, the complement and the inverting are involutory operations, and the complement commutes with the composition. In other words,

$$\overline{(\overline{\sigma})} = \sigma, \quad (\sigma^{\mathrm{I}})^{\mathrm{I}} = \sigma, \quad \overline{\sigma \circ \tau} = \overline{\sigma} \circ \overline{\tau}, \quad \overline{\sigma \circ H} = \overline{\sigma} \circ \overline{H} \quad \text{for} \quad \sigma, \tau \in \Sigma,\ H \in \Gamma.$$

Furthermore, $(\sigma \circ \tau)^{\mathrm{I}} = \tau^{\mathrm{I}} \circ \sigma^{\mathrm{I}}$. Hence, both complement and inverting preserve indecomposability.

2.2.2 Main Result for Unigraphs

The decomposition theorem implies the following:

Corollary 2.10 *A graph F with canonical sequence (2.19) is a unigraph if and only if all graphs G_i and the indecomposable part F_0 are unigraphs.*

Thus, in order to obtain a description of the structure of an arbitrary unigraph, it is sufficient to describe the indecomposable unigraphs. Let us introduce five classes of graphs as follows.

(1) For $m \geq 1$, $n \geq 2$, let $U_2(m, n)$ denote the disjoint union of the perfect matching mK_2 and the star $K_{1,n}$ (see Figure 2.2):

$$U_2(m, n) = mK_2 \cup K_{1,n}.$$

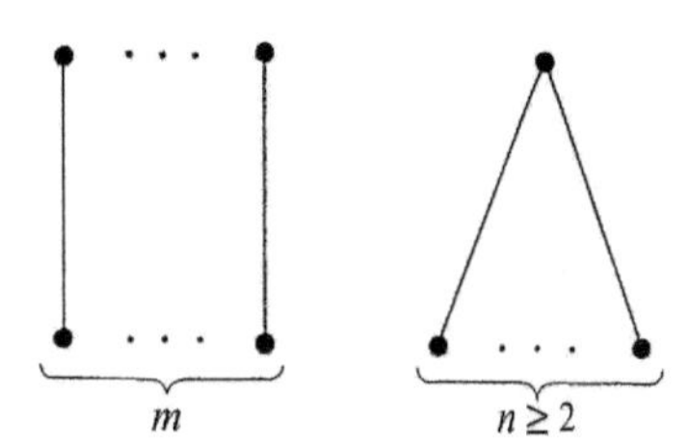

Figure 2.2 Graph $U_2(m, n)$.

(2) Let us choose a vertex in each connected component of the disjoint union of the chordless cycle C_4 and m triangles K_3, $m \geq 1$, and 'glue' together the components in such a way that the chosen vertices become a single common vertex. The resulting graph is denoted $U_3(m)$ (see Figure 2.3).

(3) Let us consider a *multistar*; that is, a split graph S with a bipartition $V(S) = A \cup B$ such that $|A| = l \geq 2$ and $\deg b = 1$ for $b \in B$. The graph S can be obtained from the disjoint union of l arbitrary stars by adding all the edges connecting the centres of the stars (only one vertex of K_2 is regarded as the central vertex). If

$$\{K_{1,p_i} : i = 1, 2, \ldots, r\}$$

is the set of non-isomorphic stars among the components of the above union and the star K_{1,p_i} occurs in the union exactly q_i times, then we set

$$S = S_2 = S_2(p_1, q_1; p_2, q_2; \ldots; p_r, q_r),$$

where

$$p_i \geq 1, \quad q_i \geq 1, \quad r \geq 1, \quad q_1 + q_2 + \cdots + q_r = l \geq 2$$

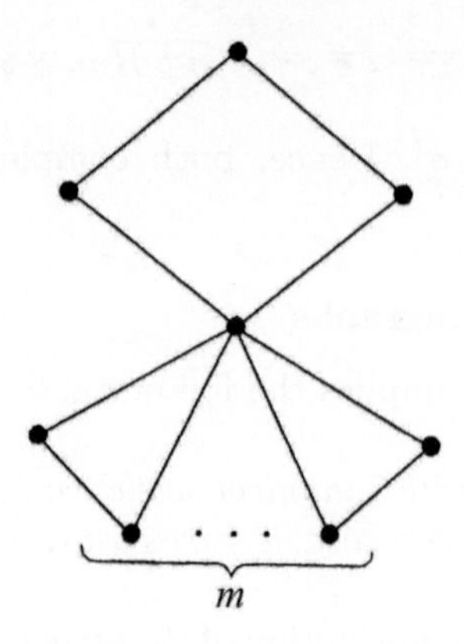

Figure 2.3 Graph $U_3(m)$.

(see Figure 2.4). In particular, for $r=1$ and $q \geq 2$, we denote $S_2(p,q)$ by $S(p,q)$ (see Figure 2.5). Note that in Figures 2.4–2.7, the edges of the upper parts are not shown.

(4) Let us consider the graph S_2 with

$$r=2, \quad p_1=p \geq 1, \quad p_2=p+1, \quad q_1 \geq 2, \quad q_2 \geq 1.$$

We add a new vertex e to the lower part B and also add the following set of q_1 edges:

$$\{ea : a \in A,\ \deg_B a = p\}.$$

The resulting graph is denoted by

$$S_3 = S_3(p, q_1, q_2)$$

(see Figure 2.6), where $p \geq 1$, $q_1 \geq 2$ and $q_2 \geq 1$.

(5) In the graph S_3, we set $q_1 = 2$ and $q_2 = q \geq 1$. Now, add a new vertex f to $S_3(p,2,q)$ and connect it by edges with all vertices except e. Denote the obtained graph by

$$S_4 = S_4(p,q)$$

(see Figure 2.7), where $p \geq 1$ and $q \geq 1$.

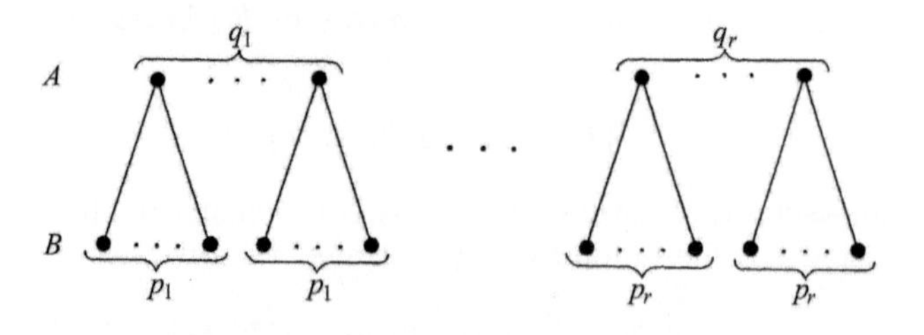

Figure 2.4 Graph $S_2(p_1, q_1; \ldots; p_r, q_r)$.

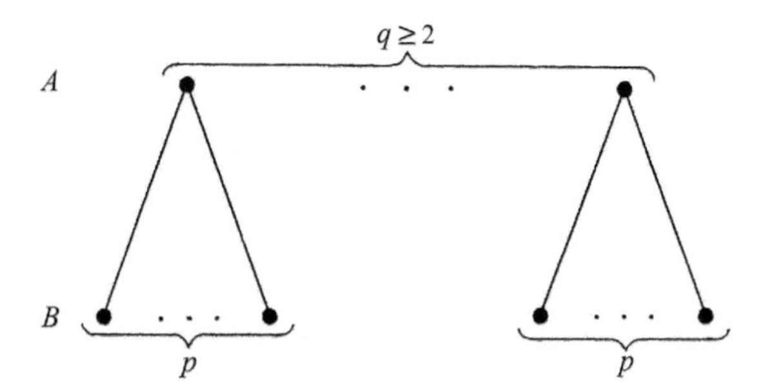

Figure 2.5 Graph $S(p, q)$.

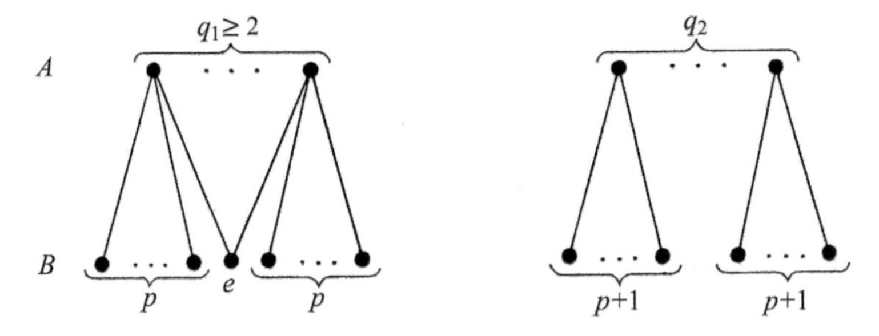

Figure 2.6 Graph $S_3(p, q_1, q_2)$.

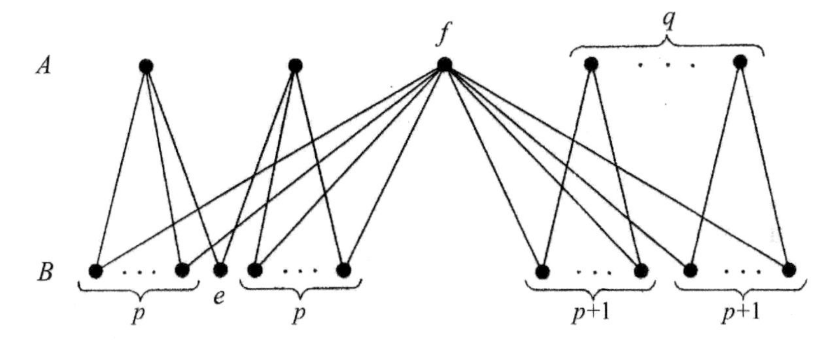

Figure 2.7 Graph $S_4(p, q)$.

Theorem 2.4 *Statements (i)–(iii) are true.*

(i) Decomposable unigraphs are graphs of the form

$$(G_1, A_1, B_1) \circ (G_2, A_2, B_2) \circ \ldots \circ (G_k, A_k, B_k) \circ G,$$

where $k \geq 1$, (G_i, A_i, B_i) *run independently from one another over the set of indecomposable splitted unigraphs and* G *runs over the set of all indecomposable unigraphs.*

(ii) An indecomposable non-split graph G *is a unigraph if and only if either* G *or* $\overline{G}$ *is contained in the following list:*

$$C_5; \ mK_2, \ m \geq 2; \ U_2(m, n); \ U_3(m). \tag{2.20}$$

(iii) An indecomposable split graph G is a unigraph if and only if at least one of the graphs G, $\overline{G}$, G^{I} or $\overline{G}^{\mathrm{I}}$ belongs to the following list:

$$K_1; \quad S(p,q); \quad S_2, \ r \geq 2; \quad S_3(p,q_1,q_2); \quad S_4(p,q). \tag{2.21}$$

Statement (i) of the theorem follows immediately from the decomposition theorem. Obviously, if G is a unigraph, then $\overline{G}$ is a unigraph too. If, in addition, G is split and indecomposable, then G^{I} is split and indecomposable as well.

Evidently, the graphs in list (2.20) are non-split, whereas those in list (2.21) are split. One easily observes that all the graphs in lists (2.20) and (2.21) are indecomposable unigraphs. This is obvious for K_1, C_5, mK_2, $U_3(m)$, $U_2(m,n)$ and S_2. Let us consider S_3. If

$$S_3 = (G, X, Y) \circ H, \tag{2.22}$$

then for $S_2 = S_3 - e$ we have $A \subseteq X$ and $B \subseteq Y$, since S_2 is indecomposable. The part A contains vertices both adjacent and non-adjacent to e, hence $e \notin V(H)$, $V(H) = \emptyset$, (2.22) is impossible and S_3 is indecomposable.

Let us show that S_3 is a unigraph. It is sufficient to prove that $S_3 \simeq tS_3$ for each 2-switch t (Fact 4). A similar statement holds for $S_2 = S_3 - e$, so one has to consider 2-switches t touching e, that is,

$$t = a_1 a_2 e b, \quad a_1, a_2 \in A, \quad b \in B.$$

Obviously, $tS_3 \simeq S_3$.

Similar arguments prove that S_4 is an indecomposable unigraph. It remains to prove the completeness of lists (2.20) and (2.21). The proof is based on the series machinery developed in Section 2.2.3 and is given in Sections 2.3.1 and 2.3.2.

Corollary 2.11 *A unigraph G is not perfect if and only if the chordless pentagon C_5 is the last indecomposable component in the canonical decomposition of G. Hence, G is perfect if and only if C_5 is not an induced subgraph of G.*

Proof: Evidently, a graph is perfect if and only if each of its indecomposable components is perfect. On the other hand, all split graphs are perfect and C_5 is the unique non-perfect graph in list (2.20). □

2.2.3 Technical Lemmas

For an arbitrary graph G and a non-negative integer s, we will define a sequence of subgraphs

$$G^0, G^1, \ldots, G^s \tag{2.23}$$

with vertex sets

$$V^0 \supseteq V^1 \supseteq \ldots \supseteq V^s, \tag{2.24}$$

respectively. Let us set

$$V^0 = V(G) \quad \text{and} \quad G^0 = G.$$

Next, let V^k and G^k be defined for all $k \leq i < s$. We denote by $|v|_i$ the degree of v in G^i. The *regular subset of rank* i in G (at a vertex $a \in V^i$) is the set

$$V_a^i = \{v \in V^i : |v|_k = |a|_k,\ k = 0, 1, \ldots, i\}.$$

Taking liberty with the language, in what follows a regular subset of rank i in G will be called a *regular subset of* G^i. Let us take some regular subset in G^i or a union of several regular subsets as V^{i+1}. We set

$$G^{i+1} = G(V^{i+1}).$$

In the following Lemmas 2.2–2.8, let G be a unigraph with a fixed sequence (2.23). In general, a subgraph of a unigraph is not a unigraph. Let us consider, for instance, the unigraph $U_3(1)$ defined in Section 2.2.2. There are two non-isomorphic graphs with the same degree sequence $(3, 2^3, 1)$ among the subgraphs of the form $U_3(1) - v$, where v is adjacent to the vertex of degree 4. Nevertheless, the following holds.

Lemma 2.2 *If* $V(F) = V^i$ *for a graph* F *and* $\deg_F v = |v|_i$ *for each vertex* $v \in V(F)$, *then* $F \simeq G^i$ *and*

$$F(V^j) \simeq G^j \text{ for } i < j \leq s.$$

Proof: Let us complement F to F^0, adding those vertices and edges that complement G^i to G. The degree of each vertex in F^0 is equal to that in G and, consequently, $G \simeq F^0$. Every isomorphism $f : G \to F^0$ preserves vertex degrees, so

$$f(V^i) = V^i.$$

The restriction of f to V^i is an isomorphism of graphs $G^i \to F^0(V^i) = F$ preserving regular subsets, $f(V^j) = V^j$ and $F(V^j) \simeq G^j$ for $j > i$. □

Corollary 2.12 *Each term of sequence (2.23) is a unigraph.*

Lemma 2.3 *If*

$$a, b \in V^{i+1} \quad \text{and} \quad |a|_i = |b|_i, \tag{2.25}$$

then

$$||a|_{i+1} - |b|_{i+1}| \leq 1.$$

Proof: Suppose that $|a|_{i+1} > |b|_{i+1}$. Then, there is $c \in V^{i+1}$ such that $c \sim a$ and $c \not\sim b$. Because (2.25) holds, there exists $d \in V^i - V^{i+1}$ for which $d \not\sim a$, $d \sim b$ and G^i admits the 2-switch $t = abcd$. Let us denote

$$F = tG^i \quad \text{and} \quad F^{i+1} = F(V^{i+1}).$$

Since $F = G^i - ac - bd + ad + bc$, we obtain $F^{i+1} = G^{i+1} - ac + bc$. Therefore, if α and β are degrees of the vertices a and b in F^{i+1}, respectively, then

$$\alpha = |a|_{i+1} - 1 \quad \text{and} \quad \beta = |b|_{i+1} + 1.$$

All other vertices have the same degrees in F^{i+1} as in G^{i+1}. However, by Lemma 2.2, $G^{i+1} \simeq F^{i+1}$, hence $|b|_{i+1} = \alpha = |a|_{i+1} - 1$. □

In what follows, we write aib if the vertices a and b are contained in the same regular subset of G^i.

Lemma 2.4 *If G^i admits the 2-switch abcd, then*

$$aib \ \text{ or } \ cid. \tag{2.26}$$

Proof: Let G^i admit the 2-switch $t = abcd$. First, we prove that one of the following holds:

$$aib, \ bic, \ cid, \ dia. \tag{2.27}$$

Let us set

$$tG^i = F, \quad F(V^i_a \cup V^i_c) = F^{i+1}, \quad G(V^i_a \cup V^i_c) = H. \tag{2.28}$$

If none of the statements in (2.27) holds, then $F^{i+1} = H - ac$. By Lemma 2.2, $F^{i+1} \simeq H$. Taking d instead of c in (2.28), we observe that one of the following holds:

$$aib, \ bid, \ dic, \ cia. \tag{2.29}$$

Comparing (2.27) and (2.29), we obtain (2.26). □

Lemma 2.5 *If G^i admits the 2-switch abcd and aib, then the neighbourhoods in $G(V^0 - V^i)$ of the vertices in the regular subset V^i_a coincide.*

Proof: Let us denote by $N_i(v)$ the neighbourhood of v in $G(V^0 - V^i)$ and show that

$$N_i(v) \subseteq N_i(a) \tag{2.30}$$

for $v \in V^i_a$. On the contrary, let there exist $v \in V^i_a$ and $e \in V^0 - V^i$ such that $v \sim e$, $a \not\sim e$; and denote $j = \max\{k : e \in V^k\}$. It is clear that $j < i$. There are three cases to consider:

(1) $v \neq c$ and $v \not\sim c$. In this case, G^j admits the 2-switch $t = avce$. Taking Lemma 2.2 into account, we obtain

$$F = tG^j = G^j - ac - ve + ae + vc, \quad F(V^{j+1}) = G^{j+1} - ac + vc \simeq G^{j+1}.$$

The latter contradicts Lemma 2.3 because $a(j+1)v$. Hence, (2.30) holds.

(2) $v \sim c$. The inclusion (2.30) holds for $v = b$, since this situation is descried by case (1). Hence, $v \neq b$ and $b \not\sim e$ as $a \not\sim e$. Because vib, $v \sim c$ and $b \not\sim c$, there exists $f \in V_i$ such that $f \neq v$, $f \sim b$ and $f \not\sim v$. The graph G^j admits the 2-switch $bvfe$ and

$$b, f, v \in V^{j+1}, \quad e \notin V^{j+1}, \quad b(j+1)v.$$

As above, the latter yields a contradiction. Hence, (2.30) holds.

(3) $v = c$. We have $e \in V^j - V^{j+1}$ and $c \sim e$, $a \not\sim e$, $b \not\sim e$. Let us consider the closed neighbourhoods $N[a]$ and $N[c]$ of a and c in G^{j+1}. If $N[a] \subseteq N[c]$, then $N[a] = N[c]$ as $|a|_{j+1} = |c|_{j+1}$. Hence, $c \not\sim d$ because $a \not\sim d$. Again, we have a contradiction as G^j admits the 2-switch

$$cbed, \quad b, c, d \in V^{j+1}, \quad e \notin V^{j+1}, \quad b(j+1)c.$$

Consequently, $N[a] \not\subseteq N[c]$. But $a \sim c$, so there exists $g \in V^{j+1}$ such that $g \neq a$, $g \neq c$, $g \sim a$ and $g \not\sim c$. Now, we have a contradiction because G^j admits the 2-switch

$$acge, \quad a, c, g \in V^{j+1}, \quad e \notin V^{j+1}, \quad a(j+1)c.$$

Inclusion (2.30) is proved.

The complement $\overline{G}$ is a unigraph as well as G. The sequence

$$\overline{G}^0, \overline{G}^1, \ldots, \overline{G}^s$$

has the same meaning for $\overline{G}$ as (2.23) for G, and sequence (2.24) remains valid. The graph $\overline{G}^i$ admits the 2-switch $abdc$ and aib holds as above. Therefore, according to (2.30),

$$\overline{N}_i(v) \subseteq \overline{N}_i(a),$$

where $\overline{N}_i(v)$ is the neighbourhood of v in $\overline{G}(V^0 - V^i)$. Hence, $N_i(a) \subseteq N_i(v)$ and so $N_i(a) = N_i(v)$ for any vertex $v \in V^i_a$. □

If a graph H admits a 2-switch $t = abcd$ where $a \in A \subseteq V(H)$, then we shall say that t is a *2-switch of type Abcd.* 2-Switches of types $ABcd$, $ABCd$ etc., where $B, C, D \subseteq V(H)$, are defined analogously.

Lemma 2.6 *If G^i admits the 2-switch abcd and aib, then there exists a partition of V^{i-1}:*

$$V^{i-1} = V^i \cup B \cup C, \quad B \sim V^i_a, \quad C \not\sim V^i_a. \tag{2.31}$$

Furthermore, if the regular subset V^i_a has two non-adjacent vertices, then B is a clique; if V^i_a has two adjacent vertices, then C is an independent set.

Proof: The existence of partition (2.31) is guaranteed by Lemma 2.5. A 2-switch of type $BV_a^iV_a^iB$ is excluded by Lemma 2.4, so B is a clique if V_a^i has non-adjacent vertices. The arguments for C are similar taking into account that 2-switches of type $CV_a^iCV_a^i$ are excluded. □

Below, $V_a = V_a^0$ is the regular subset of G^0 at a vertex a.

Lemma 2.7 *Assume that every regular subset of a unigraph G is either a clique or an independent set.*

(i) If the regular subsets V_a and V_b are independent and

$$|V_a| > 1, \ \ |V_b| > 1, \tag{2.32}$$

then $V_a \cup V_b$ is independent too.

(ii) If the regular subsets V_a and V_b are cliques and (2.32) holds, then $V_a \cup V_b$ is a clique as well.

Proof: (i) Assume that $a \sim b$ and $\deg a > \deg b$, and take $a' \in V_a$, $a' \neq a$. There exists $c \in V(G)$ such that $c \sim a'$ and $c \not\sim b$ because $\deg a' > \deg b$, $a' \not\sim a$ and $b \sim a$. So, G admits the 2-switch $acba'$, but $\deg a \neq \deg c$ as V_a is independent and $\deg b \neq \deg a'$. The latter contradicts Lemma 2.4. Thus, (i) is proved. To obtain (ii), it is enough to consider the complement $\overline{G}$. □

A vertex a is called *singular* if the regular subset V_a is not a clique; in particular, $|V_a| > 1$. The following corollary is obvious.

Corollary 2.13 *The set of singular vertices of the unigraph in Lemma 2.7 is independent.*

Lemma 2.8 *A unigraph is split if and only if each of its regular subsets is either a clique or an independent set.*

Proof: Suppose that each regular subset in a unigraph G is either a clique or an independent set. Taking Corollary 2.3 into account, it is necessary to prove that G is split. Let C be a maximum clique in G meeting the minimal number of regular subsets. First, let us show that if $a \in C$ and the regular subset V_a is a clique, then

$$V_a \subseteq C. \tag{2.33}$$

Indeed, let there exist

$$a_1 \in V_a, \ \ c \in C, \ \ a_1 \not\sim c. \tag{2.34}$$

Then, V_c is independent; otherwise, by Lemma 2.7, the union $V_a \cup V_c$ would be a clique contradicting (2.34). Hence, $V_c \cap C = \{c\}$. If $a_1 \sim C - \{c\}$, then $C - \{c\} \cup \{a_1\}$ is a maximum clique meeting fewer regular subsets than C. This contradicts the choice of C. So, there exists $c_1 \in C$ such that $c_1 \neq c$, $c_1 \not\sim a_1$. Similar to V_c, the set V_{c_1} is independent. Since $c \sim c_1$, at least one of the sets V_c and V_{c_1} is a one-vertex set by Lemma 2.7. Suppose that

$$|V_{c_1}| = 1. \tag{2.35}$$

Since (2.34) holds and $a \sim c$, there exists $b \in V(G)$ such that $b \sim a_1$ and $b \not\sim a$. The graph G admits the 2-switch ba_1ca for $b \sim c$ and the 2-switch ca_1c_1b for $b \not\sim c$. This contradicts Lemma 2.4. Indeed, $\deg c \neq \deg a_1 = \deg a$ by (2.34), $\deg b \neq \deg c_1$ by (2.35) and $\deg a_1 \neq \deg b$ because $b \not\sim a$. Inclusion (2.33) is proved.

Now, let G be non-split. We will prove that every maximum clique C of G meeting the minimal number of regular subsets has just one singular vertex. Let us denote $B = V(G) - C$. Since G is non-split, B contains two adjacent vertices u and v. Obviously, C contains no vertex c such that $u \sim C - \{c\}$ and $v \sim C - \{c\}$ simultaneously because the contrary would contradict the choice of C. Hence, there exist $c_1, c_2 \in C$ such that $c_1 \neq c_2$, $c_1 \not\sim u$ and $c_2 \not\sim v$. Therefore, G admits the 2-switch uc_2vc_1. Taking into account Lemma 2.4 and the symmetry of 2-switch, suppose that $\deg u = \deg c_2$. We have $u \notin C$, so u and c_2 are singular by (2.33). By Corollary 2.13, no clique has more than one singular vertex. According to the same corollary, v is not singular, so $V_{c_2} \neq V_v$. Therefore, if $v \sim C - \{c_2\}$, then $C' = C - \{c_2\} \cup \{v\}$ is a maximum clique intersecting the same number of regular subsets as C but not containing a singular vertex. This is impossible as stated above. Hence, C has a vertex $c_3 \neq c_2$ satisfying $v \not\sim c_3$. Now, using the same argument as before, we obtain that c_1 or c_3 is singular, contradicting the fact that c_2 is the only singular vertex in C. □

2.3 Indecomposable Unigraphs

2.3.1 Indecomposable Non-split Unigraphs

For graphs G and H, we shall write $G \equiv H$ if $G \simeq H$ or $\overline{G} \simeq H$. Let G be an arbitrary indecomposable non-split unigraph. It is proved in this section that $G \equiv H$, where H is a graph from list (2.20). In the case of regular graphs, this follows immediately from Fact 2.

Let G be an arbitrary unigraph. Suppose that, in each step when constructing (2.23), we take a regular subset of G^i as V^{i+1}. Then, this sequence of graphs stabilizes on a regular graph; that is, we obtain the following sequence:

$$G^0, G^1, \ldots, G^l, \tag{2.36}$$

where $G^{i+1} \neq G^i$ and, for $i < l$, all G^i are not regular whereas G^l is regular. We shall call such a sequence a *series* of G.

Lemma 2.3 implies the following result:

Corollary 2.14 *If (2.36) is a series of a unigraph G, then for $1 \leq i \leq l-1$ all G^i are biregular. Moreover, the difference of any two vertex degrees in G^i is at most 1.*

We shall call the number l the *length* of series (2.36). The maximal length of all such series is called the *step* $l(G)$. The equality $l(G) = 0$ means that G is regular. All such unigraphs are described in [7, 11] (Fact 2). Below, we consider indecomposable non-split unigraphs, which are not regular. Series (2.36) of length $l(G)$ in which G^l admits a 2-switch will be called *good.* If G has no such series, then all its series of length $l(G)$ will be called *good.*

For an indecomposable non-split non-regular unigraph G, let us fix a good series (2.36). We have $l \geq 1$. The following notation will be used, unless otherwise stated: x, x_k or x^k is a vertex that belongs to a subset X, X_k or X^k, respectively.

Lemma 2.9 *Statements (i)–(iii) are true.*

(i) $l \geq 2$.

(ii) $G^{l-2} \equiv U_2(m,n)$ *for appropriate* m *and* n.

(iii) $l(G) = 2$ *for an indecomposable non-split unigraph* G *if and only if* $G \equiv U_2(m,n)$.

Proof:

(i) First, let us consider G^{l-1}. There are two possibilities: (a) G^l admits a 2-switch; (b) G^l has no 2-switches.

(a) Since G^l is regular, by Lemma 2.6, we have the following partition for V^{l-1}:

$$V^{l-1} = V^l \cup B \cup C, \quad B \sim V^l, \quad C \not\sim V^l,$$

where B is a clique and C is an independent set. Obviously, $G^{l-1} = (G - V^l, B, C) \circ G^l$. Hence, it is a decomposable graph and, consequently, $l \geq 2$.

(b) Let

$$V^{l-1} = A_1 \cup A_2 \cup \ldots \cup A_k$$

be a partition of V^{l-1} into regular subsets. One can choose each of A_i, $i = 1, 2\ldots, k$ as V^l. Since series (2.36) is good, $G_i = G(A_i)$ is regular and admits no 2-switch. By Fact 2, G_i is either complete or empty. Then, by Lemma 2.8, G^{l-1} is split. Since G^0 is non-split, we obtain $l \geq 2$ and (i) is proved.

(ii) By Corollary 2.14, G^{l-1} is biregular. In case (a), $B \cup C$ is a regular subset in G^{l-1} because V^l is a regular subset. So, one of the sets B and C is empty. Without loss of generality, let $C \neq \emptyset$. We obtain

$$|c|_{l-1} = 0, \quad |v^l|_{l-1} = |v^l|_l = 1, \quad G^l = mK_2, \quad G^{l-1} = mK_2 \cup O_n, \qquad (2.37)$$

where $m \geq 2$ and $n \geq 1$. In case (b), suppose that A and B are the upper and lower parts of G^{l-1}, respectively, $|A| = p$ and $|B| = q$. Then, according to Corollary 2.4,

$$|b|_{l-1} \in \{0, 1, p-1, p\} \quad \text{or} \quad |a|_{l-1} \in \{p-1, p, p+q-2, p+q-1\}.$$

Further, $|a|_{l-1} - |b|_{l-1} = 1$. Therefore, when constructing the complement, there are two possibilities:

$$|a|_{l-1} = 1, \quad |b|_{l-1} = 0, \quad G^{l-1} = K_2 \cup O_n; \tag{2.38}$$

or

$$|a|_{l-1} = 2, \quad |b|_{l-1} = 1, \quad G^{l-1} = P_4.$$

In fact, the second possibility does not occur. Indeed, for $G^{l-1} = P_4$, one can assume that

$$V^{l-1} = V^l \cup D, \quad |v^l|_{l-1} = 2, \quad |d|_{l-1} = 1.$$

The graph G^{l-1} admits a 2-switch of type V^lV^lDD, and V^l and D are regular subsets of G^{l-1}. By Lemma 2.6, V^{l-2} has the following partition:

$$V^{l-2} = V^{l-1} \cup B \cup C, \quad B \sim V^l, \quad C \not\sim V^l,$$

and C is independent. In a similar way, there exists the partition

$$V^{l-2} = V^{l-1} \cup E \cup F, \quad E \sim D, \quad F \not\sim D,$$

where E is a clique. Let us set $E = B_1 \cup C_1,\ B_1 \subseteq B,\ C_1 \subseteq C$.

Since E is a clique and C is independent, we have $|C_1| \leq 1$. Now,

$$2 + |B| = |v^l|_{l-2} = |d|_{l-2} = 1 + |E|,$$

$$|E| = |B| + 1 = |B_1| + |C_1|.$$

So, $|C_1| = 1$, $B_1 = B$ and $c_1 \sim B$. We obtain

$$|c_1|_{l-2} = |D| + |B| = 2 + |B| = |v^l|_{l-2}.$$

This yields a contradiction as c_1 and v^l belong to different regular subsets of G^{l-2}.

Thus, it remains to consider situations (2.37) and (2.38). We have

$$G^l = mK_2, \quad G^{l-1} = mK_2 \cup O_n, \ m, n \geq 1, \quad V^{l-1} = V^l \cup A, \quad A \not\sim V^l$$

and A is an independent regular subset in G^{l-1}.

Let us consider the partition

$$V^{l-2} = V^l \cup A \cup B \cup C \cup D, \tag{2.39}$$

where B and C are the subsets in $V^{l-2} - V^{l-1}$ maximal with respect to the conditions $B \sim V^l$, $C \not\sim V^l$. The set C is independent, since otherwise G^{l-2}

would admit a 2-switch of type V^lCV^lC contradicting Lemma 2.4 ($i = l - 2$). We have

$$|v^l|_{l-1} > |a|_{l-1} \quad \text{and} \quad |v^l|_{l-2} = |a|_{l-2}.$$

Hence, there exists

$$x \in C \cup D, \quad x \sim a, \quad x \not\sim v^l. \tag{2.40}$$

Now, let us consider the following two cases separately: (1) $m \geq 2$ and (2) $m = 1$. In case (1), by Lemma 2.5 ($i = l - 1$),

$$D = \emptyset, \tag{2.41}$$

so $x \in C$. In case (2), let us put

$$V^l = \{v_1, v_2\}, \quad D = D_1 \cup D_2, \quad D_i \sim v_i, \; i = 1, 2.$$

It follows from the definition of partition (2.39) that $D_1 \cap D_2 = \emptyset$. Hence, $D_i \not\sim v_j$ for $i \neq j$.

Assume that $a_0 \not\sim C$ for some $a_0 \in A$. Condition (2.40) implies that there exist $d_i \sim a_0$, $i = 1, 2$. The graph G^{l-2} admits the 2-switch $d_1v_1d_2v_2$ for $d_1 \sim d_2$ and the 2-switch $d_1v_2a_0d_2$ for $d_1 \not\sim d_2$. The latter contradicts Lemma 2.4 because

$$|a|_{l-2} \neq |d|_{l-2} \neq |v^l|_{l-2}.$$

So, for each vertex a there exists $c \sim a$.

Let us return to the general situation. The sets A and C are independent and the 2-switches of type $ACCA$ are forbidden by Lemma 2.4 ($i = l - 2$). Therefore, $n = |A| = 1$ or there exists $c_1 \in C$ such that $c_1 \sim A$ and $A \not\sim C - \{c_1\}$. In both cases, we have the following partition of C:

$$C = C_1 \cup C_2, \quad C_1 \sim A, \quad C_2 \not\sim A, \quad C_1 \neq \emptyset,$$

and either $|C_1| = 1$ or $n = 1$. The graph G^{l-2} admits a 2-switch of type $c_1V^ldV^l$ for $c_1 \sim d$ and a 2-switch of type dAV^lc_1 for $c_1 \not\sim d$. Both situations contradict Lemma 2.4, hence (2.41) holds in both cases. The same lemma excludes the 2-switches of type V^lC_1BA, so $B \sim C_1$. We have

$$|v|_{l-2} = 1 + |B| \neq |c_1|_{l-2} = n + |B|.$$

Hence, $n \neq 1$ and, therefore, $|C_1| = 1$.

Let us fix a vertex a and represent B in the form $B = B_1 \cup B_2$, $B_1 \sim a$, $B_2 \not\sim a$. Now,

$$1 + |B_1| = |a|_{l-2} = |v|_{l-2} = 1 + |B|,$$

and we have $B_1 = B$, $B_2 = \emptyset$ and $B \sim A$. Therefore, B is a clique because the 2-switches of type BV^lAB would contradict Lemma 2.4. Hence,

$$V^{l-2} = V^l \cup A \cup B \cup \{c_1\} \cup C_2,$$

where B is a clique, A and $\{c_1\} \cup C_2$ are independent sets and

$$B \sim V^l, \quad B \sim A, \quad B \sim c_1, \quad C_2 \not\sim V^l, \quad C_2 \not\sim A.$$

If $B \cup C_2 \neq \emptyset$, then

$$G^{l-2} = (G(B \cup C_2), B, C_2) \circ (G^{l-2} - B - C_2).$$

Hence, $l \geq 3$ because G is indecomposable. We have

$$|a|_{l-2} = |B| + 1, \quad |b|_{l-2} \geq |B| - 1 + 2m + n + 1 = |B| + 2m + n \geq |B| + 4.$$

This contradicts Lemma 2.3 unless $B = \emptyset$. Furthermore,

$$|a|_{l-2} = 1, \quad |c_1|_{l-2} = n \geq 2 \quad \text{and} \quad |c_2|_{l-2} = 0.$$

Hence, $C_2 = \emptyset$ by Lemma 2.3, and

$$V^{l-2} = V^l \cup A \cup \{c\}, \quad A \not\sim V^l, \quad c \not\sim V^l, \quad c \sim A,$$

A is independent, $n = |A| \neq 1$. Thus, it is proved that

$$G^{l-2} = mK_2 \cup K_{1,n} = U_2(m,n).$$

(iii) It was observed in Section 2.2.2 that $U_2(m,n)$ is an indecomposable unigraph for any $m \geq 1$ and $n \geq 2$. Evidently, $l\,(U_2(m,n)) = 2$. □

Lemma 2.10 *If $l(G) \geq 3$, then*

(i) $l(G) = 3$;
(ii) $G \equiv U_3(m)$;
(iii) $l(U_3(m)) = 3$ *for any* $m \geq 1$.

Proof: (i) Taking into account Lemma 2.9, we may assume without loss of generality that

$$G^{l-2} = mK_2 \cup K_{1,n}, \quad G^{l-1} = mK_2 \cup O_n, \quad G^l = mK_2, \quad m \geq 1,\ n \geq 2,$$

$$V^{l-2} = V^{l-1} \cup \{c\}, \quad V^{l-1} = V^l \cup A, \quad |A| = n, \quad A \not\sim V^l,$$

$$|V^l| = 2m, \quad c \not\sim V^l, \quad c \sim A, \quad A \text{ is independent.}$$

We observe that $|c|_{l-2} = n$ and $|v^l|_{l-2} = 1$. Now, by Lemma 2.3, the difference between these two degrees is equal to 1 because $l \geq 3$. Therefore, $n = 2$.

Next, G^{l-2} admits a 2-switch of type $V^l A V^l c$, and V^{l-1}(equal to $V^l \cup A$) is a regular subset in G^{l-2}, so one can apply Lemma 2.6 with $i = l-2$. Obviously, V^{l-1}

has both adjacent and non-adjacent vertices. Consequently, there exists a partition of V^{l-3} in the form

$$V^{l-3} = V^{l-2} \cup E \cup F, \quad E \sim V^{l-1}, \quad F \not\sim V^{l-1},$$

where E is a clique and F is an independent set. We observe that

$$|c|_{l-2} = 2 > 1 = |v^l|_{l-2} \qquad \text{and} \qquad |c|_{l-3} = |v^l|_{l-3},$$

so there exists a vertex e with $e \sim v^l$ and $e \not\sim c$. We have

$$|e|_{l-3} \geq 2m + 2 + |E| - 1 \geq 3 + |E| \qquad \text{and} \qquad |a|_{l-3} = 1 + |E|.$$

Consequently, $l = 3$, since otherwise Lemma 2.3 would yield a contradiction.

(ii) Let us consider the following partition:

$$E = E_1 \cup E_2, \quad E_1 \sim c, \quad E_2 \not\sim c.$$

We showed above that $E_2 \neq \emptyset$. If $e_2 \sim f$, then $G^{l-3} = G$ admits the 2-switch e_2afc. But

$$a, c \in V^{l-2}, \quad e_2, f \notin V^{l-2}$$

and V^{l-2} is a regular subset in G. The latter contradicts Lemma 2.4, hence $E_2 \not\sim F$. On the other hand, a 2-switch of type E_2cV^lF contradicts the same lemma, hence $c \not\sim F$.

Therefore, E_1 is a clique, F is an independent set and

$$E_1 \sim V^{l-2}, \quad E_1 \sim E_2, \quad F \not\sim V^{l-2}, \quad F \not\sim E_2.$$

We obtain

$$G = (G(E_1 \cup F), E_1, F) \circ (G - E_1 - F)$$

if $E_1 \cup F$ is non-empty. Since G is indecomposable, this union is empty.

We observe that $\deg a = 1 + |E_2|$ and $\deg c = 2$. But the vertices a and c belong to the same regular subset V^{l-2} of G, hence $\deg a = \deg c$ and $|E_2| = 1$. Thus,

$$V(G) = V^{l-3} = V^{l-1} \cup \{c, e\}, \quad G - e = G^{l-2} = mK_2 \cup K_{1,2}$$

and $e \sim V^{l-1}$, $e \not\sim c$. This means that $G = U_3(m)$.

(iii) Obviously, $l(U_3(m)) = 3$ for any $m \geq 1$. □

The last two lemmas together with Fact 2 imply statement (ii) of Theorem 2.4.

2.3.2 Indecomposable Split Unigraphs

According to Corollary 2.3, the step of any graph considered in this section is at most 1. Therefore, the study of the series, which played such an important role in Section 2.3.1, is useless here because all non-trivial information concerning the original graph disappears already in the first step. In order to exploit the techniques developed in Section 2.2.3, we modify the definition of series.

Until the end of this section, G is a non-trivial indecomposable split unigraph. By Lemma 2.1, G has a unique splitting (G, A, B), where $A \neq \emptyset$, $B \neq \emptyset$ and each regular subset of G is entirely contained either in A or in B. When constructing sequence (2.23) for G, let us put

$$G^0 = G, \quad A^0 = A, \quad B^0 = B, \quad V^i = A^i \cup B^i,$$

where A^i and B^i are regular subsets of G^{i-1} contained in A^{i-1} and B^{i-1}, respectively. Then, we obtain a sequence of split unigraphs

$$G^0, G^1, \ldots, G^s \tag{2.42}$$

with splittings (G^i, A^i, B^i); for $i < s$, the splittings (G^i, A^i, B^i) are not biregular, and the splitting (G^s, A^s, B^s) is biregular. In similar situations, we will write 'G^s is biregular' or 'G^i is not biregular'. The obtained sequence will be called a (*splitted*) *series* of G, and s will be called the *length* of this series. Let us denote by $s(G)$ the maximal length of all (splitted) series of G. If $s = s(G)$ and G^s admits a 2-switch, then series (2.42) is called *good*. If there is no such series, then any series of length $s(G)$ is called *good*.

When we pass from G to the complement $\overline{G}$, the splitting (G, A, B) is replaced by $(\overline{G}, B, A)$, and series (2.42) is replaced by the following one:

$$\overline{G^0}, \overline{G^1}, \ldots, \overline{G^s}, \qquad \overline{G^i} = \overline{G}(A^i \cup B^i), \tag{2.43}$$

where A^i and B^i are the same as above, but the upper and lower parts are interchanged. Analogously, when we pass from G to G^{I}, series (2.42) is replaced by

$$(G^0)^{\mathrm{I}}, (G^1)^{\mathrm{I}}, \ldots, (G^s)^{\mathrm{I}}, \qquad (G^i)^{\mathrm{I}} = G^{\mathrm{I}}(A^i \cup B^i). \tag{2.44}$$

Hence, $s(\overline{G}) = s(G^{\mathrm{I}}) = s(G)$ and all three series (2.42)–(2.44) are either simultaneously good or not good.

In what follows, we write $G \equiv H$ for non-trivial indecomposable split graphs G and H if one of the following holds: $G \simeq H$, $\overline{G} \simeq H$, $G^{\mathrm{I}} \simeq H$ or $\overline{G}^{\mathrm{I}} \simeq H$.

The equality $s(G) = 0$ means biregularity. All biregular unigraphs are described with the help of Fact 3, which immediately implies the following result.

Corollary 2.15 *An indecomposable non-regular split unigraph* G *is biregular if and only if* $G \equiv S(p, q)$*,* $p \geq 1$*,* $q \geq 2$*.*

We shall assume that $s(G) \geq 1$ and (2.42) is a good series of G.

Lemma 2.11 *The following statements are true:*

(i) *For* $i \in [2, s]$*, we have*

$$A^{i-1} = A^i \quad \text{or} \quad B^{i-1} = B^i. \tag{2.45}$$

(ii) *For* $i \in [1, s-1]$*, the equalities* $A^{i-1} = A^i$ *and* $A^i = A^{i+1}$ *cannot hold simultaneously.*

Proof: (i) Assume to the contrary that none of equalities (2.45) holds. Then, by Lemma 2.3, A^{i-1} and B^{i-1} are partitioned in G^{i-1} into two regular subsets:

$$A^{i-1} = C \cup D, \quad B^{i-1} = E \cup F, \quad |c|_{i-1} > |d|_{i-1}, \quad |e|_{i-1} > |f|_{i-1}.$$

According to Fact 1, there exists a pair of adjacent vertices c and e as well as a pair of non-adjacent vertices d and f. We have

$$|c|_{i-2} = |d|_{i-2} \quad \text{and} \quad |e|_{i-2} = |f|_{i-2}.$$

Hence, there exist

$$b \in B^{i-2} - B^{i-1}, \quad a \in A^{i-2} - A^{i-1}, \quad b \sim d, \quad b \not\sim c, \quad a \sim f, \quad a \not\sim e.$$

Then, G^{i-2} admits the 2-switch $caeb$ or $dabf$ for $a \sim b$ and $a \not\sim b$, respectively. This contradicts Lemma 2.4.

(ii) If $A^{i+1} = A^i = A^{i-1}$, then

$$B^i = B^{i+1} \cup C, \quad B^{i-1} = B^{i+1} \cup C \cup D, \quad C, D \neq \emptyset.$$

Consequently, the vertices b^{i+1} and c belong to the same regular subset of rank $i-1$, but to different regular subsets of rank i. On the other hand, $|v|_{i-1} = |v|_i$ for $v \in B^{i-1}$. □

Remark Both equalities in (2.45) cannot hold simultaneously for any $i \in [1, s]$ because this would imply $s = i - 1$.

Lemma 2.12 *Assume that* $A^{i-1} = A^i$ *for some* $i \in [1, s]$, *and denote* $C = B^{i-1} - B^i$. *Then,* $C \neq \emptyset$. *Furthermore, if* $C = C_0 \cup C_1 \cup C_2$, *where* C_1 *and* C_2 *are maximal subsets in* C *with respect to the conditions* $C_1 \sim A^i$ *and* $C_2 \not\sim A^i$, *and* $C_0 = C - (C_1 \cup C_2)$, *then*

$$C = \begin{cases} C_0 & \text{for } i < s, \\ C_0, C_1 \text{ or } C_2 & \text{for } i = s. \end{cases}$$

The lemma holds if one interchanges A and B.

Proof: Obviously, $C \neq \emptyset$ because G^{i-1} is not biregular. Let us consider the partition of C determined in the statement of the lemma.

First, let $i = 1$. Then,

$$A = A^0 = A^1, \qquad B = B^0 = B^1 \cup C_0 \cup C_1 \cup C_2.$$

If $C_1 \neq \emptyset$, then

$$G = (G - C_1, A, B - C_1) \circ G(C_1)$$

is a decomposable graph. For $C_2 \neq \emptyset$, the graph G is decomposable as well:

$$G = (G(C_2), \emptyset, C_2) \circ (G - C_2).$$

Hence, $C_1 = C_2 = \emptyset$ and $C = C_0$.

For $i > 1$, the set C is a regular subset in G^{i-1} by Lemma 2.3. But

$$|c_2|_{i-1} < |c_0|_{i-1} < |c_1|_{i-1},$$

so only one of the sets C_k, $k = 0, 1, 2$, is not empty.

Let $i < s$. The set A^{i-1} is a regular subset in G^{i-1} because $A^i = A^{i-1}$. If $C = C_1$ or C_2, then A^i is a regular subset in G^i, that is, $A^{i+1} = A^i$. The latter contradicts Lemma 2.11, so $C = C_0$. This proves the lemma for A. Passing to G^{I}, we obtain the same for B. □

Lemma 2.13 *Assume that for some $i \in [1, s]$ there exist $u_1, u_2 \in A^i$ and $v_1, v_2 \in B^i$ such that $u_1 \sim v_1$ and $u_2 \not\sim v_2$. Let us represent A^{i-1} and B^{i-1} in the form*

$$A^{i-1} = A^i \cup D_1 \cup D_2 \cup D \quad \textit{and} \quad B^{i-1} = B^i \cup E_1 \cup E_2 \cup E,$$

where D_1 and D_2 (E_1 and E_2) are the maximal subsets in A^{i-1} (B^{i-1}) with respect to the conditions

$$D_1 \sim B^i, \quad D_2 \not\sim B^i, \quad E_1 \sim A^i, \quad E_2 \not\sim A^i.$$

Then,

$$D_1 \sim E_1 \quad \textit{and} \quad D_2 \not\sim E_2. \tag{2.46}$$

Proof: If (2.46) does not hold, then G^{i-1} admits a 2-switch of type

$$D_1 A^i B^i E_1 \quad \text{or} \quad D_2 A^i E_2 B^i.$$

This contradicts Lemma 2.4 because the degrees of the vertices d_1 and a^i, b^i and e_1, d_2 and a^i, e_2 and b^i in G^{i-1} are not equal. □

Corollary 2.16 *Statements (i)–(iii) are true.*

(i) $s(G) \geq 2$.

(ii) For some of the graphs G, $\overline{G}$, G^{I} or $\overline{G}^{\mathrm{I}}$, the following holds:

$$A^{s-1} = A^s, \quad B^{s-1} = B^s \cup C, \quad C \neq \emptyset, \quad C \cap B^s = \emptyset, \quad C \not\sim A^s; \tag{2.47}$$

$$|b^s|_s = |b^s|_{s-1} = 1. \tag{2.48}$$

(iii) If the formulae in (2.47) hold for G, then $G^s = S(p,q)$ and G^{s-1} is the disjoint union of the graph $S(p,q)$ and the empty graph $O(C)$:

$$G^{s-1} = S(p,q) \cup O(C).$$

Proof: (i) Let us distinguish the following two cases: (1) G^s admits a 2-switch, (2) G^s has no 2-switches.

(1) Evidently, a 2-switch admitted by G^s has type $A^s A^s B^s B^s$. Because A^s and B^s are regular subsets in G^s, by Lemma 2.5, there exist the partitions

$$A^{s-1} = A^s \cup A_1 \cup A_2, \quad A_1 \sim B^s, \quad A_2 \not\sim B^s,$$

$$B^{s-1} = B^s \cup B_1 \cup B_2, \quad B_1 \sim A^s, \quad B_2 \not\sim A^s.$$

By Lemma 2.13, $A_1 \sim B_1$ and $A_2 \not\sim B_2$. So, we have

$$G^{s-1} = \begin{cases} (G(A_1 \cup B_2), A_1, B_2) \circ (G^{s-1} - A_1 - B_2) & \text{for } A_1 \cup B_2 \neq \emptyset, \\ (G^{s-1} - A_2 - B_1, A^s \cup A_1, B^s \cup B_2) \circ G(A_2 \cup B_1) & \text{for } A_2 \cup B_1 = \emptyset. \end{cases}$$

Hence, $s = s(G) \geq 2$ because G is indecomposable.

(2) Let

$$A^{s-1} = A_1 \cup A_2 \cup \ldots \cup A_k, \qquad B^{s-1} = B_1 \cup B_2 \cup \ldots \cup B_l$$

be the partitions into the regular subsets of G^{s-1}, and

$$|a_1|_{s-1} > |a_2|_{s-1} > \ldots > |a_k|_{s-1}, \qquad |b_1|_{s-1} > |b_2|_{s-1} > \ldots > |b_l|_{s-1}.$$

Since series (2.42) is good, the graph $G(A_i \cup B_j)$ is biregular and admits no 2-switches for all $i = 1, 2, \ldots, k$, $j = 1, 2, \ldots, l$. By Fact 5,

$$A_i \sim B_j \quad \text{or} \quad A_i \not\sim B_j. \tag{2.49}$$

If $|b_l|_{s-1} \neq 0$, it follows from (2.49) and Fact 1 that $A_1 \sim B^{s-1}$. Thus, G^{s-1} contains either a dominating vertex or an isolated vertex and, therefore, it is decomposable and $s \geq 2$.

(ii) Since $s \geq 2$, we have by Lemma 2.11:

$$A^{s-1} = A^s \quad \text{or} \quad B^{s-1} = B^s.$$

Replacing G by G^{I} if necessary, one may assume that $A^{s-1} = A^s$. By Lemma 2.12,

$$B^{s-1} = B^s \cup C, \quad C = C_0, C_1 \text{ or } C_2.$$

In case (1), C_0 is excluded by Lemma 2.5. In case (2), as shown above,

$$A^s \sim B^s, \ A^s \not\sim C \quad \text{or} \quad A^s \not\sim B^s, \ A^s \sim C;$$

that is, C_0 is excluded as well. In any situation for $C = C_1$, we obtain $C = C_2$ replacing G by $\overline{G}^{\mathrm{I}}$. Formulae (2.47) are proved.

Obviously, (2.47) implies (2.48). Indeed, by (2.47), $|c|_{s-1} = 0$, and Lemma 2.3 implies (2.48).

(iii) Let formulae (2.47) and, consequently, (2.48) hold for G. Then, G^s is a biregular split graph such that all degrees of vertices in the lower part are equal to 1. Obviously, $G^s = S(p, q)$, $p, q \geq 1$. By (2.47) and (2.48), $G^{s-1} = S(p, q) \cup O(C)$. □

Lemma 2.14 *Assume that G^s admits the 2-switch $t = a_1 a_2 b_1 b_2$, where $a_1, a_2 \in A^s$, $b_1, b_2 \in B^s$. Also, suppose that $A^{i-1} = A^i$ and $B^{i-1} = B^i \cup C$ for some $i \in [2, s]$, where C is defined in Lemma 2.12 and $C \not\sim A^s$. Then,*

$$B^{i-2} = B^{i-1}, \quad A^{i-2} = A^i \cup D, \quad D \neq \emptyset, \quad D \cap A^i = \emptyset. \tag{2.50}$$

Moreover, $D \not\sim B^i$, $\deg_D c = 1$ *and* $\deg_C d \neq 0$ *if*

$$|b^i|_{i-1} > |c|_{i-1}, \tag{2.51}$$

and $D \sim B^i$ *otherwise.*

Proof: Let us consider the partition

$$A^{i-2} = A^{i-1} \cup D \cup D_1 \cup D_2, \tag{2.52}$$

where D_1 and D_2 are the maximal sets satisfying the following:

$$D_1 \sim B^{i-1}, \quad D_2 \not\sim B^{i-1}. \tag{2.53}$$

We observe that there are no vertices $d, d' \in D$, b^i and c such that

$$d \sim b^i, \quad d \not\sim c, \quad d' \not\sim b^i, \quad d' \sim c. \tag{2.54}$$

Indeed, suppose that such a quadruple exists. The graph G^{i-1} as well as G^s admits the 2-switch

$$t = a_1 a_2 b_1 b_2, \quad a_1, a_2 \in A^i, \quad b_1, b_2 \in B^i,$$

and B^i is a regular subset in G^{i-1}. By Lemma 2.5, the neighbourhoods in the set D of all vertices from B^i coincide. Hence, (2.54) implies

$$d \sim B^i, \quad d' \not\sim B^i. \tag{2.55}$$

Thus, G^{i-2} admits the 2-switch $t_1 = a_1 d' b_1 c$. If $H = t_1 G^{i-2}$ and $H^{i-1} = H(A^{i-1} \cup B^{i-1})$, then, by Lemma 2.2, $H \simeq G^{i-2}$ and $H^{i-1} \simeq G^{i-1}$. But

$$H^{i-1} = G^{i-1} - a_1 b_1 + a_1 c.$$

The part B^{i-1} must be partitioned into two regular subsets in H^{i-1} as well as in G^{i-1}. By Lemma 2.3, the vertex degrees of these regular subsets must differ by 1. Consequently, c lies in the same regular subset with $B^i - \{b_1\}$ in H^{i-1}. The graph H^{i-1} admits the 2-switch $a_1 a_2 c b_2$ and, by Lemma 2.5 for the graph H, the neighbourhoods of b_2 and c in D must coincide. However, $b_2 \sim d$ by (2.55), and $c \not\sim d$ by (2.54). It is proved that (2.54) does not hold.

In what follows, we consider the following situations: (1) (2.51) holds, (2) (2.51) does not hold.

(1) Since

$$|b^i|_{i-2} = |c|_{i-2}, \tag{2.56}$$

for each pair b^i and c there exists $d' \in D$ such that $d' \not\sim b^i$ and $d' \sim c$. But (2.54) is impossible, so the implication $d \sim b^i \Rightarrow d \sim C$ is true. Let us recall that $d \sim b^i$ implies $d \sim B^i$. Consequently, $d \sim B^{i-1}$, which contradicts the definition of D (see (2.52)). Hence, $D \not\sim B^i$.

Lemma 2.3 implies that $|b^i|_{i-1} = |c|_{i-1} + 1$ and by the definition of D, $\deg_D c = 1$ and $\deg_C d \neq 0$.

(2) We have $|b^i|_{i-1} < |c|_{i-1}$. So, by (2.56), there exists $d \in D$ such that $d \sim b^i$ and $d \not\sim c$. As above, we conclude that $D \sim B^i$.

It remains to prove (2.50). For $i > 2$, these equalities follow from Lemmas 2.11 and 2.12. Suppose that $i = 2$. Since G^1 admits the 2-switch t and A^1 is a regular subset in G^1, by Lemma 2.5, there exists a partition

$$B = B^1 \cup E_1 \cup E_2, \quad E_1 \sim A^1, \quad E_2 \not\sim A^1. \tag{2.57}$$

By Lemma 2.13,

$$D_1 \sim E_1, \quad D_2 \not\sim E_2. \tag{2.58}$$

Next, we deduce that

$$D \sim E_1, \quad D \not\sim E_2, \tag{2.59}$$

otherwise G^{i-2} would admit a 2-switch of type A^sDE_1C, $A^2DB^2E_2$ (case (1)), $A^sDE_1B^s$ or A^2DCE_2 (case (2)). The latter contradicts Lemma 2.4.

It follows from (2.52), (2.53) and (2.57)–(2.59) that

$$G = \begin{cases} (G(D_1 \cup E_2), D_1, E_2) \circ (G - D_1 - E_2) & \text{for } D_1 \cup E_2 \neq \emptyset, \\ (G - D_2 - E_1, A - D_2, B - E_1) \circ G(D_2 \cup E_1) & \text{for } D_2 \cup E_1 \neq \emptyset. \end{cases}$$

Since G is indecomposable, we obtain

$$D_1 = D_2 = E_1 = E_2 = \emptyset.$$

This proves (2.50). □

Lemma 2.15 *Let $s(G) \geq 2$, G^s admit no 2-switches and $G^{s-1} = S(p,q) \cup O(C)$; that is, (2.47) and (2.48) hold for G. Then,*

(i) $G = S_2(p_1, 1; p_2, 1; \ldots; p_r, 1)$, $r \geq 2$;

(ii) $s(G) = 2$.

Proof: Since G^s admits no 2-switches, we have $q = 1$ and

$$A^{s-1} = A^s = \{a\}, \quad B^{s-1} = B^s \cup C, \quad a \sim B^s, \quad a \not\sim C.$$

Let us consider the partitions

$$A^{s-2} = A^{s-1} \cup D_1 \cup D_2 \cup D \quad \text{and} \quad B^{s-2} = B^{s-1} \cup E_1 \cup E_2,$$

where D_1 and D_2 (E_1 and E_2) are the maximal subsets such that

$$D_1 \sim B^{s-1}, \quad D_2 \not\sim B^{s-1}, \quad E_1 \sim a, \quad E_2 \not\sim a.$$

By Lemma 2.13, $D_1 \sim E_1$ and $D_2 \not\sim E_2$.

Assume that there exist vertices b^s, c and d such that $b^s \sim d$ and $c \not\sim d$. Since

$$|b^s|_{s-1} > |c|_{s-1} \quad \text{and} \quad |b^s|_{s-2} = |c|_{s-2},$$

there exists $d' \in D$ such that $b^s \not\sim d'$ and $c \sim d'$, and G^{s-2} admits the 2-switch $t = b^scdd'$. Let us set $H = G^{s-2} - a$. The graph H can be constructed by the union of some regular subsets in G^{s-2} and it is a unigraph by Corollary 2.12. Also, H admits t. Since $b^s \sim a$ and $c \not\sim a$, by Lemmas 2.4 and 2.5 the vertices d and d' belong to the same regular subset in H and, consequently, in G^{s-2}. Let D_0 be the regular subset in G^{s-2} containing d and d', and

$$H^{s-1} = G(D_0 \cup B^{s-1}).$$

Similar to H, the graph H^{s-1} admits t, and B^{s-1} is not a regular subset in H^{s-1}; that is, $(H^{s-1}; D_0, B^{s-1})$ is not biregular. Hence, one can complement the sequence

$$G^0, G^1, \dots, G^{s-2}, H^{s-1}$$

to the series

$$G^0, G^1, \dots, G^{s-2}, H^{s-1}, H^s.$$

Since the initial series (2.42) is good and G^s does not admit 2-switches, H^s does not admit them either. By Corollary 2.16, H^{s-1} cannot admit 2-switches. This contradiction proves that the following implication is true:

$$d \sim b^s \Rightarrow d \sim C.$$

Consequently, both $d \not\sim C$ and $d \sim B^s$ are impossible because this would contradict the definition of D. Now, we observe that $D \sim E_1$ and $D \not\sim E_2$, since otherwise G^{s-2} would admit a 2-switch of type aDE_1C or aDB^sE_2, thus contradicting Lemma 2.4.

For $s = 2$, we have

$$D_1 = E_2 = D_2 = E_1 = \emptyset,$$

otherwise

$$G^{s-2} = (G(D_1 \cup E_2), D_1, E_2) \circ (G^{s-2} - D_1 - E_2)$$

or

$$G^{s-2} = (G^{s-2} - D_2 - E_1, A^{s-2} - D_2, B^{s-2} - E_1) \circ G(D_2 \cup E_1).$$

But $G^{s-2} = G$ and G is indecomposable. Hence,

$$A^{s-2} = \{a\} \cup D \quad \text{and} \quad B^{s-2} = B^{s-1} = B^s \cup C. \tag{2.60}$$

For $s > 2$, the equalities in (2.60) also hold; they follow from Lemmas 2.11–2.12. We have $\deg_C d \neq 0$. The sets $\{a\}$, D and B^{s-2} are regular subsets in G^{s-2}. Because

$$|c|_{s-1} = |b|_{s-1} - 1 \quad \text{and} \quad |b|_{s-2} = |b|_{s-1},$$

we have $\deg_D c = 1$. It follows immediately from above that

$$G^{s-2} = S_2(p_1, 1; p_2, g_2; \dots; p_r, q_r), \quad r \geq 2.$$

We can show that

$$q_i = 1, \quad i = 2, 3, \dots, r.$$

Indeed, G^{s-2} is a multistar obtained from the disjoint union of the stars K_{1,p_i} by adding all edges that connect the centres of these stars. Here, the star K_{1,p_i} is included in the above union q_i times. We took the multistar $S_2(p_1, 1) = K_{1,p_1}$ as G^s. Obviously, each multistar $S_2(p_i, q_i)$ can be taken as G^s. Since the initial series (2.42) is good, such a multistar cannot admit 2-switches; that is, q_i must be equal to 1. Part (i) is proved.

(ii) Let $s(G) \geq 3$. By Lemma 2.3, D is a regular subset in G^{s-2}, so $r=2$. Therefore,

$$G^{s-2} = S_2(p_1, 1; p_2, 1), \quad p_1 = |B^s|, \quad p_2 = |C|.$$

We have $A^s = \{a\}$ and $D = \{d\}$. Without loss of generality, one may assume that $p_1 < p_2$ and, consequently,

$$|a|_{s-2} < |d|_{s-2}.$$

We set $E = B^{s-3} - B^{s-2}$. The degrees of vertices a and d in G^{s-3} are equal because $a, d \in A^{s-2}$. Hence, there exists a vertex e such that

$$e \sim a, \quad e \not\sim d. \tag{2.61}$$

If $s(G) > 3$, we obtain $A^{s-3} = A^{s-2}$ by Lemma 2.11. Hence,

$$|e|_{s-3} = 1 = |c|_{s-3}.$$

But $e \notin B^{s-2}$ and $c \in B^{s-2}$. Therefore,

$$s(G) = 3. \tag{2.62}$$

Let us put $F = A - A^1$. Observe that the implication $f \sim e \Rightarrow f \sim C$ is true, since otherwise G would admit a 2-switch of type *eCfd*, which is forbidden by Lemma 2.4. Taking also into account (2.61), we conclude that $\deg e \leq \deg c$. But the equality is impossible by the definition of a regular subset, hence $\deg e < \deg c$ and there exists $f' \in F$ such that $f' \sim c$ and $f' \not\sim e$. The graph G admits the 2-switch *ecaf*, thus contradicting Lemma 2.4. This proves that (2.62) is impossible and, hence, $s(G) = 2$. □

Corollary 2.17 *Statements (i) and (ii) are true.*

(i) $2 \leq s(G) \leq 4$.

(ii) For $s(G) = 2, 3$ *or* 4, *the following formulae hold, respectively:*

$$G \equiv S_2(p_1, q_1; \ldots; p_r, q_r), \quad G \equiv S_3(p, q_1, q_2), \quad G \equiv S_4(p, q).$$

Proof: By Corollary 2.16, $s(G) \geq 2$. If G^s admits no 2-switches, then by Lemma 2.15, $s(G) = 2$ and

$$G = S_2(p_1, 1; p_2, 1; \ldots; p_r, 1), \quad r \geq 2.$$

Let G^s admit a 2-switch. By Corollary 2.16, we may assume that

$$G^s = S(p, q), \quad q \geq 2, \quad G^{s-1} = G^s \cup O(C),$$

$$A^{s-1} = A^s, \quad B^{s-1} = B^s \cup C, \quad C \not\sim A^s. \tag{2.63}$$

Hence, for $i = s$, Lemma 2.14 (case (1)) can be applied; that is,

$$B^{s-2} = B^{s-1}, \ A^{s-2} = A^s \cup D, \ D \not\sim B^s, \ \deg_D c = 1, \ \deg_C d \neq 0. \tag{2.64}$$

Obviously,

$$G^{s-2} = S_2(p_1, q_1; p_2, q_2; \ldots; p_r, q_r) = S_2, \quad r \geq 2.$$

In particular, $G = S_2$ for $s(G) = 2$.

Suppose that $s(G) \geq 3$. By Lemma 2.3, D is a regular subset in G^{s-2}, so $r = 2$. Let us consider the graph $H = G^{\mathrm{I}}$. Its parts will be interchanged; that is, the sets A^i will become the regular subsets of the lower part and B^i will become the regular subsets of the upper part. For *H*, formulae (2.63) and (2.64) are valid, so the assumptions of Lemma 2.14 ($i = s - 1$) hold. By this lemma,

$$A^{s-3} = A^{s-2}, \quad B^{s-3} = B^{s-2} \cup E, \quad E \neq \emptyset, \quad B^{s-2} \cap E = \emptyset. \tag{2.65}$$

Next, for each vertex e, we obtain

$$e \sim D \text{ or } e \not\sim D. \tag{2.66}$$

Indeed, for $q_2 \geq 2$, the graph G^{s-2} admits a 2-switch of type *DDCC*. Since D is a regular subset in G^{s-2}, Lemma 2.5 implies (2.66). For $q_2 = 1$, condition (2.66) is trivial.

Now, if $p_1 > p_2$, we obtain the case (1) of Lemma 2.14:

$$E \not\sim A^s, \quad \deg_E d = 1, \quad \deg_D e \neq 0, \tag{2.67}$$

and (2.66) and (2.67) imply that $E \sim D$ and $|E| = 1$.

For $p_1 < p_2$, we obtain the case (2) of Lemma 2.14 ($E \sim A^s$). As

$$|a^s|_{s-3} = |d|_{s-3} = p_1 + 1,$$

we have $|E| = 1$. By Lemma 2.12 ($i = s - 3 < s$), there exists d such that $d \not\sim e$ and (2.66) implies $e \not\sim D$.

Let us return to G. It was proved above that, in any case, G^{s-3} can be obtained from $S_2(p_1, q_1; p_2, q_2)$ by adding the vertex e to the lower part. Here, for $p_1 < p_2$, we have

$$e \sim A^s, \quad e \not\sim D, \quad p_2 = p_1 + 1;$$

that is, $G^{s-3} = S_3(p, q_1, q_2)$ and $G = G^{s-3}$ for $s(G) = 3$.

Now, suppose that $s(G) \geq 4$. Formulae (2.64) and (2.65) hold for G. If $p_1 > p_2$, then (2.67) is true, so $e \not\sim A^s$ and $e \sim D$. The degrees of c and e in G^{s-3} must differ by 1, hence

$$q_2 = |D| = |e|_{s-3} = 2.$$

Therefore, the graph $G' = G(D \cup C)$ admits a 2-switch, and replacing G^s by G' in (2.42) yields a good series. Hence, one may assume without loss of generality that

$$p_1 < p_2, \quad e \sim A^s, \quad e \not\sim D, \quad q_1 = |e|_{s-3} = 2 > |c|_{s-3} = 1, \quad p_2 = p_1 + 1.$$

Let us set $H = \overline{G}^{\mathrm{I}}$. For H, we obtain

$$e \not\sim A^s, \quad e \sim D, \quad |c|_{s-3} > |e|_{s-3}.$$

So, one can apply Lemma 2.14 ($i = s - 2$, case (1)). We have

$$B^{s-4} = B^{s-3}, \quad A^{s-4} = A^{s-3} \cup F, \quad F \neq \emptyset,$$

$$F \cap A^{s-3} = \emptyset, \quad F \not\sim B^{s-2}, \quad F \sim e.$$

For G, we obtain $F \sim B^{s-2}$ and $F \not\sim e$. Since the vertices c and e have equal degrees in G^{s-4}, and the degrees differ from each other by just 1 in G^{s-3}, we have $|F| = 1$. Thus, $G^{s-4} = S_4(p, q)$ and $G = S_4(p, q)$ for $s(G) = 4$. Next, we have

$$|f|_{s-4} - |d|_{s-4} = 2 + (p + 1)(q - 1) \geq 2.$$

Hence, according to Lemma 2.3, $s = 4$. □

Statement (iii) of Theorem 2.4 follows from the previous corollary. Theorem 2.4 is proved.

2.4 References

[1] M. Aigner and E. Triesch, Realizability and uniqueness in graphs, *Discrete Mathematics*, **136** (1–3)(1994), 3–20.

[2] A. Brandstädt, *Special Graph Classes: A Survey* (preliminary version), Friedrich–Schiller Universität Jena, Forschungsergebnisse, Nr: N/90,6, 1991.

[3] V. Chvatal and P. L. Hammer, Aggregation of inequalities in integer programming, *Annals of Discrete Mathematics*, **1** (1977), 145–162.

[4] S. Földes and P. L. Hammer, Split graphs, *Congressus Numerantium*, **19** (1977), 311–315.

[5] D. R. Fulkerson, A. J. Hoffman and M. H. McAndrew, Some properties of graphs with multiple edges, *Canadian Journal of Mathematics*, **17** (1965), 166–177.

[6] P. L. Hammer and B. Simeone, The splittence of a graph, *Combinatorica*, **1** (3)(1981), 275–284.

[7] R. H. Johnson, Simple separable graphs, *Pacific Journal of Mathematics*, **56** (1)(1975), 143–158.

[8] R. H. Johnson, Properties of unique realizations—a survey, *Discrete Mathematics*, **31** (2)(1980), 185–192.

[9] D. J. Kleitman and S.-Y. Li, A note on unigraphic sequences, *Studies in Applied Mathematics*, **54** (4)(1975), 283–287.

[10] M. Koren, Pairs of sequences with a unique realization by bipartite graphs, *Journal of Combinatorial Theory, Ser. B*, **21** (3)(1976), 224–234.

[11] M. Koren, Sequences with unique realization, *Journal of Combinatorial Theory, Ser. B*, **21** (3)(1976), 235–244.

[12] S.-Y. Li, Graphic sequences with unique realization, *Journal of Combinatorial Theory, Ser. B*, **19** (1)(1975), 42–68.

[13] F. Maffray and M. Preissmann, Linear recognition of pseudosplit graphs, *Discrete Applied Mathematics*, **52** (3)(1994), 307–312.

[14] N. V. R. Mahadev and U. N. Peled, *Threshold Graphs and Related Topics*, Annals of Discrete Mathematics, Vol. 56, Amsterdam: Elsevier, 1995.

[15] R. I. Tyshkevich, Canonical decomposition of a graph, *Doklady Akademii Nauk BSSR*, **24** (8)(1980), 677–679.

[16] R. I. Tyshkevich, Once more on matrogenic graphs, *Discrete Mathematics*, **51** (1984), 91–100.

[17] R. I. Tyshkevich and A. A. Chernyak, Unigraphs I, *Izvestiya Akademii Nauk BSSR*, Ser. fiz.-mat. nauk, (5)(1978), 5–11.

[18] R. I. Tyshkevich and A. A. Chernyak, Unigraphs II, *Izvestiya Akademii Nauk BSSR*, Ser. fiz.-mat. nauk, (1)(1979), 5–12.

[19] R. I. Tyshkevich and A. A. Chernyak, Unigraphs III, *Izvestiya Akademii Nauk BSSR*, Ser. fiz.-mat. nauk, (2)(1979), 5–11.

[20] R. I. Tyshkevich and A. A. Chernyak, Canonical decomposition of a unigraph, *Izvestiya Akademii Nauk BSSR*, Ser. fiz.-mat. nauk, (5)(1979), 14–26.

[21] R. I. Tyshkevich, and A. A. Chernyak, Box-threshold graphs: the structure and the enumeration, Proceedings of the 30th International Scientific Colloquium *'Graphen und Netzwerke—Theorie und Anwendungen'*, Ilmenau, Germany, 1985, 119–121.

[22] R. I. Tyshkevich and A. A. Chernyak, Decomposition of graphs, *Kibernetika*, **21** (2)(1985), 67–74.

[23] R. I. Tyshkevich, O. I. Melnikov and V. M. Kotov, On graphs and degree sequences, *Kibernetika*, **6** (1981), 5–8.

3

Matrogenic, Matroidal and Threshold Graphs

T. Calamoneri, R. Petreschi, I. Sciriha & V. Zverovich

In this chapter, we consider some important subclasses of unigraphs such as matrogenic, matroidal and threshold graphs. The last class of graphs is well studied and motivated by many applications. First, we will discuss some important properties of those graphs as well as their decomposition, enumeration and recognition. Next, we will study a variant of the general problem of assigning channels to the stations in a wireless network when the graph representing the possible interferences is a matrogenic graph. In this problem, channels assigned to adjacent vertices must be at least 2 apart, while channels assigned to vertices at distance 2 must be different. Exploiting the decomposition of a graph into degree boxes, we will discuss algorithms for this problem in the aforementioned classes of graphs. Further, we will study the spectrum of a threshold graph and explore its nullspace. In particular, we show that the spectrum of a connected threshold graph G and its underlying antiregular graph have common characteristics, provided that G has an equitable partition with the minimal number of parts. The graphic appeal of the Ferrers–Young diagram, with rows representing the degree sequence of a threshold graph, was instrumental in exploring the nullity and structure of graphs.

3.1 Subclasses of Unigraphs

This chapter is devoted to some subclasses of *unigraphs*; that is, graphs uniquely determined by their own degree sequence up to isomorphism. In [4], unigraphs are presented as a superclass including matrogenic graphs, matroidal graphs, split matrogenic graphs and threshold graphs as shown in Figure 3.1. In this section, we give definitions of those graphs and formulate some interesting results.

Let $G = (V, E)$ be a finite, simple, loopless graph, where V is the vertex set of G with cardinality n and E is the edge set of G with cardinality m. A vertex $x \in V$ is called *dominating* if it is adjacent to all other vertices of V; such a vertex has degree $d(x) = n - 1$. A vertex $x \in V$ is called *isolated* if it is adjacent to no other

T. Calamoneri, R. Petreschi, I. Sciriha & V. Zverovich, *Matrogenic, Matroidal and Threshold Graphs.* In: *Methods of Graph Decompositions.*
Edited by: Vadim Zverovich and Pavel Skums, Oxford University Press. © Vadim Zverovich & Pavel Skums (2024).
DOI: 10.1093/oso/9780198882091.003.0003

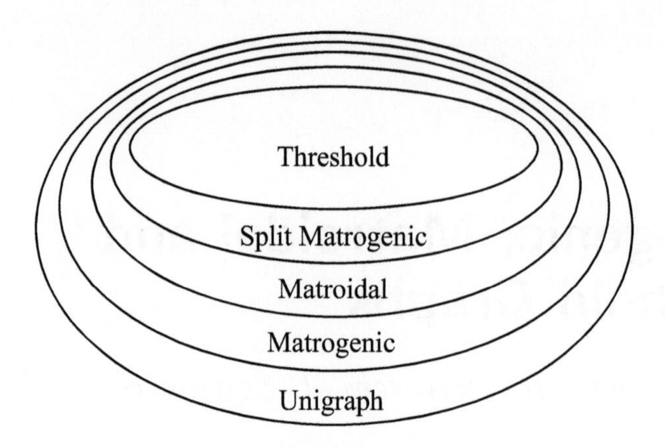

Figure 3.1 Relationships of inclusion among the subclasses of unigraphs.

vertex in V; such a vertex has degree $d(x) = 0$. A graph $I = (V_I, E_I)$, where $V_I \subseteq V$ and $E_I = E \cap (V_I \times V_I)$ is said to be *induced* by V_I. A graph G is an *empty graph* if its edge set is empty, irrespective of the size of the vertex set. In this chapter, the complement of a graph G is denoted G^{c} for convenience of presentation of large formulae.

Denote by $(\rho_1, \rho_2, \ldots, \rho_n)$ the degree sequence of a graph G sorted in non-increasing order:

$$\rho_1 \geq \rho_2 \geq \ldots \geq \rho_n \geq 0.$$

The equivalence classes of vertices in G under equality of degree are called *boxes.* In terms of boxes, the degree sequence can be compressed as

$$(d_1^{m_1}, d_2^{m_2}, \ldots, d_r^{m_r}), \quad d_1 > d_2 > \ldots > d_r \geq 0, \tag{3.1}$$

where d_i is the degree of the m_i vertices contained in the box $B_i(G)$, $1 \leq m_i \leq n$. We call a box *isolated (dominating)* if it contains only isolated (dominating) vertices.

The graph I induced by a subset $V_I \subseteq V$ is called

- *complete* or a *clique* if any two distinct vertices in V_I are adjacent in G; the set V_I may be called a *clique* too;
- *empty* if no two vertices in V_I are adjacent in G; the set V_I is *independent.*

A graph G is said to be *split* if there is a partition $V = V_K \cup V_S$ of its vertices such that the induced subgraphs K and S are complete and empty, respectively. For any graph G, let $N(x)$ be the set of neighbours of a vertex x. We define the *vicinal preorder* $\preceq$ on V as follows:

$$x \preceq y \quad \text{if and only if} \quad N(x) - y \subseteq N(y) - x.$$

Let us briefly describe the classes of graphs we will use in this chapter. For more details, the interested reader can consult [4, 21].

3.1.1 Threshold Graphs

Definition 3.1 [12, 25] *A graph $G = (V, E)$ is a threshold graph if and only if the vicinal preorder on V is total; that is, for any pair $x, y \in V$, either $x \preceq y$ or $y \preceq x$.*

In Figure 3.2a, an example of a threshold graph with nine vertices is depicted; its degree sequence is $(8, 6, 5, 4, 4, 3, 2, 1, 1)$. Because threshold graphs are split graphs, in Figure 3.2b the same graph is represented highlighting the vertices in the clique K and the vertices in the empty graph S, and the adjacencies in K are represented by a rectangle for the sake of clarity. Finally, in Figure 3.2c the same graph is represented in terms of boxes with compressed degree sequence $(8^1, 6^1, 5^1, 4^2, 3^1, 2^1, 1^2)$. Observe that if G is connected, the box of maximum degree contains all dominating vertices of G.

Threshold graphs have a rich structure. Here we highlight only some properties, for more details see the book by Mahadev and Peled [29].

Property 1T G is a threshold graph if and only if its adjacencies are determined by the following rule: for all $x \in B_i(G)$, $y \in B_j(G)$, $x \neq y$, we have $(x, y) \in E$ if and only if

$$r + 1 \geq i + j,$$

where r is the number of boxes in G. Thus, the structure of G is completely described by its degree sequence, implying that threshold graphs are unigraphs.

Property 2T Thresholdness is a hereditary property; that is, all induced subgraphs of a threshold graph are threshold graphs.

Property 3T A threshold graph G does not contain the configuration in Figure 3.3a; that is, G contains neither P_4, nor chordless C_4 nor $2K_2$ as induced subgraphs. It follows that threshold graphs are a subclass of cographs, since they cannot contain P_4 as an induced subgraph.

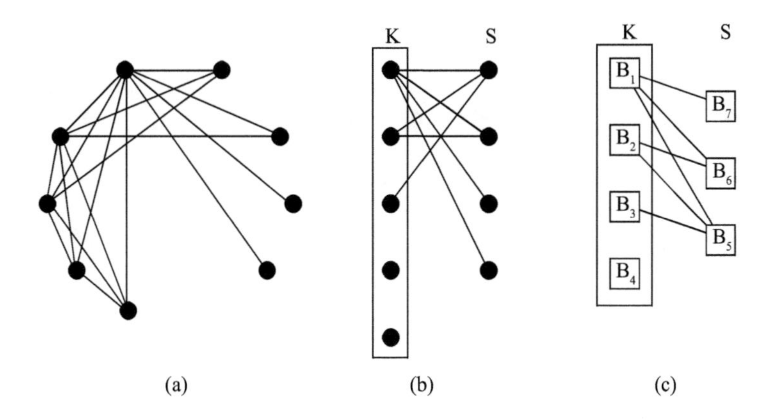

Figure 3.2 Threshold graph.

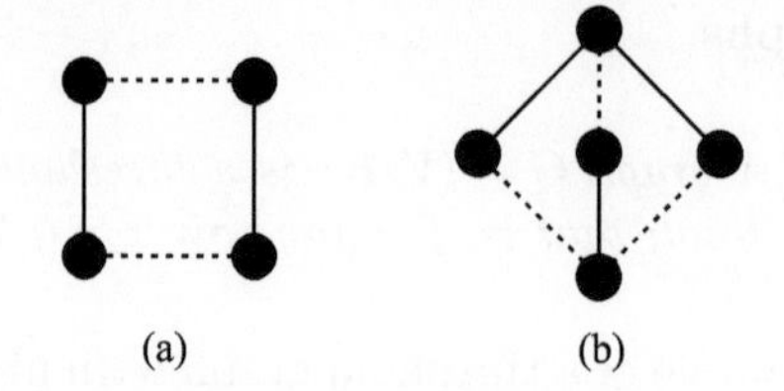

Figure 3.3 The forbidden configurations of a threshold graph (a) and of a matrogenic graph (b): — shows a present edge, --- shows an absent edge.

Property 4T In a connected threshold graph G, there always exists at least one dominating vertex, hence G has diameter 2.

3.1.2 Matrogenic and Matroidal Graphs

Before introducing matrogenic and matroidal graphs, let us recall some definitions. A set M of edges is a *perfect matching of dimension* h of X onto Y if and only if X and Y are disjoint subsets of vertices with the same cardinality h and each edge is incident to exactly one vertex $x \in X$ and to one vertex $y \in Y$, and different edges must be incident to different vertices. We say that x and y are *dually correlated* (see Figure 3.4a).

An *antimatching of dimension* h of X onto Y is a set A of edges such that

$$M(A) = X \times Y - A$$

is a perfect matching of dimension h of X onto Y. $M(A)$ is an *uncorrelated matching* and two endpoints of any edge in $M(A)$ are *dually uncorrelated* (see Figure 3.4b). Observe that, by definition, a non-trivial antimatching of dimension h must have $h \geq 3$. Indeed, $h = 1$ implies that the unique vertex in X (Y) is isolated, while $h = 2$

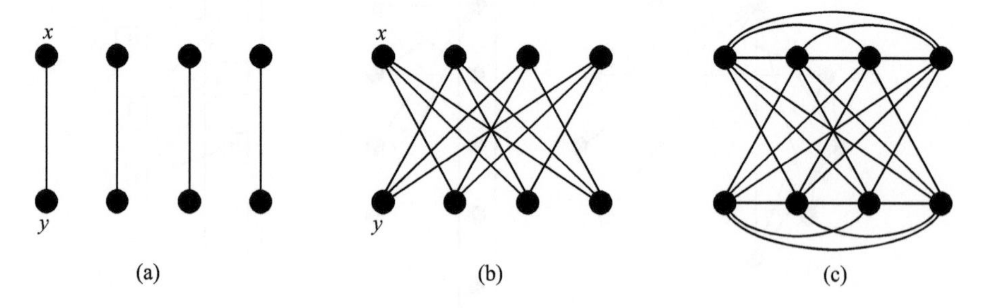

Figure 3.4 (a) A matching of dimension 4; (b) an antimatching of dimension 4; (c) a 4-hyperoctahedron.

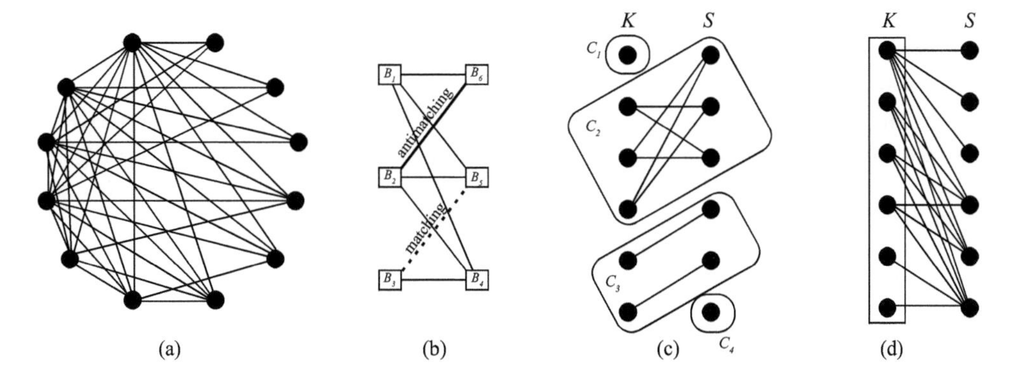

Figure 3.5 A split matrogenic graph (a) and its representation in terms of boxes (b); (c) its red graph; (d) its black graph.

implies that the antimatching coincides with a matching. In the following, we will visualize each vertex in X in front of its dually uncorrelated mate in Y in some order. With respect to this order, we naturally define the *leftmost* and *rightmost* vertices in X and Y. An *h-hyperoctahedron* is the complement of a perfect matching of dimension h, see Figure 3.4c.

In the following, we first show how to build split matrogenic graphs and then use them to construct all matrogenic graphs. Namely, the following theorem presents split matrogenic graphs. They are obtained by a superposition of a *black graph B* and a *red graph R*, where B is a threshold graph and R is the union of vertex-disjoint perfect matchings, antimatchings and empty graphs according to specific rules.

Theorem 3.1 [30] *A split graph G with clique V_K and independent set V_S is matrogenic if and only if the edges of G can be coloured red and black so that:*

(a) The red subgraph is the union of vertex-disjoint pieces $C_1, C_2, \ldots, C_z$. Each piece is either an empty graph N_j belonging to either K or S; or a matching M_r of dimension h_r of $K_r \subseteq V_K$ onto $S_r \subseteq V_S$, $r = 1, 2, \ldots, \mu$; or an antimatching A_t of dimension h_t of $K_t \subseteq V_K$ onto $S_t \subseteq V_S$, $t = 1, 2, \ldots, \alpha$ (Figure 3.5c). Exactly ν_K vertices of K and exactly ν_S vertices of S belong to the red empty graph, so

$$\sum_j |V_{N_j}| = \nu_K + \nu_S.$$

(b) The linear ordering $C_1, C_2, \ldots, C_z$ is such that each vertex in V_K belonging to C_i is not linked to any vertex in V_S belonging to C_j, $j = 1, 2, \ldots, i-1$, but is linked by a black edge to every vertex in V_S belonging to C_j, $j = i+1, i+2, \ldots, z$. Furthermore, any two vertices in V_K are linked by a black edge (Figure 3.5d).

Figure 3.5a depicts a split matrogenic graph having degree sequence $(11, 10, 10, 10, 7, 7, 6, 5, 5, 3, 3, 3)$. Figures 3.5c and 3.5d illustrate items (a) and (b) of the previous theorem, respectively.

By the definition of perfect matching and Theorem 3.1, it is not restrictive to assume that any matching M_r has dimension at least 2. Indeed, if a matching M_r of $K_r = \{v_r\}$ onto $S_r = \{w_r\}$ has dimension 1, it is possible to colour edge (v_r, w_r) black instead of red, and transform v_r and w_r into empty graphs.

Now, we classify split matrogenic graphs considering their representation in terms of boxes in the following way:

Theorem 3.2 [30] *If G is a split matrogenic graph, then one of the following four cases must occur:*

(a) *$B_r(G)$ consists of vertices of degree 0.*
(b) *$B_1(G)$ consists of vertices of degree $n-1$.*
(c) *$B_r(G)$ consists of h vertices of degree 1 and $B_1(G)$ consists of h vertices of degree $n-h$, and $B_1(G) \cup B_r(G)$ induces a perfect matching of dimension h.*
(d) *$B_r(G)$ consists of h vertices of degree at least 2 and $B_1(G)$ consists of h vertices of degree $n-2$, and $B_1(G) \cup B_r(G)$ induces an antimatching of dimension h.*

Figure 3.5b depicts the matrogenic graph of Figure 3.5a in terms of boxes with compact degree sequence $(11^1, 10^3, 7^2, 6^1, 5^2, 3^3)$. From the previous theorem, the following definition naturally arises: two boxes $B_i(G)$ and $B_j(G)$ are *partially connected* if the graph induced by $B_i(G) \cup B_j(G)$ in the red graph is either a perfect matching or an antimatching. Generalizing, a box $B_i(G)$ is partially connected to itself whenever it is neither partially connected to any other box nor is a clique nor is an independent set; that is, $B_i(G)$ induces a crown—this concept is explained in the next definition.

Now, we can introduce the classes of *matrogenic* and *matroidal graphs.*

Definition 3.2 [17] *A graph is matrogenic if and only if its vertex set V can be partitioned into three disjoint sets V_K, V_S and V_C such that:*

(1) *$V_K \cup V_S$ induces a split matrogenic graph in which V_K is a clique and V_S is an independent set.*
(2) *V_C induces a* crown; *that is, either a perfect matching or an h-hyperoctahedron (see Figure 3.4c) or a chordless cycle C_5.*
(3) *Every vertex in V_C is adjacent to every vertex in V_K and to no vertex in V_S.*

Notice that split matrogenic graphs are matrogenic graphs in which $V_C = \emptyset$. A result in [17] states that a graph $G = (V, E)$ is *matrogenic* if and only if it does not contain the configuration in Figure 3.3b.

Definition 3.3 [33] *A graph $G = (V, E)$ is matroidal if and only if it contains neither the configuration in Figure 3.3b nor a chordless cycle C_5.*

In other words, a matroidal graph differs from a matrogenic one only by the fact that its crown cannot be a chordless C_5. Therefore, split matrogenic and split matroidal graphs coincide, so in the following we will only use the name of split matrogenic graphs.

Property 1M Matrogenicity and matroidality are hereditary properties.

3.1.3 Decomposition, Enumeration, Recognition

As described in Chapter 1, every graph admits the unique decomposition into indecomposable components; it is called the *canonical decomposition.* Note that if a graph G is split, then the triple (G, A, B) should be decomposed when constructing the corresponding canonical decomposition, where A and B induce complete and empty graphs, respectively.

Theorem 3.3 [50] *The following statements are true:*

(i) *Every matrogenic graph is a unigraph.*

(ii) *A graph is threshold, matroidal, matrogenic if and only if all its indecomposable components are threshold, matroidal, matrogenic, respectively.*

(iii) *The 1-vertex graph K_1 is the only indecomposable threshold graph.*

(iv) *The following graphs are all indecomposable matroidal graphs:*

- *K_1, mK_2 and $(mK_2)^c$ (in both cases $m \geq 2$);*
- *Split graphs (G, A, B) and (G^c, A, B) such that*

$$|A| = |B|, \quad G = mK_2 \cup K(A)$$

and the matching mK_2 $(m \geq 2)$ consists of edges (a, b) where $a \in A$, $b \in B$. Here, $K(A)$ is the complete graph with the vertex set A.

(v) *The only indecomposable matrogenic graphs are all the indecomposable matroidal graphs and the chordless cycle C_5.*

Let s_n denote the number of n-vertex pairwise non-isomorphic matroidal graphs, and l_n be the same for matrogenic graphs. Consider the enumeration series for matroidal and matrogenic graphs, respectively:

$$s(x) = \sum_{n=1}^{\infty} s_n x^n \quad \text{and} \quad l(x) = \sum_{n=1}^{\infty} l_n x^n.$$

Theorem 3.3 implies (see [50]) that

$$s(x) = \frac{2p(x) - 3x + x^4}{1 - p(x)} \quad \text{and} \quad l(x) = \frac{2p(x) - 3x + x^4 + x^5}{1 - p(x)},$$

where

$$p(x) = 2x + x^4 + 2\sum_{n=3}^{\infty} x^{2n}.$$

In fact, $p(x)$ is the enumeration series for indecomposable split matroidal graphs. Now, one can obtain the recurrent formulae for s_n and l_n.

The following theorem gives an efficient recognition algorithm for matrogenic graphs; for threshold graphs such an algorithm is known [20, 24]. The theorem would also imply a recognition algorithm for matroidal graphs if the word 'matrogenic' is replaced by 'matroidal' and condition (7) is removed. A graphical sequence d in form (3.1) is called *matroidal* (*matrogenic*) if it has a matroidal (matrogenic) realization; let us denote $n = \sum_{i=1}^{r} m_i$.

Theorem 3.4 [50] *A sequence d is matrogenic if and only if one of conditions (1)–(7) holds:*

(1) $d_1 = n - 1$ *and for* $r \geq 2$ *the following sequence is matrogenic:*

$$\left((d_2 - m_1)^{m_2}, (d_3 - m_1)^{m_3}, \ldots, (d_r - m_1)^{m_r}\right);$$

(2) $d_r = 0$ *and for* $r \geq 2$ *the following sequence is matrogenic:*

$$\left(d_1^{m_1}, d_2^{m_2}, \ldots, d_{r-1}^{m_{r-1}}\right);$$

(3) $m_1 = m_r,\ d_r = 1,\ d_1 = n - m_1$ *and for* $r \geq 3$ *the following sequence is matrogenic:*

$$\left((d_2 - m_1)^{m_2}, (d_3 - m_1)^{m_3}, \ldots, (d_{r-1} - m_1)^{m_{r-1}}\right); \qquad (3.2)$$

(4) $m_1 = m_r,\ d_r = m_1 - 1,\ d_1 = n - 2$ *and for* $r \geq 3$ *the sequence (3.2) is matrogenic;*

(5) $d = \left(1^{2k}\right)$;

(6) $d = \left((2k-2)^{2k}\right)$;

(7) $d = \left(2^5\right)$.

3.2 λ-Colouring Matrogenic Graphs

Graph colouring is one of the most fertile and widely studied areas in graph theory, as is evident from browsing through the list of solved and unsolved problems in a comprehensive book on graph colouring problems [26]. The most general problem in this field is vertex colouring, consisting in assigning values (colours) to the vertices of a graph so that adjacent vertices have distinct colours; the objective is to minimize the number of colours used. The decision version of this problem, and of most of its modifications and generalizations, is NP-complete [19]. Generalizations of graph colouring arise in the design of wireless communication systems [22], where radio channels must be assigned to transmitters. The graph has a vertex for each transmitter and two vertices are joined by an edge if assigning them channels which are too close together could cause interference. The variant of the vertex colouring problem that we focus on in this section, the *λ-colouring problem*, consists in an assignment of colours from the integer set $0, 1, \ldots, \lambda$ to the vertices of a graph G such that vertices at distance at most 2 have different colours and adjacent vertices have colours which are at least 2 apart. The aim is to minimize λ.

For some special classes of graphs—such as paths, cycles, wheels and tilings—tight bounds for the number of colours necessary for a λ-colouring are known and such a colouring can be computed efficiently [1, 8, 11, 22]. Nevertheless, in general, both determining the minimum number of necessary colours [22] and deciding if this number is less than k for any fixed $k \geq 4$ [16] are NP-complete. Therefore, for many classes of graphs—such as chordal graphs [37], interval graphs [11], split graphs [3], outerplanar and planar graphs [3, 9]—approximate bounds have been looked for. For a complete survey, see [7].

In this section, we consider the λ-colouring problem restricted to some subclasses of unigraphs: matrogenic graphs, matroidal graphs, split matrogenic graphs and threshold graphs (see Figure 3.1 and Section 3.1 for definitions). We will focus on each of these subclasses, while it remains an open problem to extend our algorithm paradigm to the whole class of unigraphs.

In the following, we will discuss linear time algorithms to λ-colour those graphs taking advantage of the decompositions of degree sequences. Namely, a general algorithm paradigm is provided first, and then it is modified for threshold, split matrogenic, matrogenic and matroidal graphs. For threshold graphs, the algorithm is exact, while it approximates the optimal solution for the other classes. Note that threshold graphs are a subclass of *cographs*; that is, graphs not containing P_4 as an induced subgraph. For cographs, Chang and Kuo [11] proved that the λ-colouring problem is polynomially solvable, but they did not provide an explicit algorithm. The algorithm given in this section presents a simple method for generating an optimal solution.

Moreover, we will discuss some upper bounds for λ that are linear in the maximum degree Δ of a graph. For example, the upper bound

$$\lambda \leq \Delta^{1.5} + 2\Delta + 2$$

proved by Bodlaender et al. [3] for split graphs can be improved when the problem is restricted to both threshold and split matrogenic graphs. Griggs and Yeh [22] showed that graphs with diameter 2 satisfy $\lambda \leq \Delta^2$. Connected threshold graphs are of diameter 2, however they admit an upper bound on λ that is linear in Δ.

3.2.1 An Algorithm Paradigm for λ-Colouring Matrogenic Graphs

In the following, we will deal with the representation of a graph $G = (V, E)$ in terms of boxes with degree sequence

$$(d_1^{m_1}, d_2^{m_2}, \ldots, d_r^{m_r}), \quad d_1 > d_2 > \ldots > d_r.$$

In the previous section, we underlined how it is possible to identify the structure of a graph from our specific classes by analysing only its degree sequence. The idea of decomposition into boxes will be exploited further. For our specific graph classes, those boxes are either dominating, or isolated, or a crown, or induce in pairs either a matching or an antimatching. Since a disconnected matrogenic graph consists of a connected graph and an isolated box, we can assign the same colour to all vertices in the isolated box and run the algorithm for λ-colouring the non-trivial connected component.

The λ-colouring algorithm paradigm proceeds by labelling the boxes according to their structure. Namely, it works on the current graph (G at the beginning) and colours vertices belonging to the extremal boxes $B_1(G)$ and $B_r(G)$. Then, the *pruned graph* G_p is derived, where G_p is the subgraph induced by:

- $V - B_1(G) - B_r(G)$ if $B_1(G)$ and $B_r(G)$ are partially connected;
- $V - B_1(G)$ if $B_1(G)$ is dominating;
- $V - B_r(G)$ if $B_r(G)$ is isolated.

The pruning procedure works on compressed degree sequences and transforms the degree sequence of G into the degree sequence of G_p by eliminating either $d_1^{m_1}$ or $d_r^{m_r}$ according to the definition of the pruned graph. When $d_1^{m_1}$ is eliminated, the degree of each other box B_i is decreased by m_1; when $d_r^{m_r}$ is eliminated, the degree of box B_1 is decreased by d_r. In the algorithm paradigm, this procedure is called $\texttt{Prune}(G, imax, imin)$, where $imax$ and $imin$, initialized to 1 and r respectively, indicate at each step the indices of the boxes to be considered. From the definition of the pruned graph, observe that the third argument is equal to 0 if $B_{imax}(G)$ is a dominating box while the second one is 0 when $B_{imin}(G)$ is an isolated box. In the case of the crown, we can denote $c = imax = imin$. Notice that G_p can be disconnected but the vertices in its isolated boxes must anyway receive different colours in view of the connections eliminated during the pruning operation.

Before describing the algorithm paradigm for λ-colouring our specific classes of graphs, let us partition the colours $0, 1, \ldots, \lambda$ into *even* and *odd* in the obvious way. Furthermore, we say that the colour k is the *first odd (even) available* colour if we have already used all odd (even) colours from 0 to $k-1$. Finally, we say that the colour k is *thrown out* if we decide not to use it. After the colour k has been thrown out, it is not available anymore.

The λ-colouring algorithm paradigm is the following:

Algorithm Paradigm `colour-Matrogenic` [10]

Input: Degree sequence $d_1^{m_1}, d_2^{m_2}, \ldots, d_r^{m_r}$ of a matrogenic graph G.

Output: λ-Colouring for G.

`Initialize-Queue` $Q = \emptyset$; $G_p \leftarrow G$;

```
IF d_r = 0
   THEN colour the m_r vertices in B_r(G_p) with 0;
            G_p ← Prune(G_p, 0, r);
            r ← r − 1;
imax ← 1; imin ← r;
REPEAT
   Consider the boxes B_imax(G_p) and B_imin(G_p);
   Step 1 (B_imax(G_p) dominating)
   IF d_imax = Σ_{j=imax}^{imin} m_j − 1
      THEN colour the m_imax vertices of B_imax(G_p) with the first m_imax
               available even colours;
         FOR each i = 1, 2, ..., m_imax DO
            IF Queue-is-empty
               THEN throw the first available odd colour out
               ELSE v ← Extract-from-Queue
                     Colour v with the first available odd colour;
         G_p ← Prune(G_p, imax, 0); imax ← imax + 1;
      ELSE
      Step 2 (B_imin(G_p) isolated)
      IF d_imin = 0
         THEN Enqueue(Q, B_imin(G_p));
               G_p ← Prune(G_p, 0, imin); imin ← imin − 1;
         ELSE
         Step 3 (Matching, antimatching or crown)
         Handle the configuration induced by B_imin ∪ B_imax appropriately
            (details in Sections 3.2.3 and 3.2.4);
         G_p ← Prune(G_p, imax, imin);
         imax ← imax + 1;  imin ← imin − 1;
UNTIL G_p is empty;
Step 4
IF Queue-is-not-empty
   THEN colour the k vertices in Q with the first k available
            consecutive colours (both odd and even);
```

Theorem 3.5 [10] *The algorithm paradigm* `colour-Matrogenic` *correctly* λ*-colours a matrogenic graph* G *in* $O(n)$ *time.*

Proof: First, we show that the algorithm is correct. Observe that iteratively running the pruning procedure is a well-defined method, since each subclass we consider is closed under the pruning operation in view of Properties 2T and 1M.

Furthermore, the colouring found by the algorithm is feasible. Indeed, the vertices in V_K are even coloured and, therefore, have colours at distance at least 2. Moreover, by the nature of the algorithm, each vertex in V_S cannot be coloured with a colour at distance at most 1 to the colours of all its adjacent vertices (in V_K).

The time complexity depends on two phases: λ-colouring the box $B_i(G)$ and the pruning procedure. The first phase takes $O(m_i)$ time, while the second one is a constant. Indeed, consider the compressed degree sequence and the two indices *imax* and *imin* initialized to 1 and r, respectively. After pruning G, *imax* goes forward to $imax + 1$ if the maximum degree box has been removed, *imin* goes backward to $imin - 1$ if the minimum degree box has been removed, and another pruning step can be run considering the new degree sequence of G_p. In view of the fact that the algorithm is iterated at most r times, the required time is

$$O(n) = \sum_{i=1}^{r} (O(m_i) + O(1)) .$$

□

In what follows, we modify the previous algorithm paradigm according to the properties of considered graphs.

3.2.2 λ-Colouring Threshold Graphs

In this section, we will exploit the following constructive characterization:

Fact 3.1 [29] *Any threshold graph can be derived from a one-vertex graph by repeatedly adding either an isolated or a dominating vertex. In other words, if* $G_1 = (\{v_1\}, \emptyset)$*, then* G_k *is the threshold graph obtained by adding a vertex* v_k *(either isolated or dominating) to the threshold graph* G_{k-1}*.*

This fact implies that when the algorithm paradigm `colour-Matrogenic` is adjusted to threshold graphs, Step 3 never occurs.

Now, we want to prove that the number of colours used by the algorithm paradigm `colour-Matrogenic` when the input graph G is threshold is linear in Δ and that the resulting λ-colouring is optimal. Notice that Chang and Kuo [11] proved that the λ-colouring problem is polynomially solvable for cographs, a superclass of threshold graphs, but they did not provide an explicit algorithm. Our algorithm presents a simple method for generating an optimal solution for the class in question.

As discussed earlier, the following upper bound holds for split graphs [3]: $\lambda \le \Delta^{1.5} + 2\Delta + 2$. Also, $\lambda \le \Delta^2$ for graphs with diameter 2 [22]. The following theorem shows that $\lambda \le 2\Delta + 1$ for threshold graphs.

Theorem 3.6 [10] *The algorithm paradigm* `colour-Matrogenic` *without Step 3 λ-colours a threshold graph* G *with at most* $2\Delta + 1$ *colours.*

Proof: For the graph G, we have $\Delta = |V_K| - 1 + |V_S|$. Step 1 assigns exactly $|V_K|$ even colours, i.e. $2|V_K| - 1$ colours, and Step 4 assigns at most $|V_S|$ additional colours. It follows that the largest colour assigned to a vertex is at most

$$2|V_K| - 1 + |V_S| = 2(|V_K| + |V_S| - 1) + 1 - |V_S| \le 2\Delta + 1.$$

□

Let us restrict our attention to a connected threshold graph G. Indeed, if G is not connected, then according to the algorithm paradigm all isolated vertices are coloured with the same colour 0.

Theorem 3.7 [10] *The algorithm paradigm* `colour-Matrogenic` *without Step 3 λ-colours a connected threshold graph* G *with the minimum number of colours.*

Proof: First, at least n colours are necessary to λ-colour an n-vertex threshold graph, since connected threshold graphs have diameter 2. Further, it is easy to see that colouring the clique of G needs at least $2|V_K| - 1$ colours. Finally, all dominating vertices are at distance 1 from any other vertex (both in V_K and in V_S), therefore the colours at distance at most 1 from those chosen for the dominating vertices must be thrown out. For these reasons, during the proof, we need to consider a λ-colouring where each colour is used at most once. This new labelling is called *λ′-colouring.*

The proof is by induction on the number of vertices, and it is based on the constructive characterization given in Fact 3.1, although it is worth noticing that the sequence of vertices defined in Fact 3.1 is different from the sequence in which vertices are coloured by the algorithm.

Basis: Graph G_1 is trivially a threshold graph and $\lambda'_1 = \lambda_1 = 0$ (minimum).

The graph G_2 is λ'-coloured with the minimum number of colours. Indeed, G_2 is constituted either by two isolated vertices v_1 and v_2 and $\lambda'_2 = 1$, or by the edge (v_1, v_2) and $\lambda'_2 = \lambda_2 = 2$. Notice that $\lambda' = \lambda$ when the threshold graph is connected, while it holds $\lambda' > \lambda$ when the graph is not connected.

Inductive step: Let us assume that the threshold graph G_k with the vertex set $\{v_1, v_2, \dots, v_k\}$ is optimally λ'-coloured. Let us denote by Q_k the subset of colours in $0, 1, \dots, \lambda'_k$ not used to colour G_k. As an example, $Q_2 = \{1\}$ when G_2 is an edge.

Moving from k to $k+1$ vertices, v_{k+1} can be added either as an isolated vertex or as a dominating one. In the first case, any colour c in Q_k is suitable to colour v_{k+1} because v_{k+1} is not connected to the vertices $v_1, v_2, \dots, v_k$ and Fact 3.1 guarantees that the edges (v_i, v_{k+1}), $1 \le i \le k$, will never be introduced. It follows that $\lambda'_{k+1} = \lambda'_k$ and Q_{k+1} is obtained by eliminating c from Q_k. Notice that if $Q_k = \emptyset$ then $\lambda'_{k+1} = \lambda'_k + 1$ because the first available colour has to be added.

In the second case, the new vertex v_{k+1} requires a colour different from all the previous ones and two apart. Then, λ'_{k+1} is incremented by two colours with respect to λ'_k; that is, $\lambda'_{k+1} = \lambda'_k + 2$ because one colour is used for v_{k+1}, while the other is added to Q_k to obtain Q_{k+1}. In this case, $\lambda'_{k+1} = \lambda_{k+1}$ and G_{k+1} is optimally λ_{k+1}-coloured. Because $G_n = G$ is connected, the result follows. □

3.2.3 λ-Colouring Split Matrogenic Graphs

If the input graph of the algorithm paradigm `colour-Matrogenic` is split matrogenic, then Step 3 can deal with both perfect matchings and antimatchings. In the following, we will detail how these structures can be coloured. Furthermore, we prove that $3\Delta + 1$ colours are enough to λ-colour split matrogenic graphs, thus improving for this class the general result for split graphs due to Bodlaender et al. [3].

Lemma 3.1 [10] *Let $G = (V_K \cup V_S, E)$ be a split matrogenic graph and let $K_r \subseteq V_K$ and $S_r \subseteq V_S$, $K_t \subseteq V_K$ and $S_t \subseteq V_S$ induce a perfect matching M_r and an antimatching A_t, respectively. Then:*

(a) The perfect matching M_r of dimension $h_r \geq 3$ can be λ-coloured with $2h_r$ consecutive colours, and all the vertices in K_r are coloured with even colours and all the vertices in S_r are coloured with odd colours.

(b) The perfect matching M_r of dimension $h_r = 2$ can be λ-coloured with 5 consecutive colours.

(c) The antimatching A_t of dimension $h_t \geq 3$ can be λ-coloured with at most

$$2h_t + \lceil h_t/2 \rceil - 1$$

consecutive colours, and the leftmost vertex in K_t and the rightmost vertex in S_t are coloured with the minimum and maximum used colours, respectively.

Proof: To prove item (a), first consider the perfect matching M_r of dimension $h_r \geq 3$. We can colour vertices in K_r from left to right with h_r consecutive even colours:

$$2k,\ 2k + 2,\ \dots\ , 2k + 2h_r.$$

It is not difficult to assign the remaining odd colours to the vertices in S_r in such a way that dually correlated vertices have colours at distance at least 2:

$$2k + 3,\ 2k + 5,\ \dots\ , 2k + 2h_r - 1,\ 2k + 1 \quad \text{(see Figure 3.6a)}$$

If $h_r = 2$, it is enough to use four consecutive colours but we cannot use the successive colour that must be either thrown out or used for the enqueued vertices (see Figure 3.6b). So, we need at least five colours.

Now, let us consider the antimatching A_t of dimension $h_t \geq 3$, and let us group the involved vertices four by four putting together two consecutive vertices in K_t and their dually uncorrelated vertices in S_t. As the antimatching is determined, each group can be coloured with four consecutive colours, but—before starting to

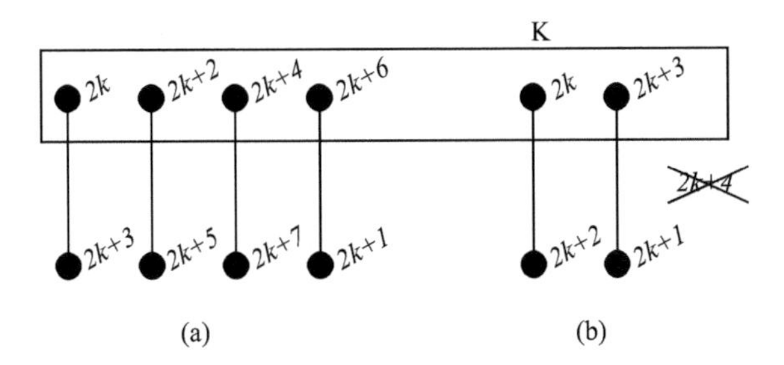

Figure 3.6 Some coloured matchings.

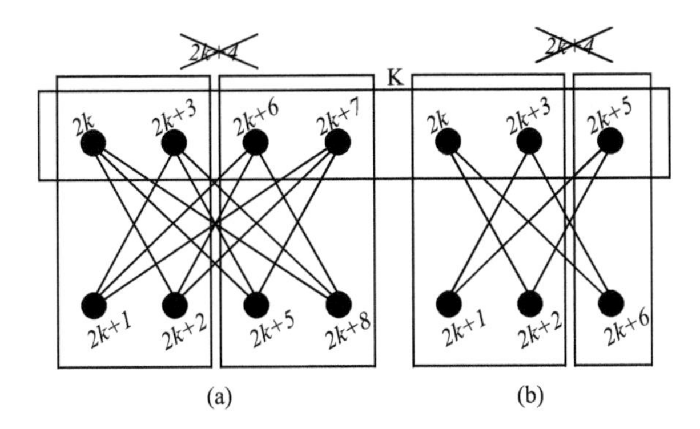

Figure 3.7 Some coloured antimatchings.

colour another group—one colour c must be thrown out (see $2k+4$ in Figure 3.7a and 3.7b). Indeed, c cannot colour any vertex in K or in S_t because it would be connected to the vertex in K coloured $c-1$. Hence, exactly $2h_t + \lceil h_t/2 \rceil - 1$ colours are used, and it is always possible to impose the constraint that the colours assigned to the leftmost vertex in K_t and to the rightmost vertex in S_t are the minimum and the maximum, respectively. □

It is worth pointing out that the vertex of C_i ($C_i = M_t$ or A_t, $t \geq 3$) with the largest colour is in S and not in K. This ensures the correctness of the following detailed Step 3 in the algorithm paradigm `colour-Matrogenic` for split matrogenic graphs:

Step 3 (`Split Matrogenic`)

IF $d_{imin} = 1$ AND $m_{imin} \geq 3$
(i.e. matching M of dimension $h = m_{imin} \geq 3$)
THEN colour M (cf. Lemma 3.1 (a))with the first $2\,h$ available colours;
$G_p \leftarrow$ `Prune`$(G_p, imax, imin)$;
$imax \leftarrow imax + 1$; $imin \leftarrow imin - 1$;

```
ELSE
IF d_imin = 1 AND m_imin = 2 (i.e. matching M of dimension 2)
  THEN colour M (cf. Lemma 3.1 (b)) with the first 5 colours, leaving
           1 unused colour (it is 2k + 4 in Figure 3.6b);
       IF Queue-is-empty
         THEN throw the available colour out
         ELSE v ← Extract-from-Queue
                Colour v with the first available colour;
 ELSE
 IF d_imin = h − 1 ≥ 2 (i.e. antimatching A of dimension h = m_imin)
   THEN colour A (cf. Lemma 3.1)with the first 2h + ⌈h/2⌉ − 1
            available colours leaving ⌈h/2⌉ − 1 unused colours;
        FOR each i = 1, 2, ..., ⌈h/2⌉ − 1 DO
          IF Queue-is-empty
            THEN throw the first unused colour out (it is 2k + 4 in
                   Figure 3.7a and 3.7b)
            ELSE v ← Extract-from-Queue
                   Colour v with the first available colour;
 G_p ← Prune(G_p, imax, imin);
 imax ← imax + 1;   imin ← imin − 1;
```

The next theorem bounds the number of used colours and shows that this number is linear in Δ. To do that, we need to prove some results about the maximum degree Δ of a split matrogenic graph. Without loss of generality, we assume that $|V_S| > 0$.

Lemma 3.2 [10] *If $G = (V_K \cup V_S, E)$ is a connected split matrogenic graph, then its maximum degree Δ is bounded as follows:*

$$|V_K| \leq \Delta \leq |V_K| + |V_S| - 1 \tag{3.3}$$

and

$$\Delta \geq |V_K| + \nu_S - 1. \tag{3.4}$$

Proof: Notice that Δ has the smallest value if the red graph of G is constituted by a unique perfect matching. In this case, the maximum degree is $|V_K|$, where $|V_K| - 1$ is the contribution of the clique and 1 is the contribution of the perfect matching. On the contrary, G has the largest maximum degree if it contains a dominating vertex; in such a case, $\Delta = |V_K| + |V_S| - 1$.

Concerning the second inequality, consider the first box B_1. Each vertex inside B_1 has degree Δ and receives a contribution of

$$|V_K| - 1 + |V_S| - |C_1 \cap V_S|$$

from the black graph. The contribution of the red graph is 0 if B_1 is the set of dominating vertices, it is 1 if B_1 is involved in a perfect matching, and it is more than 1 if B_1 is involved in an antimatching. Then,

$$\Delta \geq |V_K| - 1 + |V_S| - |C_1 \cap V_S| \geq |V_K| + \nu_S - 1.$$

□

Theorem 3.8 [10] *The algorithm paradigm* `colour-Matrogenic` *with Step 3* `(Split Matrogenic)` *λ-colours a connected split matrogenic graph G with at most* $3\Delta + 1$ *colours.*

Proof: First of all, note that the following equalities hold in view of Theorem 3.1:

$$\nu_K + \sum_{r=1}^{\mu} h_r + \sum_{t=1}^{\alpha} h_t = |V_K| \tag{3.5}$$

and

$$\nu_S + \sum_{r=1}^{\mu} h_r + \sum_{t=1}^{\alpha} h_t = |V_S|. \tag{3.6}$$

Furthermore, from the nature of the algorithm, each empty graph of K in the red graph of G contributes to the number of colours twice the number of its vertices, and each empty graph of S in the red graph contributes to the number of colours no more than the number of its vertices. Finally, each perfect matching and each antimatching in the red graph contribute according to Lemma 3.1. Therefore, the algorithm uses the number of colours bounded by

$$\tau = 2\nu_K + \nu_S + \sum_{r=1}^{\mu}(2h_r + 1) + \sum_{t=1}^{\alpha}\left(2h_t + \left\lceil \frac{h_t}{2} \right\rceil - 1\right).$$

Using Lemma 3.2 and the values of $|V_K|$ and $|V_S|$ in this proof, we obtain

$$\begin{aligned}
\tau &\overset{(3.5)}{=} 2|V_K| + \nu_S + \mu + \sum_{t=1}^{\alpha}\left\lceil \frac{h_t}{2} \right\rceil - \alpha - 1 + 1 \\
&\overset{(3.4)}{\leq} \Delta + |V_K| + \mu + \sum_{t=1}^{\alpha}\left\lceil \frac{h_t}{2} \right\rceil - \alpha + 1 \\
&\overset{(3.3)}{\leq} 2\Delta + \sum_{r=1}^{\mu} h_r + \sum_{t=1}^{\alpha} h_t - \alpha + 1 \\
&\overset{(3.5)}{\leq} 2\Delta + |V_K| + 1 \\
&\overset{(3.3)}{\leq} 3\Delta + 1.
\end{aligned}$$

□

3.2.4 λ-Colouring Matrogenic and Matroidal Graphs

Matrogenic and matroidal graphs differ from split matrogenic graphs only by the fact that they have a non-empty crown induced by the vertices in V_C. Hence, we have to specialize Step 3 further in order to guarantee that the crown is handled correctly. Notice that no colour already used to λ-colour $V_K \cup V_S$ can be reused for the crown, since each vertex in V_C is adjacent to each vertex in V_K and hence it is connected by a path of length 2 to each vertex of V_S.

A crown that is a perfect matching of dimension $h \geq 2$ can be λ-coloured with $2h$ consecutive colours in such a way that dually correlated vertices receive colours with difference at least 2 (see Figure 3.8a and Lemma 3.1 (a)). Note that if $h = 2$, we do not need to add a further colour, as the crown is the last box we consider. A crown that is a chordless C_5 (which never occurs if G is matroidal) can be trivially λ-coloured with five consecutive colours (see Figure 3.8b). Now, it remains to consider only the case in which the crown is a hyperoctahedron. Notice that a hyperoctahedron H of dimension h, $h \geq 2$, of X onto Y can be obtained from the clique induced by $X \cup Y$ when a perfect matching of dimension h of X onto Y is eliminated.

Lemma 3.3 [10] *A h-hyperoctahedron H, $h \geq 2$, can be optimally λ-coloured with $3h-1$ consecutive colours.*

Proof: Consider a hyperoctahedron of dimension h. Each vertex can receive a colour adjacent to the colour of its dually uncorrelated vertex, but at distance at least 2 from the colour of any other vertex. Therefore, we can colour each pair of uncorrelated vertices with two consecutive colours and—before starting to colour another pair—throw out one colour (see Figure 3.8c). It follows that $3h-1$ colours are sufficient.

On the other hand, if a vertex v has colour c, then the two adjacent colours $c+1$ and $c-1$ cannot be assigned to any vertex different from its dually uncorrelated

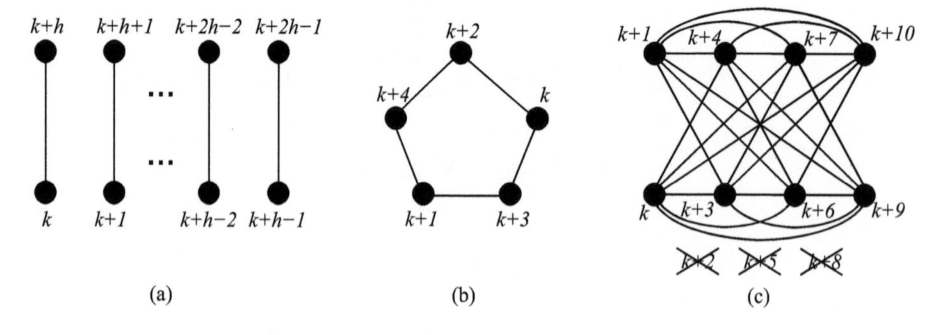

Figure 3.8 All possible crowns with a feasible λ-colouring: (a) a perfect matching; (b) a chordless C_5; (c) a hyperoctahedron.

one. Therefore, the elimination of one colour from each triple generates an optimal λ-colouring of the hyperoctahedron. $\square$

It remains to complete the algorithm `colour-Matrogenic` by adding the following code to Step 3 described in the previous section:

Step 3 (`Crown`)

```
...
IF imin = imax = c (i.e. the crown is reached)
  THEN
    IF d_c = 1 (i.e. crown is a matching M of dimension h = m_c)
      THEN colour M with the first 2h available colours;
      ELSE
        IF d_c = 2 (and m_c = 5, i.e. crown is a chordless C_5)
          THEN colour C_5 with the first 5 available colours;
          ELSE
            IF d_c = m_c - 2 (i.e. crown is a hyperoctahedron H of
                     dimension h = m_c/2)
              THEN colour H (cf. Lemma 3.3) with the first 3h - 1
                         available colours;
                     FOR each unused colour DO
                       IF Queue-is-empty
                         THEN throw the current colour out
                         ELSE v <- Extract-from-Queue
                                 Colour v with the first available colour;
```

Theorem 3.9 [10] *The algorithm paradigm* `colour-Matrogenic` *with Step 3* (`Split Matrogenic`) *completed with Step 3* (`Crown`) *λ-colours a matrogenic or matroidal graph G with at most 3Δ colours.*

Proof: The proof is very similar to the proof of Theorem 3.8 if we add the contribution of the crown to the inequalities of Lemma 3.2, to the bound on the number of colours and to the bounds on Δ. Namely, the contribution of the crown to degrees is $|V_C|$ and to the number of colours is at most $\frac{3}{2}|V_C|$. $\square$

3.3 On the Spectrum of Threshold Graphs

A graph $G = G(\mathcal{V}, \mathcal{E})$ of order n has a labelled vertex set $\mathcal{V} = \{1, 2, \ldots, n\}$ containing n vertices and a set $\mathcal{E}$ of m edges consisting of unordered pairs of vertices. When a subset $\mathcal{V}_1$ of $\mathcal{V}$ is deleted, the edges incident to $\mathcal{V}_1$ are also deleted. The subgraph $G - \mathcal{V}_1$ of G is said to be an *induced* subgraph of G. The subgraph of G obtained by deleting a particular vertex v is simply denoted by $G - v$. The *cycle* and the *complete graph* on n vertices are denoted by C_n and K_n, respectively.

We use boldface, say $\boldsymbol{G}$, to denote the $(0,1)$-*adjacency matrix* of the graph bearing the same name G, where the ij-th entry of the symmetric matrix $\boldsymbol{G}$ is 1 if $(i,j) \in \mathcal{E}$ and 0 otherwise. We note that the graph G is determined, up to isomorphism, by $\boldsymbol{G}$. The adjacency matrix $\boldsymbol{G}^{\mathrm{c}}$ of the complement G^{c} of G is $\boldsymbol{J} - \boldsymbol{I} - \boldsymbol{G}$, where each entry of $\boldsymbol{J}$ is one and $\boldsymbol{I}$ is the identity matrix. The degree of a vertex i is the number of non-zero entries in the i-th row of $\boldsymbol{G}$.

The disconnected graph with two components G_1 and G_2 is their *disjoint union*, denoted by $G_1 \sqcup G_2$. For $r \geq 2$, the graph rG is the disconnected graph with r components where each component is isomorphic to G. The *join* $G_1 \nabla G_2$ of two graphs G_1 and G_2 is $(G_1^{\mathrm{c}} \sqcup G_2^{\mathrm{c}})^{\mathrm{c}}$.

For the linear transformation $\boldsymbol{G}$, the n real numbers $\{\lambda\}$ satisfying $\boldsymbol{G}\boldsymbol{x} = \lambda \boldsymbol{x}$ for some non-zero vector $\boldsymbol{x} \in \mathbb{R}^n$ are said to be *eigenvalues* of $\boldsymbol{G}$ and form the *spectrum* of G. They are the solutions of the *characteristic polynomial* $\phi(G, \lambda)$ of $\boldsymbol{G}$ defined as the polynomial $\det(\lambda \boldsymbol{I} - \boldsymbol{G})$ in λ. The subspace $\ker \boldsymbol{G}$ of $\mathbb{R}^n$ that maps to zero under $\boldsymbol{G}$ is said to be the *nullspace* of $\boldsymbol{G}$. A graph G is said to be *singular of nullity* η if the dimension of $\ker(\boldsymbol{G})$ is η. The non-zero vectors $\boldsymbol{x} \in \mathbb{R}^n$ in the nullspace, termed *kernel eigenvectors* of G, satisfy $\boldsymbol{G}\boldsymbol{x} = \boldsymbol{0}$. We note that the multiplicity of the eigenvalue zero is η. If there exists a kernel eigenvector of $\boldsymbol{G}$ with no zero entries, then G is said to be a *core graph.* The cycle C_4 on four vertices is a core graph of nullity two with a kernel eigenvector $(1,1,-1,-1)^{\mathrm{T}}$ for the usual labelling of the vertices round the cycle. A core graph of nullity one is said to be a *nut graph* [45]. The *minimal configuration* for a particular core, to be defined formally in Section 3.3.4, is intuitively a graph of nullity one with the minimal number of vertices and edges for that core.

Let $\boldsymbol{j}$ be the vector with each entry equal to one. The distinct eigenvalues

$$\mu_1, \mu_2, \ldots, \mu_p, \quad 1 \leq p \leq n,$$

which have an associated eigenvector *not* orthogonal to $\boldsymbol{j}$, are said to be *main.* We denote the remaining distinct eigenvalues by $\mu_{p+1}, \ldots, \mu_s$, $s \leq n$, and refer to them as *non-main.* By the Perron–Frobenius Theorem [18, p. 6], the maximum eigenvalue of the adjacency matrix of a connected graph has an associated eigenvector (termed the *Perron vector*) with all its entries positive. Therefore, at least one eigenvalue of a graph is main.

A *cograph*, or a complement-reducible graph, is a graph that can be generated from the single-vertex graph K_1 by complementation and disjoint union. Threshold graphs, discussed in the previous section, are a subclass of cographs. There are several equivalent definitions of threshold graphs; the most useful for our purposes is one in terms of their degree sequences [20, 29]. For this definition, which is given below, one needs the *graph partition* Π of $2m$ into *parts* equal to the vertex degrees $\{\rho_1, \rho_2, \ldots, \rho_n\}$. The array of boxes $F(\Pi)$, known as a *Ferrers–Young diagram* for the monotonic non-increasing sequence $\Pi = \{\rho_1, \rho_2, \ldots, \rho_n\}$, consists of n rows of ρ_i boxes as i runs successively from 1 to n. Threshold graphs are characterized by a particular shape of the Ferrers–Young diagram (see Figure 3.12), which will be described in Section 3.3.1. Also, as discussed in Section 3.1.1, a threshold graph is a graph with no induced subgraphs isomorphic to any of the following graphs on four vertices: the path P_4, the cycle C_4 and the graph $2K_2$ (see Figure 3.3).

Definition 3.1 [32] *If the monotonic non-increasing degree sequence* $\Pi = \{\rho_1, \rho_2, \ldots, \rho_n\}$ *of a graph* G *is represented by the rows of a Ferrers–Young diagram* $F(\Pi)$, *where the length of the principal square of* $F(\Pi)$ *is* $f(\Pi)$ *and the lengths* $\{\pi_k^* : 1 \leq k \leq f(\Pi)\}$ *of the columns of* $F(\Pi)$ *satisfy* $\pi_k^* = \rho_k + 1$, *then* G *is said to be a threshold graph.*

If the parts of a threshold graph partition of $2m$ are all equal, then the graph is regular and corresponds to the complete graph. If, on the other hand, there are as many distinct sizes of the parts of a threshold graph partition of $2m$ as possible, then the graph is antiregular. Recall that at least two vertices in a graph have the same degree.

Definition 3.2 *An antiregular graph on* r *vertices is defined as a threshold graph whose vertex degrees take as many different values as possible; that is,* $r-1$ *distinct non-negative integral values.*

Definition 3.3 *The partition* $\mathcal{V}_1 \sqcup \mathcal{V}_2 \sqcup \ldots \sqcup \mathcal{V}_r$ *of the vertex set* $\mathcal{V}$ *of a graph* G *is said to be an equitable partition if, for all* $i, j \in \{1, 2, \ldots, r\}$, *the number of neighbours in* $\mathcal{V}_j$ *of a vertex in* $\mathcal{V}_i$ *depends only on the choice of* i *and* j.

The overall aim of this section is to explore the spectrum of the adjacency matrix of an antiregular connected graph and show common properties with connected threshold graphs, having an equitable partition with the minimal number r of parts.

Now, let us discuss cographs. They will be used in Section 3.3.1 to determine a particular representation of a threshold graph that has earned it the name of *nested split graph.* In addition, various other representations will be presented and selectively used to simplify our proofs.

A cograph is the union or the join of subgraphs of the form

$$(\ldots((r_1K_1)^{\mathrm{c}} \sqcup (r_2K_1))^{\mathrm{c}} \sqcup \ldots \sqcup (r_sK_1))^{\mathrm{c}},$$

where $r_i \in \mathbb{Z}^+ \cup \{0\}$ for all i. Therefore, the family of cographs is the smallest class of graphs that includes K_1 and is closed under complementation and disjoint union. It is well known that no cograph on at least four vertices has P_4 as an induced subgraph [4]. In fact, cographs can be characterized as P_4-free graphs. Cographs have received much attention since the 1970s, and they were discovered independently by many authors [27, 28, 46, 48]. For a more detailed treatment of cographs, see [4].

Connected graphs, which are $\{2K_2, P_4, C_4\}$-free, necessarily have a *dominating vertex*; that is, a vertex adjacent to all other vertices of the graph. Thus, all connected threshold graphs have a dominating vertex. By construction, a connected cograph also has a dominating vertex. Therefore, its complement has at least one isolated vertex. A necessary condition for a connected non-trivial graph to have a connected complement is that it has P_4 as an induced subgraph [32, Theorem 1.19].

The set of cographs and the class of non-trivial graphs with a connected complement are disjoint as sets. However, if the graph H is $P_4 \sqcup K_1$, then both H and H^c have an induced P_4. Thus, there exist connected graphs that are neither P_4-free nor have a connected complement.

Recall that

$$G_1 \nabla G_2 = (G_1^c \sqcup G_2^c)^c.$$

Hence, cographs are also characterized as the smallest class of graphs that includes K_1 and is closed under join and disjoint union. This definition of cographs is the basis of the proof in [2] that cographs are polynomial reconstructible from the deck of characteristic polynomials of one-vertex-deleted subgraphs.

A cograph can be represented uniquely by a *cotree*, as explained in [13] and later in [2]. Figure 3.9 shows the cotree T_G of the cograph G. The vertices $\oplus$, $\otimes$ and $\bullet$ of the cotree represent the disjoint union, the join and the vertices of the cograph, respectively. For simplicity, we say that the terminal vertices of T_G are vertices of G. The cotree T_G is a rooted tree and only the terminal vertices represent the cograph vertices. An interior vertex $\oplus$ or $\otimes$ of T_G represents the subgraph of G induced by its terminal successors. The immediate successors of $\oplus$ can be cograph vertices $\bullet$ or $\otimes$. Similarly, the immediate successors of $\otimes$ can be cograph vertices $\bullet$ or $\oplus$. Therefore, the interior vertices of T_G on an (oriented) path descending from the root to a terminal vertex of T_G are a sequence of alternating $\otimes$ and $\oplus$.

3.3.1 Representations of Threshold Graphs

In this section, we present some of the various representations of threshold graphs. Collectively, they provide a wealth of information that determines combinatorial properties of these graphs. We start with the cotree representation discussed above. There are certain restrictions on the structure of a cotree in the case when a cograph is a threshold graph.

We give a proof of the following result quoted in [2].

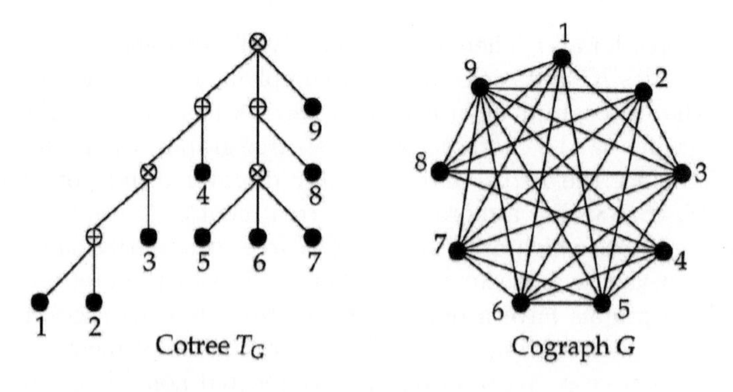

Figure 3.9 Cotree T_G for the cograph G.

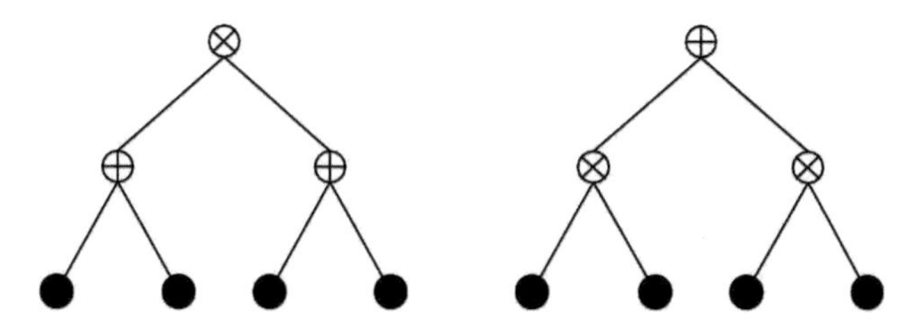

Figure 3.10 Representations of C_4 (left) and $2K_2$ (right) in a cotree.

Lemma 3.4 [2] *If a cograph G is also a threshold graph, then each interior vertex of T_G has at most one interior vertex as an immediate successor.*

Proof: A threshold graph G is P_4-free and, therefore, is a cograph which can be represented by a cotree T_G. Note that P_4 cannot be represented as a cotree. In a threshold graph, there are no induced subgraphs isomorphic to C_4 or $2K_2$. Therefore, the configurations in Figure 3.10 representing C_4 and $2K_2$ as cotrees are not allowed in the cotree T_G corresponding to a threshold graph G. We deduce that the number of interior vertices that are immediate successors of an interior vertex is less than two, as required. □

We now present various other representations of threshold graphs that are used in the proofs that follow.

Cotrees of Nested Split Graphs

A *caterpillar* is a tree in which the removal of all terminal vertices (i.e. those of degree 1) gives a path. The following result follows immediately from Lemma 3.4.

Corollary 3.1 *The cotree of a threshold graph is a caterpillar.*

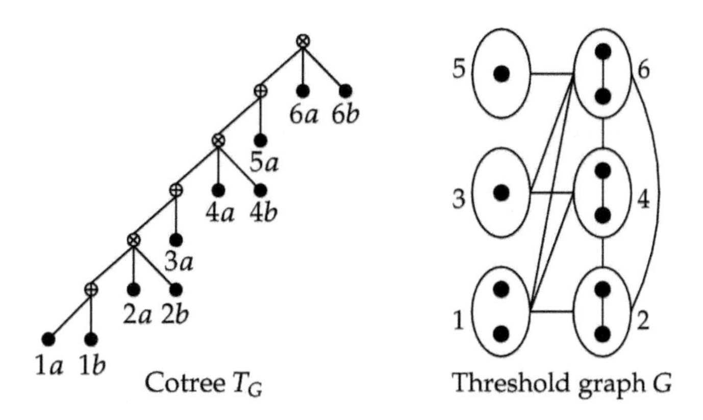

Figure 3.11 Cotree T_G and the nested structure of the threshold graph G.

The vertex set of a *split graph* is partitioned into two subsets, one of which is a *clique* (inducing a complete subgraph) and the other is a *coclique* or an independent set (inducing an empty graph with no edges). Because of its structure, a threshold graph is also referred to as a *nested split graph.*

The first vertex labelling of a threshold graph is according to its construction; this labelling will be called *Lab1.* Starting from K_1 (vertex 1a), the graph in Figure 3.11 is

$$((((((((K_1 \sqcup K_1)\nabla K_1)\nabla K_1) \sqcup K_1)\nabla K_1)\nabla K_1) \sqcup K_1)\nabla K_1)\nabla K_1)$$

coded as

$$((((((K_1 \sqcup K_1)\nabla_2 K_1) \sqcup K_1)\nabla_2 K_1) \sqcup K_1)\nabla_2 K_1)$$

to avoid repetitions of successive joins or unions. Therefore, according to the vertex labelling in Figure 3.11, G is

$$((((((((1a \sqcup 1b)\nabla 2a)\nabla 2b) \sqcup 3a)\nabla 4a)\nabla 4b) \sqcup 5a)\nabla 6a)\nabla 6b).$$

The cotree T_G represents the threshold graph G drawn next to it in a way so as to emphasize the nested split graph structure of G. In the graph G, the circumscribed vertices labelled 1 form the subgraph (coclique) induced by the vertices 1a and 1b, and similarly for the other circumscribed subsets of vertices. Thus, circumscribed vertices in G represent either a clique or a coclique. A line in G joining a clique and a coclique (or another clique) means that each vertex of the former is adjacent to each vertex of the latter. In the cotree T_G, the terminal vertices ($\bullet$) which are immediate successors of a vertex $\otimes$ form a clique, whereas those immediately succeeding a vertex $\oplus$ form a coclique.

Minimal Equitable Partition of the Vertex Set

Our labelling of the r parts in the equitable partition of the vertices of a connected threshold graph $C(a_1, a_2, \ldots, a_r)$ follows the addition of the vertices in the construction in order, namely:

$$(\ldots((\sqcup_{a_1} K_1 \nabla_{a_2} K_1) \sqcup_{a_3} K_1)\nabla \ldots \sqcup_{a_r} K_1)$$

according to the coded representation of the graph in Figure 3.11. Then, the nested structure of the threshold graph becomes clear. The parts are cliques or cocliques of size a_i for $1 \leq i \leq r$. For the minimal value of r, Π is said to be a *non-degenerate* equitable partition for the *non-degenerate* representation $C(a_1, a_2, \ldots, a_r)$. All other equitable partitions of the vertex set are refinements of Π with a larger number of parts, when an equitable partition and the corresponding representation $C(a_1, a_2, \ldots, a_r)$ are said to be *degenerate.* Unless otherwise stated, we will assume that equitable vertex partitions and representations are non-degenerate. In particular, $a_1 \neq 1$.

According to our labelling convention (Lab1) for $C(a_1, a_2, \ldots, a_r)$ as in Figure 3.11, a threshold graph G whose cotree T_G has root $\otimes$ is connected. If

r is even, then a_1 is associated with a coclique, whereas for r odd, a_1 is associated with a clique. It follows that the monotonic non-increasing vertex degree sequence of G will be associated with

$$a_r, a_{r-2}, \ldots, a_2, a_1, a_3, \ldots a_{r-1}$$

in that order if r is even, and

$$a_r, a_{r-2}, \ldots, a_1, a_2, a_4, \ldots a_{r-1}$$

in that order if r is odd. By convention, therefore, for a non-degenerate equitable partition, $a_i \geq 1$ for $2 \leq i \leq r-1$ and $a_1 \geq 2$. According to this representation, the graph of Figure 3.11 has the non-degenerate representation $C(2, 2, 1, 2, 1, 2)$.

The Binary Code of a Threshold Graph

For the purposes of inputting an n-vertex threshold graph to be processed in a computer program, the graph is encoded as a string of $n-1$ bits. The graph is represented as a sequence of 0 and 1 entries where 0 represents the *addition* of an isolated vertex and 1 represents the *addition* of a dominating vertex in the construction of the graph, starting from K_1, as described previously. For example, the graph of Figure 3.11 is encoded as (011011011).

Degree Sequence

The last representation of a threshold graph that we now give is constructed from its degree sequence. Following Definition 3.1, let $F(\Pi)$ be the Ferrers–Young diagram (Figure 3.12) for the non-increasing degree sequence giving a vertex partition $\Pi = \{\rho_1, \rho_2, \ldots, \rho_n\}$ of $2m$ for an n-vertex graph. The largest principal square of boxes in $F(\Pi)$ is termed the *Durfee square* and $f(\Pi)$ denotes the size of the Durfee square; that is, the length of a side of the Durfee square. The diagram is graphical if and only if $\pi_i^* \geq \rho_i + 1$ for $1 \leq i \leq f(\Pi)$ [36].

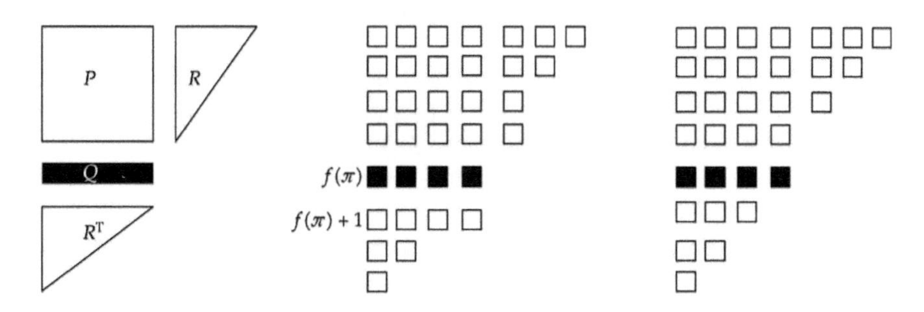

Figure 3.12 Ferrers–Young diagram for a threshold graph.

We will need the following result, which was already discussed in Section 3.1.1 (see Property 1T).

Lemma 3.5 [32] *A threshold graph is a unigraph.*

The degree sequence Π of a threshold graph also produces a particular structure of the Ferrers–Young diagram $F(\Pi)$ shown in Figure 3.12.

Lemma 3.6 [32] *For a threshold graph, $F(\Pi)$ consists of four blocks P, Q, R and its transpose R^{T}, where P is the Durfee square, Q is the $(f(\Pi)+1)$-th row of $F(\Pi)$ of length $f(\Pi)$ and R is the array of boxes left after removing the Durfee square from the first $f(\Pi)$ rows of $F(\Pi)$.*

3.3.2 The Structure of Threshold Graphs

An interesting algorithm was presented in [36] to construct a threshold graph. The adjacency list *adjList* of the graph (i.e. the lists of neighbours of each vertex) is, in fact, obtained by filling in the boxes of the i-th row in $F(\Pi)$ with consecutive integers starting from 1, but skipping i. By Lemma 3.5, $F(\Pi)$ gives a unique threshold graph up to isomorphism and, therefore, provides a canonical vertex labelling. We now present a procedure to produce the adjacency matrix of the labelled threshold graph corresponding to *adjList* from $F(\Pi)$. Note that this gives us the second labelling, Lab2, in order of the non-increasing degree sequence and, therefore, different from Lab1 used for $C(a_1, a_2, \ldots, a_r)$.

Theorem 3.10 [43] *The $n \times n$ adjacency matrix $\mathbf{G}$ of a threshold graph G is obtained from its Ferrers–Young diagram $F(\Pi)$, representing the degree sequence of an n-vertex graph, as follows. The i-th box is inserted in each i-th row and filled with entry 0. The rest of the existing boxes are filled with entry 1. Boxes are now inserted so that an $n \times n$ array of boxes is obtained. Each of the remaining empty boxes is filled with zero. The obtained $n \times n$ array of $(0,1)$-numbers is the adjacency matrix $\mathbf{G}$.*

The rows and columns of the adjacency matrix constructed in Theorem 3.10 are indexed according to the non-increasing degree sequence. If, for a threshold graph, each of the boxes of the i-th row in $F(\Pi)$ is filled with i to obtain $H(\Pi)$, then the adjacency list *adjList* of the graph is just a rearrangement of the entries of $H(\Pi)$ since, by Definition 3.1, $\pi_k^* = \rho_k + 1$. Due to the shape of the non-zero part, the adjacency matrix is said to have a '*stepwise form*' [5, 6].

The Antiregular Graph

The antiregular graph A_r may be considered to be the smallest threshold graph for an equitable vertex partition having a given number $(r-1)$ of parts.

Definition 3.4 *An antiregular graph A_r on r vertices is a graph whose vertex degrees take the values of $r-1$ distinct (non-negative) integers.*

We shall use the r-vertex connected antiregular graph A_r with the largest number $(r-1)$ of parts in its equitable partition, having degenerate representation

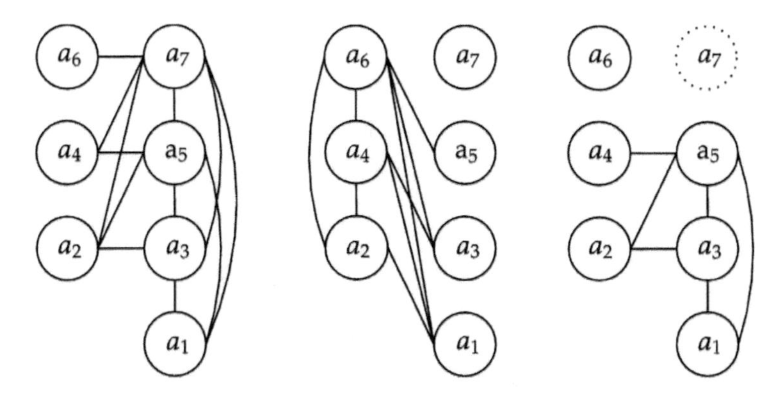

Figure 3.13 The threshold graph $G = C(a_1, a_2, \ldots, a_7)$, G^c and $G - v$ for $a_7 = 1$. (From left to right.)

$C(1, 1, \ldots, 1)$ using Lab1. Any part can be expanded to produce a threshold graph $C(a_1, a_2, \ldots, a_r)$, taking care to preserve the nested split structure. The connected antiregular graph A_r with degenerate equitable partition into r parts is adopted as the *underlying graph* of a connected threshold graph for an equitable vertex partition with r parts.

Lemma 3.7 [43] *An induced subgraph H of $G = C(a_1, a_2, \ldots, a_r)$ is $C(b_1, b_2, \ldots, b_r)$, where $0 \leq b_i \leq a_i$ for $1 \leq i \leq r$.*

Proof: When $a_i \neq b_i$ for at least one value of i to produce H, vertices are deleted from the part of size a_i in the equitable partition of $\mathcal{V}_G$. This procedure produces an induced subgraph at each stage and it is repeated until b_i is reached for each i. □

The threshold graph $C(1, 1, \ldots, 1)$ having r parts, where each part is of size 1, is the degenerate form of A_r. Its non-degenerate form, consistent with the cotree representations of threshold graphs, is $C(2, 1, \ldots, 1, 1)$; it has $r - 1$ parts with only the first part of size 2. As an immediate consequence of Lemma 3.7, we have the following.

Corollary 3.2 [43] *The connected antiregular graph $C(2, 1, 1, \ldots, 1)$, having $r - 1$ parts, with degenerate representation $C(1, 1, 1, \ldots, 1)$, having r parts, is an induced subgraph of $C(a_1, a_2, \ldots, a_r)$, where $1 \leq a_i$ for $2 \leq i \leq r$ and $a_1 \geq 2$.*

On taking the complement of $C(a_1, a_2, \ldots, a_r)$ or on deleting a dominating vertex when $a_r = 1$, a disconnected graph is obtained (see Figures 3.13 and 3.14).

Proposition 3.1 [43] *Let v be the dominating vertex of A_r. Then,*

(i) $A_r - v$ *is* $K_1 \sqcup A_{r-2}$ *and*
(ii) $A_r^c = K_1 \sqcup A_{r-1}$.

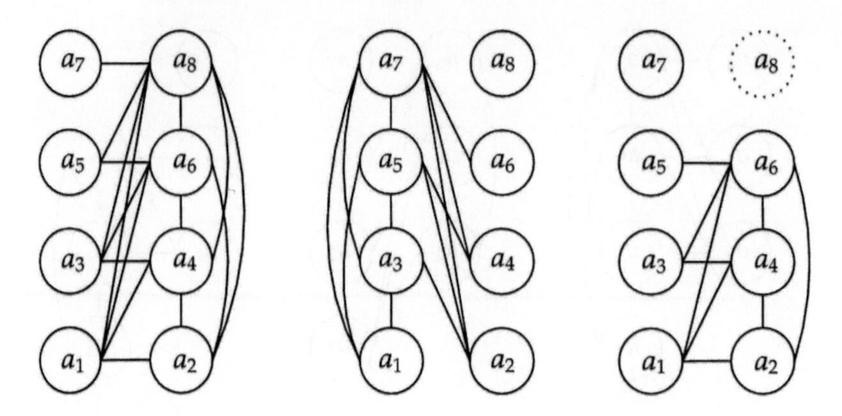

Figure 3.14 The threshold graph $G = C(a_1, a_2, \ldots, a_8)$, G^c and $G - v$ for $a_8 = 1$ (Lab1). (From left to right.)

Figures 3.13 and 3.14, respectively, show the threshold graphs with underlying A_7 and A_8, their complements and the v-deleted subgraphs when v is the only dominating vertex. The corresponding representations of A_7 and A_8 are $C(2,1,1,1,1,1)$ and $C(2,1,1,1,1,1,1)$, respectively.

Proposition 3.2 [43] *For the connected antiregular graphs A_{2k} and A_{2k+1}, the binary codes are, respectively, the $(2k-1)$-string* $(1\ 0\ 1\ 0\ \ldots\ 1)$ *and the $2k$-string* $(0\ 1\ 0\ \ldots\ 1)$ *with alternating* 0 *and* 1 *entries.*

Since the binary code follows the construction of A_r algorithmically, we obtain:

Corollary 3.3 [43] *For $k \in \mathbb{Z}^+$, the construction of connected antiregular graphs is as follows:*

$$A_{2k} = (\ldots((K_1 \nabla K_1) \sqcup K_1)\nabla \ldots)\nabla K_1,$$
$$A_{2k+1} = (\ldots((K_1 \sqcup K_1)\nabla K_1) \sqcup \ldots)\nabla K_1.$$

The Complement of a Threshold Graph

The complement of a connected threshold graph $C(a_1, a_2, \ldots, a_r)$ is disconnected and is denoted by $D(a_1, a_2, \ldots, a_r)$. The following result is deduced from the construction of the complement.

Proposition 3.3 [47] *The cotree T_{G^c} of the complement $G^c = D(a_1, a_2, \ldots, a_r)$ of $G = C(a_1, a_2, \ldots, a_r)$ is obtained from T_G by changing the interior vertices from $\otimes$ to $\oplus$ and vice-versa.*

Corollary 3.4 *The complement G^c of the connected threshold graph $C(a_1, a_2, \ldots, a_r)$ is the disconnected threshold graph $D(a_1, a_2, \ldots, a_r)$ isomorphic to $C(a_1, a_2, \ldots, a_{r-1}) \sqcup a_r K_1$.*

Proof: Since $C(a_1, a_2, \ldots, a_r)$ is connected, its cotree has $\otimes$ as a root. Hence, by Proposition 3.3, the cotree $D(a_1, a_2, \ldots, a_r)$ has $\oplus$ as a root and, therefore, it has coclique K_{a_r}. □

Proposition 3.4 [43] *The binary string coding of the threshold graph $C(a_1, a_2, \ldots, a_{2k})$ with underlying graph A_{2k} is the $2k$-string $(0^{a_1-1}\, 1^{a_2} \ldots 1^{a_{2k}})$ of entries 0 and 1. (The superscripts stand for repetition; e.g. 1^{a_i} denotes the substring 111... with 1 repeated a_i times.)*

Similarly, the binary string coding of the threshold graph $C(a_1, a_2, \ldots, a_{2k+1})$ with underlying graph A_{2k+1} is the $(2k+1)$-string $(1^{a_1-1}\, 0^{a_2} \ldots 1^{a_{2k+1}})$.

3.3.3 The Nullity of Threshold Graphs

A pair of *duplicate* vertices of a graph are non-adjacent and have the same neighbours, whereas a pair of *coduplicate* vertices are adjacent and have the same neighbours. The rows of the adjacency matrix corresponding to duplicate vertices are identical and for those of coduplicate vertices k and h, the k-th and h-th rows differ only in the k-th and h-th entries. It follows that both duplicates and coduplicates produce the eigenvector with only two non-zero entries, namely, 1 and -1 at positions corresponding to the pair of vertices, with the corresponding eigenvalues of 0 and -1, respectively. A graph with duplicates is often considered as having repeated vertices and, therefore, redundant properties. The induced subgraph of a graph obtained by removing repeated vertices is called *canonical*.

In this subsection, we adopt the vertex labelling Lab2 of a threshold graph induced by the Ferrers–Young diagram in accordance with the procedure to form the stepwise adjacency matrix presented in Theorem 3.10.

Theorem 3.11 [43] *An upper bound for the nullity $\eta(G)$ of the adjacency matrix of a threshold graph is $n - f(\Pi) - 1$.*

Proof: When the adjacency matrix $\boldsymbol{G}$ is obtained from *adjList*, the first $f(\Pi)$ rows are shifted so that none of them is repeated. The first $f(\Pi) + 1$ labelled vertices form a clique, and hence the rank of the adjacency matrix $\boldsymbol{G}$ of the n-vertex graph G, which is $n - \eta(G)$, is at least $f(\Pi) + 1$. □

The bound in Theorem 3.11 is reached, for instance, for the threshold graphs $C(f(\Pi) + 1)$ (complete graphs) and for $C(f(\Pi) + 1, f(\Pi))$.

Theorem 3.12 [43] *Let G be a threshold graph on n vertices with Durfee square size $f(\Pi)$ and nullity $\eta(G)$. If $n > 2f(\Pi)$, then G has duplicate vertices.*

Proof: The last $n - f(\Pi)$ rows of $F(\Pi)$ are not affected by the introduction of the zero diagonal when constructing $\boldsymbol{G}$ as in Theorem 3.10. Hence, duplicates may only occur among the last $n - f(\Pi)$ labelled vertices. If G were to have no duplicate vertices, then the last $n - f(\Pi)$ rows of $\boldsymbol{G}$ need to be all different. Because the $f(\Pi)$-th row is $f(\Pi)$ long, by the pigeon-hole principle, the largest number n of vertices possible for the graph without duplicates is $2f(\Pi)$. Therefore, if $n > 2f(\Pi)$, G has at least one pair of duplicate vertices. □

A threshold graph may have duplicate vertices even if $n < 2f(\Pi)$. We note again that a kernel eigenvector corresponding to duplicate vertices has only two non-zero entries. This prompts the question: can the kernel eigenvector of a threshold graph have more than two non-zero entries? The answer is in the negative as we will now see.

Theorem 3.13 [43] *The nullity $\eta(G)$ of a threshold graph G is the number of vertices removed to obtain the canonical graph.*

Proof: Let H be the canonical graph obtained from G by removing all the duplicate vertices, and let the number of vertices removed be t. Because the reflection in the first column of the adjacency matrix $\boldsymbol{H}$ of H is in row echelon form, the rows of $\boldsymbol{H}$ after the $f(\Pi)$-th is in the strict stepwise form. Hence, the columns of $\boldsymbol{H}$ are linearly independent. Now, if the t vertices are added to H in turn to obtain G again, then the nullity increases by one at each stage, contributing to the nullspace of the obtained graph and a kernel eigenvector (with exactly two non-zero entries) while preserving the existing ones. We deduce that there are only t linear combinations among the rows of $\boldsymbol{G}$ arising from the repeated rows in the last $n - f(\Pi)$ rows. Therefore, the nullity of G is t. Moreover, the kernel eigenvector cannot have more than two non-zero entries. □

In the proof of Theorem 3.13, the following result becomes evident.

Corollary 3.5 [43] *If a threshold graph is singular, then no kernel eigenvector has more than two non-zero entries.*

Note that any repeated rows in the first $f(\Pi)$ rows of $F(\Pi)$ give coduplicates. Also, $f(\Pi)$ is the degree of a vertex in the first part of the equitable partition of the threshold graph represented by $C(a_1, a_2, \ldots, a_r)$ for Lab1. For A_r, this corresponds to the $\lceil \frac{r+1}{2} \rceil$-th degree in the monotonic non-increasing sequence of distinct degrees (the $\lceil \frac{r+1}{2} \rceil$-th vertex for labelling Lab2). The construction of an antiregular graph implies that it has exactly one pair of either duplicates or coduplicates.

Theorem 3.14 [43] *The following statements are true:*

(i) An antiregular graph A_{2k-1} on an odd number of vertices has a duplicate vertex.

(ii) An antiregular graph A_{2k} on an even number of vertices has a coduplicate vertex.

Proof: The graph A_r is $C(2, 1, 1, \ldots, 1)$. Therefore, if r is even, it has a clique of two and hence a pair of coduplicate vertices. On the other hand, if r is odd, then it has a coclique of two, producing a pair of duplicate vertices. □

To calculate the number of duplicate and coduplicate vertices in a threshold graph, we count the number of vertices to be removed from G and G^c, respectively, to obtain the canonical graph.

Theorem 3.15 [43] *A threshold graph with non-degenerate representation $C(a_1, a_2, \ldots, a_r)$, where r is even, has*

(i) $\sum_{k=1}^{\frac{r}{2}}(a_{2k-1} - 1)$ *duplicate vertices;*

(ii) $\sum_{k=1}^{\frac{r}{2}}(a_{2k} - 1)$ *coduplicate vertices.*

For odd r, $C(a_1, a_2, \ldots, a_r)$ has

(iii) $\sum_{k=1}^{\frac{r-1}{2}}(a_{2k} - 1)$ *duplicate vertices;*

(iv) $\sum_{k=1}^{\frac{r+1}{2}}(a_{2k-1} - 1)$ *coduplicate vertices.*

3.3.4 Minimal Configurations

Most of the information to determine the grounds for a labelled graph G to be singular is encoded in the nullspace $\ker(\boldsymbol{G})$ of its adjacency matrix $\boldsymbol{G}$; that is, in

$$\ker(\boldsymbol{G}) = \{\boldsymbol{x} : \boldsymbol{G}\boldsymbol{x} = \boldsymbol{0}\}.$$

The support of a kernel eigenvector $\boldsymbol{x}$ in $\ker(\boldsymbol{G})$ is the set of vertices corresponding to the non-zero entries. These vertices induce a subgraph termed *the core of G with respect to $\boldsymbol{x}$*. Therefore, the core of G with respect to $\boldsymbol{x}$ is a core graph in its own right. The size of the support is said to be the *core order* [38].

Definition 3.5 [38] *Let F be a core graph on at least two vertices with nullity $s \geq 1$ and a kernel eigenvector $\boldsymbol{x}_F$ having no zero entries. Suppose that N is a graph of nullity one having $\boldsymbol{x}_F$ as non-zero part of the kernel eigenvector. If the graph N is obtained by adding $s - 1$ independent vertices whose neighbours are vertices of F, then N is said to be a minimal configuration (MC) with core $(F, \boldsymbol{x}_F)$.*

Hence, an MC with core $(F, \boldsymbol{x}_F)$ is a connected singular graph of nullity one having the minimal number of vertices and edges for the core F, which satisfies $\boldsymbol{F}\boldsymbol{x}_F = \boldsymbol{0}$. The MCs may be considered as the 'atoms' of a singular graph [38, 41]. The smallest MC is P_3 corresponding to a pair of duplicates. If core order is three, the only MC is P_3. The number of MCs increases fast for higher core order (e.g. see [40]). Figure 3.15 shows two MCs: (a) P_5^c, the only MC with core C_4; and (b) a nut graph of order seven [45].

A basis for the nullspace $\ker(\boldsymbol{G})$ of the adjacency matrix $\boldsymbol{G}$ of a graph G of nullity $\eta > 1$ can take different forms. We choose the *minimal basis* $B_{\min}$ for the

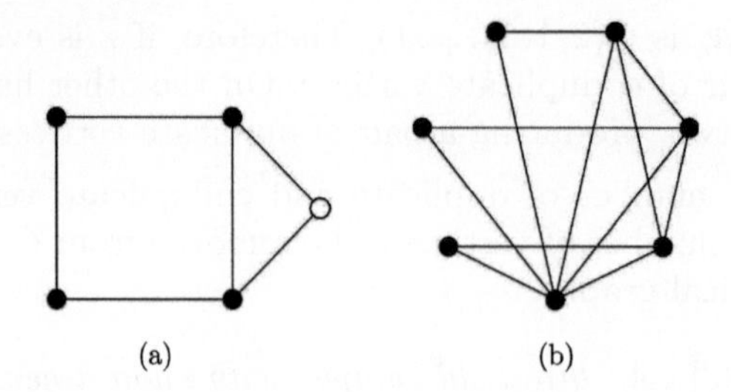

Figure 3.15 Two minimal configurations: (a) P_5^c and (b) a nut graph.

nullspace of $\boldsymbol{G}$; that is, a basis having the minimal total number of non-zero entries in its vectors [38, 44].

Such a minimal basis for $\ker(\boldsymbol{G})$ has the property that the corresponding monotonic non-decreasing sequence of core orders (termed *the core order sequence*) is unique and lexicographically placed first in the list of bases for $\ker(\boldsymbol{G})$, also ordered according to the non-increasing core orders. Moreover, for all i, the i-th entry of the core order sequence for $B_{\min}$ does not exceed the i-th entry of any other core order sequence of the graph. We say that the vectors in $B_{\min}$ determine the *fundamental system of cores* of G consisting of a collection of cores of the minimal core order corresponding to a basis of linearly independent nullspace vectors [39]. The significance of MCs can be gauged from the next result.

Theorem 3.16 [38, 41] *Let H be a singular graph of nullity η. There exist η MCs which are subgraphs of H, whose core vertices are associated with the non-zero entries of the η vectors in the minimal basis of the nullspace of $\boldsymbol{H}$.*

To give an example supporting Theorem 3.16, Figure 3.16 shows a six-vertex graph of nullity two and two MCs corresponding to the fundamental system of cores found as subgraphs.

The following result follows immediately from Theorem 3.13.

Corollary 3.6 [43] *The only MC found in a threshold graph as a subgraph is P_3.*

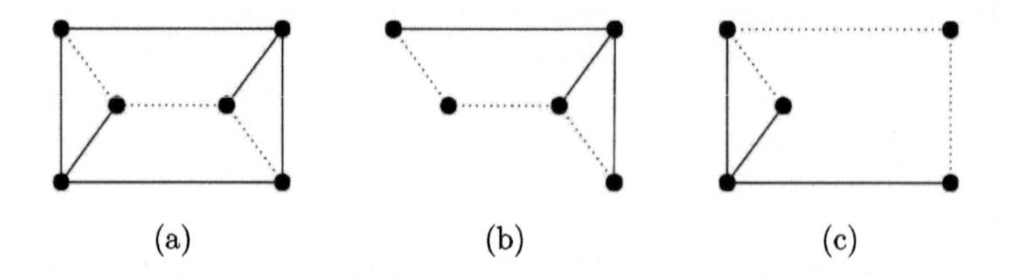

Figure 3.16 A graph of nullity two (a) with two MCs as subgraphs (b, c).

Corollary 3.6 was generalized for cographs in [35]—in cographs, only P_3 (corresponding to duplicate vertices) may be found as an MC corresponding to a vector in $B_{\min}$. Therefore, it is sufficient to have just P_4 as a forbidden subgraph for a graph to have only a core of order two contributing to the nullity.

Theorem 3.17 [43] *All MCs with core order at least three have P_4 as an induced subgraph.*

Proof: Suppose an MC is P_4-free; then it is a cograph. Therefore, the only MC to contribute to the nullity is P_3 of core order two. We deduce that all other MCs, which have core order at least three, are not cographs. □

Since P_4 is self-complementary, it follows that the complement of an MC with core order at least three also has P_4 as an induced subgraph. Figures 3.16b and 3.16c show P_4 as an induced subgraph (dotted edges) of the MC P_5^c.

The second largest eigenvalue of P_4 is the golden section $\sigma = (\sqrt{5} - 1)/2$. By interlacing, we obtain the following result:

Theorem 3.18 [43] *The second largest eigenvalue of an MC $\neq P_3$ is bounded from below by σ.*

The only MC for which the bound is known to be strict is the seven-vertex nut graph of Figure 3.15.

3.3.5 The Main Characteristic Polynomial

The main eigenvalues of a graph G are closely related to the number of walks in G. The product of those factors of the minimum polynomial of $\boldsymbol{G}$, corresponding to the main eigenvalues only, has interesting properties.

Definition 3.6 *The polynomial*

$$M(G, x) = \prod_{i=1}^{p} (x - \mu_i)$$

whose roots are the main eigenvalues of the adjacency matrix of a graph G is termed the main characteristic polynomial.

The following result was proved in several papers.

Lemma 3.8 [15, 34] *The main characteristic polynomial*

$$M(G, x) = x^p - c_0 x^{p-1} - c_1 x^{p-2} - \ldots - c_{p-2} x - c_{p-1}$$

has integer coefficients c_i for all i, $0 \leq i \leq p - 1$.

The Main Eigenvalues of Antiregular Graphs

Let us recall that A_r has exactly one pair of either duplicates or coduplicates.

Theorem 3.19 [43] *All eigenvalues of A_r other than 0 or -1 are main.*

Proof: Let Prop(r) be: All eigenvalues of A_r other than 0 or -1 are main.

We will prove Prop(r) by induction on r. Prop(2) refers to K_2 whose only non-main eigenvalue is -1. Prop(3) refers to P_3 whose only non-main eigenvalue is 0. This establishes the base cases.

Assume now that Prop(r) is true for all $r \leq k$. Therefore, for a non-main eigenvalue λ other than 0 or -1, $\boldsymbol{A}_r\boldsymbol{x} = \lambda\boldsymbol{x}$ implies $\boldsymbol{x} = \boldsymbol{0}$ for $r \leq k$.

Consider A_{k+1} and let $\boldsymbol{A}_{k+1}$ be its adjacency matrix. For the case when $k+1$ is odd and A_{k+1} is connected, let $\boldsymbol{A}_{k+1}\boldsymbol{x} = \lambda\boldsymbol{x}$ for an eigenvalue λ and $\boldsymbol{x}^{\mathrm{T}} = (x_1, x_2, \ldots, x_{k+1}) \neq \boldsymbol{0}$. It follows that

$$\sum_{i=1}^{k+2-q} x_i = (1+\lambda)x_q \quad \text{for} \quad 1 \leq q \leq f(\Pi)$$

and

$$\sum_{i=1}^{k+2-q} x_i = \lambda x_q \quad \text{for} \quad f(\Pi)+1 \leq q \leq k+1.$$

Similar equations are obtained for the case when $k+1$ is even.

The eigenvalue λ is non-main if and only if $\boldsymbol{j}^{\mathrm{T}}\boldsymbol{x} = 0$, whence $\lambda = -1$ or $\lambda = 0$ or $x_1 = x_2 = 0$. If v (labelled 1) is the dominating vertex of A_{k+1}, then, by Proposition 3.1,

$$A_{k+1} - v = K_1 \sqcup A_{k-1}.$$

If $x_1 = x_2 = 0$, then $\boldsymbol{x}$ restricted to A_{k-1} is an eigenvector for the same eigenvalue λ. Therefore, by the induction hypothesis, $\boldsymbol{x} = \boldsymbol{0}$. Hence, $\lambda = -1$ or $\lambda = 0$. The result follows by induction on r. □

The Main Eigenvalues of Threshold Graphs

By Theorem 3.19, all eigenvalues of A_r that are not 0 or -1 are main. We show that this is still the case for a threshold graph $C(a_1, a_2, \ldots, a_r)$ having $a_1 \geq 2$ and $a_i \geq 1$ for $2 \leq i \leq r$, which is obtained from the degenerate form $A_r = C(1, 1, \ldots, 1)$ by adding duplicates and/or coduplicates.

Lemma 3.9 [43] *A graph has the same number of main eigenvalues as its complement.*

Proof: Let $\boldsymbol{G}^{\mathrm{c}}$ be the adjacency matrix of the complement of a graph G and $\boldsymbol{J}$ the matrix with each entry equal to one. Then, $\boldsymbol{G} + \boldsymbol{G}^{\mathrm{c}} = \boldsymbol{J} - \boldsymbol{I}$. Now, λ is a non-main eigenvalue of G if and only if $\boldsymbol{J}\boldsymbol{x} = \boldsymbol{0}$. Hence, $\boldsymbol{G}$ and $\boldsymbol{G}^{\mathrm{c}}$ share the same eigenvectors only for non-main eigenvalues. $\square$

Theorem 3.20 [43] *Let G be a threshold graph. All its eigenvalues other than* 0 *or* -1 *are main.*

Proof: Let G be $C(a_1, a_2, \ldots, a_r)$ such that $a_1 \geq 2$ and $a_i \geq 1$ for $2 \leq i \leq r$. Let the proposition Prop(r) be: All eigenvalues of $C(a_1, a_2, \ldots, a_r)$ other than 0 or -1 are main.

We prove Prop(r) by induction on r. If $G = C(a_1, a_2)$, $a_1 \geq 2$, $a_2 \geq 1$, then G is not regular. Hence, the number of main eigenvalues is at least two. The other distinct eigenvalues, 0 and/or -1, are non-main. By Theorem 3.15, G has at least $n-2$ non-main eigenvalues equal to 0 or -1. Thus, the number of main eigenvalues of G is two. This establishes the base case, namely, Prop(2).

The induction hypothesis is as follows: assume that Prop(k) is true. We will show that this is also true for a non-degenerate $H = C(a_1, a_2, \ldots, a_{k+1})$.

The complement H^{c} of H is

$$C(a_1, a_2, \ldots, a_k) \sqcup a_{k+1}K_1.$$

By Lemma 3.9, H and H^{c} have the same number of main eigenvalues. One of the a_{k+1} isolated vertices in H^{c} contributes to the number of main eigenvalues. By the induction hypothesis, $C(a_1, a_2, \ldots, a_k)$ has k main eigenvalues and $\sum_{i=1}^{k}(a_i - 1)$ non-main eigenvalues. Hence, H has $k+1$ main eigenvalues. The result follows by induction on r. $\square$

We deduce immediately a spectral property of a threshold graph and its underlying antiregular graph.

Corollary 3.7 *The non-degenerate threshold graph $C(a_1, a_2, \ldots, a_r)$ and its underlying A_r have r and $r-1$ main eigenvalues, respectively.*

An equitable partition $\Pi = \mathcal{V}_1, \mathcal{V}_2, \ldots, \mathcal{V}_r$ of the vertex set of a graph satisfies $\boldsymbol{GX} = \boldsymbol{XQ}$, where $\boldsymbol{X}$ is the $n \times r$ indicator matrix whose i-th column is the characteristic $(0,1)$-vector associated with the i-th part, containing $|\mathcal{V}_i|$ entries equal to 1. The matrix $\boldsymbol{Q}$ turns out to be the adjacency matrix of the quotient graph $\frac{G}{\Pi}$ (also known as divisor).

Lemma 3.10 [43] *The main part of the spectrum of $\boldsymbol{G}$ is included in the spectrum of $\boldsymbol{Q}$.*

Proof: Let λ be a main eigenvalue of G. Then, $\boldsymbol{Gx} = \lambda\boldsymbol{x}$, where $\boldsymbol{j}^{\mathrm{T}}\boldsymbol{x} \neq 0$. We have $\boldsymbol{GX} = \boldsymbol{XQ}$ and $\lambda\boldsymbol{x}^{\mathrm{T}}\boldsymbol{X} = \boldsymbol{x}^{\mathrm{T}}\boldsymbol{GX} = (\boldsymbol{x}^{\mathrm{T}}\boldsymbol{X})\boldsymbol{Q}$, so that $\lambda(\boldsymbol{X}^{\mathrm{T}}\boldsymbol{x}) = \boldsymbol{Q}(\boldsymbol{X}^{\mathrm{T}}\boldsymbol{x})$. Thus, the eigenvalue λ of G is also an eigenvalue of $\boldsymbol{Q}$ provided that $\boldsymbol{X}^{\mathrm{T}}\boldsymbol{x} \neq \boldsymbol{0}$. Indeed, this is the case when λ is a main eigenvalue because $\boldsymbol{x}^{\mathrm{T}}\boldsymbol{X} = \boldsymbol{0}^{\mathrm{T}}$ would imply $\boldsymbol{j}^{\mathrm{T}}\boldsymbol{x} = 0$, and hence $\boldsymbol{x}^{\mathrm{T}}\boldsymbol{X} \neq \boldsymbol{0}^{\mathrm{T}}$. Thus, the main part of the spectrum of G is contained in the spectrum of $\boldsymbol{Q}$. $\square$

We now show that the main part of the spectrum of $G = C(a_1, a_2, \ldots, a_r)$ is precisely the spectrum of $\boldsymbol{Q}$. Let us consider the equitable vertex partition Π for $G = C(a_1, a_2, \ldots, a_r)$ as outlined in Section 3.3.1.

Theorem 3.21 [43] *Let the threshold graph $G = C(a_1, a_2, \ldots, a_r)$ have η duplicates, $\bar{\eta}$ coduplicates and an equitable partition Π corresponding to the parts $\{a_i\}$. Let $\boldsymbol{Q}$ be the adjacency matrix of the quotient graph $\frac{G}{\Pi}$. Then, $\phi(G, \lambda) = \lambda^{\eta}(1 + \lambda)^{\bar{\eta}}\phi(\boldsymbol{Q}, \lambda)$, where $\phi(\boldsymbol{Q})$ is the main characteristic polynomial $M(G, \lambda)$ of G.*

Proof: The vertex labelling Lab1 is used—the vertices are labelled in order starting from those corresponding to a_1 and followed by those for a_2 etc. Let $\boldsymbol{X}$ be the $n \times r$ indicator matrix whose i-th column is the characteristic $(0, 1)$-vector associated with a_i and containing exactly a_i non-zero entries (each equal to 1). We obtain $\boldsymbol{GX} = \boldsymbol{XQ}$ where $\boldsymbol{Q}$ is $r \times r$. In a threshold graph, by Theorem 3.20, 0 and -1 are the only non-main eigenvalues corresponding to duplicates and coduplicates, respectively. Therefore, the number of main eigenvalues of $\boldsymbol{G}$ is exactly r. Because the main spectrum of $\boldsymbol{G}$ is contained in the spectrum of $\boldsymbol{Q}$ and $\boldsymbol{Q}$ is $r \times r$, the roots of $\phi(\boldsymbol{Q})$ are the main eigenvalues of G. □

Let us give an example to clarify the procedure. Consider the threshold graph $G = C(2, 2, 1, 2, 1, 2)$ (Lab1) of Figure 3.11. We use the adjacency matrix $\boldsymbol{G}$ and the indicator matrix $\boldsymbol{X}$ indexed according to Lab2:

$$\boldsymbol{G} = \begin{pmatrix} 0 & 1 & 1 & 1 & 1 & 1 & 1 & 1 & 1 & 1 \\ 1 & 0 & 1 & 1 & 1 & 1 & 1 & 1 & 1 & 1 \\ 1 & 1 & 0 & 1 & 1 & 1 & 1 & 1 & 1 & 0 \\ 1 & 1 & 1 & 0 & 1 & 1 & 1 & 1 & 1 & 0 \\ 1 & 1 & 1 & 1 & 0 & 1 & 1 & 1 & 0 & 0 \\ 1 & 1 & 1 & 1 & 1 & 0 & 1 & 1 & 0 & 0 \\ 1 & 1 & 1 & 1 & 1 & 1 & 0 & 0 & 0 & 0 \\ 1 & 1 & 1 & 1 & 1 & 1 & 0 & 0 & 0 & 0 \\ 1 & 1 & 1 & 1 & 0 & 0 & 0 & 0 & 0 & 0 \\ 1 & 1 & 0 & 0 & 0 & 0 & 0 & 0 & 0 & 0 \end{pmatrix}, \quad \boldsymbol{X} = \begin{pmatrix} 1 & 0 & 0 & 0 & 0 & 0 \\ 1 & 0 & 0 & 0 & 0 & 0 \\ 0 & 1 & 0 & 0 & 0 & 0 \\ 0 & 1 & 0 & 0 & 0 & 0 \\ 0 & 0 & 1 & 0 & 0 & 0 \\ 0 & 0 & 1 & 0 & 0 & 0 \\ 0 & 0 & 0 & 1 & 0 & 0 \\ 0 & 0 & 0 & 1 & 0 & 0 \\ 0 & 0 & 0 & 0 & 1 & 0 \\ 0 & 0 & 0 & 0 & 0 & 1 \end{pmatrix}.$$

The rows of $\boldsymbol{Q}$ are the distinct rows of $\boldsymbol{GX}$. Therefore,

$$\boldsymbol{Q} = \begin{pmatrix} 1 & 2 & 2 & 2 & 1 & 1 \\ 2 & 1 & 2 & 2 & 1 & 0 \\ 2 & 2 & 1 & 2 & 0 & 0 \\ 2 & 2 & 2 & 0 & 0 & 0 \\ 2 & 2 & 0 & 0 & 0 & 0 \\ 2 & 0 & 0 & 0 & 0 & 0 \end{pmatrix}.$$

Its spectrum is $\{7.16, 0.892, 0.448, -1.40, -1.59, -2.50\}$, which is precisely the main part of the spectrum of G.

For $\ell \geq 0$, the entries of $\boldsymbol{G}^{\ell}\boldsymbol{j}$ give the number of walks of length ℓ from each vertex v of G. The $n \times k$ matrix whose ℓ-th column is $\boldsymbol{G}^{\ell-1}\boldsymbol{j}$ is denoted by $\boldsymbol{W}_k$. The dimension of the subspace $\mathrm{ColSp}(\boldsymbol{W}_k)$ generated by the columns of $\boldsymbol{W}_k$ is the rank of $\boldsymbol{W}_k$.

Theorem 3.22 [23] *For a graph with p main eigenvalues, the rank and* $\dim(\mathrm{ColSp}(\boldsymbol{W}_k))$ *of the $n \times k$ matrix $\boldsymbol{W}_k = \left(\boldsymbol{j}, \boldsymbol{G}\boldsymbol{j}, \boldsymbol{G}^2\boldsymbol{j}, \ldots, \boldsymbol{G}^{k-1}\boldsymbol{j}\right)$ is p for $k \geq p$.*

The columns $\boldsymbol{j}, \boldsymbol{G}\boldsymbol{j}, \boldsymbol{G}^2\boldsymbol{j}, \ldots, \boldsymbol{G}^{p-1}\boldsymbol{j}$ are a maximal set of linearly independent vectors in $\mathrm{ColSp}(\boldsymbol{W}_k)$. Thus, the first p columns provide all the information on the number of walks from each vertex of any length [42].

Definition 3.7 *The matrix $\boldsymbol{W}_p = \left(\boldsymbol{j}, \boldsymbol{G}\boldsymbol{j}, \boldsymbol{G}^2\boldsymbol{j}, \ldots, \boldsymbol{G}^{p-1}\boldsymbol{j}\right)$ of rank p is said to be the walk matrix $\boldsymbol{W}$.*

Note that $\mathbf{W}$ has the least number of columns for a walk matrix $\mathbf{W}_k$ to reach the maximum rank possible which is p. From Corollary 3.7, we see that $C(a_1, a_2, \ldots, a_r)$ has r main eigenvalues.

Theorem 3.23 [43] *The rank of the walk matrix of $C(a_1, a_2, \ldots, a_r)$ is r.*

The number of walks of length k can be expressed in terms of the main eigenvalues [14, p. 46].

Theorem 3.24 [14] *The number w_k of walks of length k starting from any vertex of G is given by*

$$w_k = \sum_{i=1}^{p} c_i' \mu_i^k,$$

where $c_i' \in \mathbb{R}\backslash\{0\}$ is independent of k for each i and $\mu_1, \mu_2, \ldots, \mu_p$ are the main eigenvalues of $\boldsymbol{G}$.

Since 0 is never a main eigenvalue of $C(a_1, a_2, \ldots, a_r)$, it follows that all the main eigenvalues of $C(a_1, a_2, \ldots, a_r)$ contribute to the number of walks.

Cases of Reducible Main Polynomial

By Theorem 3.19, only one eigenvalue of A_r is not main. Let us recall that the minimal equitable vertex partition of $G = C(a_1, a_2, \ldots, a_r)$ satisfies $\boldsymbol{G}\boldsymbol{X} = \boldsymbol{X}\boldsymbol{Q}$, where $\boldsymbol{Q}$ is the adjacency matrix of the quotient graph $\frac{G}{\Pi}$ and $\phi(\boldsymbol{Q}, \lambda)$ is the main characteristic polynomial $M(G, \lambda)$ of G.

We note that, for many threshold graphs, $\phi(\boldsymbol{Q}, \lambda)$ is irreducible over the integers. For example, the only eigenvalue of $A_8 = C(1,1,1,1,1,1,1,1)$ (in degenerate form), which is not main, is -1; and

$$M(A_8, x) = \left(1 - 7x + 9x^2 + 15x^3 - 13x^4 - 15x^5 - x^6 + x^7\right),$$

which is irreducible.

Now, we add vertices to the degenerate form $A_8 = C(1,1,1,1,1,1,1,1)$. If a vertex is added to the first part to obtain $G_1 = C(2,1,1,1,1,1,1,1)$, a negative eigenvalue (not -1) and 0 appear. The eigenvalue -1 is lost and

$$M(G_1, x) = \left(2 - 12x + 6x^2 + 40x^3 - 40x^5 - 20x^6 + x^8\right).$$

When a vertex is added to the third part to obtain $G_3 = C(2,2,1,1,1,1,1)$, the eigenvalue -1 is retained while the zero eigenvalue appears and

$$M(G_3, x) = \left(2 - 12x + 12x^2 + 22x^3 - 16x^4 - 18x^5 - x^6 + x^7\right).$$

In the last two cases, $\phi(\boldsymbol{Q}, \lambda)$ is irreducible over the integers. Now, when a vertex is added to the seventh part to obtain $G_7 = C(2,1,1,1,1,2,1)$, the eigenvalue -1 is retained while the zero eigenvalue appears. In this case, however,

$$M(G_7, x) = (x^2 + 2x - 1)(x^5 - 3x^4 - 9x^2 + 3x^3 + 8x - 2)$$

and, therefore, it is reducible over the integers.

This is also the case for some instances of the threshold graphs $C(d, 1, t)$, when the cubic polynomial $\phi(\boldsymbol{Q}, \lambda)$ has an integer as a root and, therefore, is reducible. The divisor $\boldsymbol{Q}$ is

$$\begin{pmatrix} d-1 & 0 & t \\ 0 & 0 & t \\ d & 1 & t-1 \end{pmatrix}$$

with characteristic polynomial

$$\phi(\boldsymbol{Q}, \lambda) = -t + dt + \lambda - d\lambda - 2t\lambda + 2\lambda^2 - d\lambda^2 - t\lambda^2 + \lambda^3.$$

If λ is 0, 2 or 3, there are no integral values of t and d satisfying the polynomial $\phi(\boldsymbol{Q}, \lambda)$. If $\lambda = 1$, the graph either for $t = 3$ and $d = 8$ or for $t = 4$ and $d = 6$ satisfies it. Also, for $\lambda = -2$, the graph either for $t = 3$ and $d = 5$, or for $t = 4$ and $d = 3$, or for $t = 6$ and $d = 2$ satisfies it, while for $\lambda = -3$, the graph for $t = 7$ and $d = 40$ satisfies it.

3.3.6 Sign Pattern of the Spectrum of a Threshold Graph

We conclude with a note on the distribution of eigenvalues of a threshold graph. In [31], it was remarked that an antiregular graph has a *bipartite character*; that is, the number r^- of negative eigenvalues is equal to the number r^+ of positive ones. Recall that the number of zero eigenvalues is denoted by η.

The Spectrum of A_r

For $r \geq 4$, the antiregular graph A_r is not bipartite, hence $-\lambda_{\min} \neq \lambda_{\max}$. The proof of the next result is by induction on the order of antiregular graphs. We will need the following evident fact.

Lemma 3.11 *To transform A_r to A_{r+1}, according to the labelling (Lab2) of the stepwise adjacency matrix,*

(i) *a vertex, duplicate to the $\lceil \frac{r+1}{2} \rceil$-th vertex, is added for even r;*
(ii) *a vertex, coduplicate to the $\lceil \frac{r+1}{2} \rceil$-th vertex, is added for odd r.*

Theorem 3.25 [43] *For the antiregular graph A_r, we have $r^+ = r^-$.*

Proof: The proof is by induction on r. The spectra of the three smallest antiregular graphs establish the base cases: $\mathrm{Sp}(A_1) = \{0\}$, $\mathrm{Sp}(A_2) = \{-1, 1\}$ and $\mathrm{Sp}(A_3) = \{-\sqrt{2}, 0, \sqrt{2}\}$.

Assume that the theorem is true for A_k, and let us prove it is true for A_{k+1}. If A_k is singular, then it has a duplicate vertex and k is odd. By the induction hypothesis, $r^+ = r^-$. If, on the other hand, A_k is non-singular, then A_k has a coduplicate vertex and k is even. Again, the non-zero eigenvalues satisfy $r^+ = r^-$.

We apply Lemma 3.11 using Lab2. For odd k, if a vertex w, coduplicate to the $\lceil \frac{r+1}{2} \rceil$-th vertex, is added to A_k, then only one of the duplicate vertices of A_k will have w as a neighbour in A_{k+1}. The zero eigenvalue of A_k vanishes and the eigenvalue -1 is introduced for A_{k+1}. By the Perron–Frobenius Theorem, adding edges to the graph $A_k \sqcup K_1$ increases the maximum eigenvalue. Therefore, by interlacing, the number of positive eigenvalues increases by one. Since the new coduplicate vertex w contributes the new eigenvalue -1 to the spectrum, it follows that $r^+ = r^-$ will be satisfied in A_{k+1}. By interlacing, adding a duplicate vertex to any graph retains the number of positive and negative eigenvalues and adds 0 to the spectrum. For even k, if a vertex w, duplicate to the $\lceil \frac{r+1}{2} \rceil$-th vertex, is added, then a duplicate vertex is added to the graph, retaining $r^+ = r^-$. The result follows by induction on n. □

The Spectrum of a Threshold Graph

In this subsection, we shall represent the antiregular graph A_r by the degenerate form $C(1, 1, \ldots, 1)$. As in Section 3.3.2, any its part can be expanded to produce a threshold graph $C(a_1, a_2, \ldots, a_r)$. We need the following evident facts regarding the effect on the distribution of the spectrum of the adjacency matrix when a vertex is added.

Lemma 3.12 [43] *When adding a vertex to a graph:*

(i) *If the multiplicity of an eigenvalue λ_0 of the adjacency matrix increases, then, by interlacing, the number $n^-(\lambda_0)$ of eigenvalues less than λ_0 and the number $n^+(\lambda_0)$ of eigenvalues greater than λ_0 remain the same.*

(*ii*) *If the multiplicity of an eigenvalue* λ_0 *of the adjacency matrix decreases, then, by interlacing, each of the numbers* $n^+(\lambda_0)$ *and* $n^-(\lambda_0)$ *increases by one.*

To prove the next theorem, let us apply Lemma 3.12 (i) using Lab1. We shall write n^+ for $n^+(0)$ and n^- for $n^-(0)$. For even r, if one of the even indexed a_i $(i \geq 2)$ of $C(a_1, a_2, \ldots, a_r)$ is increased, then a coduplicate of a vertex is added. This forces η and n^+ to remain unchanged, while each of n^- and the multiplicity $m(-1)$ of the eigenvalue -1 increases by one. If the odd indexed a_i for some $i \geq 1$ is increased, then a duplicate of a vertex is added forcing n^+ and n^- to remain unchanged.

Similarly, for odd r, if the even indexed a_i for some $i \geq 2$ is increased, then a duplicate of a vertex is added. This forces n^- and n^+ to remain unchanged, while η increases by one. If the odd indexed a_i for some $i \geq 3$ is increased, then a coduplicate of a vertex is added forcing n^+ and η to remain unchanged.

The case for even r and $a_1 > 1$ is the same as for odd r with $a_1 = 1$ (Lab1). Taking $C(a_1, a_2, \ldots, a_r)$ for odd r with $a_1 = 1$ and expanding to $C(a_1, a_2, \ldots, a_r)$ with $a_1 > 1$ gives the unique case where η decreases by one and $m(-1)$ increases by one. Since η decreases by one, by Lemma 3.12 (ii), each of n^+ and n^- increases by one, the latter corresponding to the increase in the multiplicity of the eigenvalue -1. We have proved the following result:

Theorem 3.26 [43] *Let the threshold graph* $C(a_1, a_2, \ldots, a_r)$ *be transformed to another threshold graph by increasing exactly one* a_i *by one.*

(*i*) *If a duplicate is added, then* n^- *and* n^+ *are unchanged and* η *increases.*

(*ii*) *Suppose that a coduplicate is added. If either* r *is even, or* r *is odd and* $a_i \geq 3$, *or* r *is odd and* $a_1 > 1$, *then* η *and* n^+ *are unchanged and* n^- *increases.*

(*iii*) *If a coduplicate is added,* r *is odd and* $a_1 = 1$, *then* n^- *and* n^+ *increase and* η *decreases.*

Acknowledgements

This chapter is based on the following publications: (A) Reprinted by permission from Elsevier: Discrete Applied Mathematics, 154, T. Calamoneri and R. Petreschi, λ-Coloring matrogenic graphs, 2445–2457, ©2006. (B) The article by I. Sciriha and S. Farrugia [43], released under the Creative Commons Attribution License.

3.4 References

[1] A. A. Bertossi, C. Pinotti and R. Tan, $L(2, 1)$- and $L(2, 1, 1)$-labeling of graphs with application to channel assignment in wireless networks, *Proceedings of ACM DIAL M*, 2000.

[2] T. Biyikoglu, S. K. Simić and Z. Stanić, Some notes on spectra of cographs, *Ars Combinatoria*, **C** (2011), 421–434.

[3] H. L. Bodlaender, T. Kloks, R. B. Tan and J. van Leeuwen, λ-Coloring of graphs, *Proceedings of the 17th International Symposium on Theoretical Aspects of Computer Science* (STACS 2000), LNCS 1770, 2000, 395–406.

[4] A. Brandstädt, V. B. Le and J. P. Spinrad, *Graph Classes: a Survey*, Philadelphia, PA: Society for Industrial and Applied Mathematics, 1999.

[5] R. A. Brualdi and A. J. Hoffman, On the spectral radius of (0,1)-matrices, *Linear Algebra and its Applications*, **65** (1985), 133–146.

[6] R. A. Brualdi and E. S. Solheid, On the spectral radius of connected graphs, *Publications de l'Institut Mathématique*, Belgrade, **39** (53)(1986), 45–54.

[7] T. Calamoneri, The $L(h,k)$-labelling problem: A survey and annotated bibliography, *The Computer Journal*, **49** (5)(2006), 585–608.

[8] T. Calamoneri and R. Petreschi, $L(2,1)$-coloring of regular tiling, *The First Cologne-Twente Workshop*, (CTW01), 2001.

[9] T. Calamoneri and R. Petreschi, The $L(2,1)$-labeling of planar graphs, *Journal on Parallel and Distributed Computing*, **64** (3)(2004), 414–426.

[10] T. Calamoneri and R. Petreschi, λ-Coloring matrogenic graphs, *Discrete Applied Mathematics*, **154** (2006), 2445–2457.

[11] G. J. Chang and D. Kuo, The $L(2,1)$-labeling problem on graphs, *SIAM Journal on Discrete Mathematics*, **9** (2)(1996), 309–316.

[12] V. Chvatal and P. Hammer, Aggregation of inequalities in integer programming, *Annals of Discrete Mathematics*, **1** (1977), 145–162.

[13] D. G. Corneil, Y. Perl and L. K. Stewart, A linear recognition algorithm for cographs, *SIAM Journal on Computing*, **14** (1985), 926–934.

[14] D. M. Cvetković, M. Doob and H. Sachs, *Spectra of Graphs*, 3rd edition, Heidelberg/Leipzig: Johann Ambrosius Barth Verlag, 1995.

[15] D. Cvetković and M. Petrić, A table of connected graphs on six vertices, *Discrete Mathematics*, **50** (1)(1984), 37–49.

[16] J. Fiala, T. Kloks and J. Kratochvíl, Fixed-parameter complexity of λ-labelings, *Proceedings of Graph-Theoretic Concepts of Computer Science*, (WG99), LNCS 1665, 1999, 350–363.

[17] S. Foldes and P. Hammer, On a class of matroid-producing graphs, *Combinatorics*, **18** (1978), 331–352.

[18] F. R. Gantmakher, *The Theory of Matrices*, New York, NY: Chelsea Publishing Company, 1960.

[19] M. R. Garey and D. S. Johnson, *Computers and Intractability – A Guide to the Theory of NP-Completeness*, New York, NY: W.H. Freeman and Co., 1979.

[20] M. C. Golumbic, *Algorithmic Graph Theory and Perfect Graphs*, New York, NY: Academic Press, 1980.

[21] M. C. Golumbic and A. N.Trenk, *Tolerance Graphs*, Cambridge: Cambridge University Press, Cambridge Studies in Advanced Mathematics, vol. 89, 2004.

[22] J. R. Griggs and R. K. Yeh, Labeling graphs with a condition at distance 2, *SIAM Journal on Discrete Mathematics*, **5** (1992), 586–595.

[23] E. M. Hagos, Some results on graph spectra, *Linear Algebra and Its Applications*, **356** (2002), 103–111.

[24] P. L. Hammer, T. Ibaraki and B. Simeone, Threshold sequences, *SIAM Journal on Algebraic Discrete Methods*, **2** (1)(1981), 39–49.

[25] P. H. Henderson and Y. Zalcstein, A graph-theoretic characterization of the PV_{chunk} class of synchronizing primitives, *SIAM Journal on Computing*, **6** (1)(1977), 88–108.

[26] T. R. Jensen and B. Toft, *Graph Coloring Problems*, New York: John Wiley & Sons, 1995.

[27] H. A. Jung, On a class of posets and the corresponding comparability graphs, *Journal of Combinatorial Theory, Ser. B*, **24** (2)(1978), 125–133.

[28] H. Lerchs, On cliques and kernels, *Technical Report*, Department of Computer Science, University of Toronto, 1971.

[29] N. V. R. Mahadev and U. N. Peled, *Threshold Graphs and Related Topics*, Annals of Discrete Mathematics, 56, North-Holland, Amsterdam: Elsevier, 1995.

[30] P. Marchioro, A. Morgana, R. Petreschi and B. Simeone, Degree sequences of matrogenic graphs, *Discrete Mathematics*, **51** (1984), 47–61.

[31] R. Merris, Antiregular graphs are universal for trees, *Publikacije Elektrotehničkog Fakulteta. Serija Matematika*, **14** (2003), 1–3.

[32] R. Merris, *Graph Theory*, New York, NY: John Wiley & Sons, 2011.

[33] U. N. Peled, Matroidal graphs, *Discrete Mathematics*, **20** (1977), 263–286.

[34] P. Rowlinson, The main eigenvalues of a graph: a survey, *Applicable Analysis and Discrete Mathematics*, **1** (2007), 445–471.

[35] G. F. Royle, The rank of a cograph, *The Electronic Journal of Combinatorics*, **10** (2003), #N11.

[36] E. Ruch and I. Gutman, The branching extent of graphs, *Journal of Combinatorics, Information and System Sciences*, **4** (4)(1979), 285–295.

[37] D. Sakai, Labeling chordal graphs: distance two condition, *SIAM Journal on Discrete Mathematics*, **7** (1994), 133–140.

[38] I. Sciriha, A characterization of singular graphs, *Electronic Journal of Linear Algebra*, **16** (2007), 451–462.

[39] I. Sciriha, Maximal core size in singular graphs, *Ars Mathematica Contemporanea*, **2** (2009), 217–229.

[40] I. Sciriha, On the construction of graphs of nullity one, *Discrete Mathematics*, **181** (1998), 193–211.

[41] I. Sciriha, On the rank of graphs, In: Y. Alavi, D. R. Lick and A. Schwenk (eds.), *Combinatorics, Graph Theory and Algorithms*, vol. 2, Kalamazoo, Michigan: New Issues Press, 1999, 769–778.

[42] I. Sciriha and D. M. Cardoso, Necessary and sufficient conditions for a Hamiltonian graph, *Journal of Combinatorial Mathematics and Combinatorial Computing*, **80** (2012), 127–150.

[43] I. Sciriha and S. Farrugia, On the spectrum of threshold graphs, *ISRN Discrete Mathematics*, 2011, Article ID 108509, 21 pages.

[44] I. Sciriha, S. Fiorini and J. Lauri, A minimal basis for a vector space, *Graph Theory Notes of New York*, The New York Academy of Sciences, **31** (1996), 21–24.

[45] I. Sciriha and I. Gutman, Nut graphs – Maximally extending cores, *Utilitas Mathematica*, **54** (1998), 257–272.

[46] D. Seinsche, On a property of the class of n-colorable graphs, *Journal of Combinatorial Theory, Ser. B*, **16** (2)(1974), 191–193.

[47] Z. Stanić, On nested split graphs whose second largest eigenvalue is less than 1, *Linear Algebra and its Applications*, **430** (2009), 2200–2211.

[48] D. P. Sumner, Dacey graphs, *Journal of the Australian Mathematical Society*, **18** (4)(1974), 492–502.

[49] R. I. Tyshkevich, Matroid decompositions of graphs, *Diskretnaya Matematika*, **1** (3)(1989), 129–138.

[50] R. I. Tyshkevich, Once more on matrogenic graphs, *Discrete Mathematics*, **51** (1984), 91–100.

[51] V. E. Zverovich, I. E. Zverovich and R. I. Tyshkevich, Graphs with the matroidal number at most 2, *Diskretnaya Matematika*, **2** (2)(1990), 82–88.

4

Further Applications of Operator Decomposition

P. Skums & V. Zverovich

The unique factorization property makes operator decomposition especially useful for graph-theoretic problems associated with graph isomorphisms. One of the most famous open problems of this type is the Kelly–Ulam reconstruction conjecture that states that every graph on at least three vertices is determined up to isomorphism by its induced subgraphs. We will demonstrate how the operator decomposition can be used to prove the reconstruction conjecture for several substantial families of graph classes. The results that we discuss are not only interesting by themselves but also in light of the prospects of further research, as they describe a general methodology that can be useful in search for the ultimate proof of the reconstruction conjecture. We will also delve into a generalization of the operator decomposition that determines a general family of associative algebraic operations on the sets of n-partitioned graphs. Each operation in this family is governed by a fixed digraph H, and as we will demonstrate, it exhibits the unique factorization property. The corresponding factorization of an n-partitioned graph into a product of prime factors is termed H-decomposition. We then introduce H-threshold graphs and discuss the associated parameter called threshold-width, which shares a natural connection with clique-width and NLC-width of graphs. Finally, we will consider properties of realizations of a degree sequence and how 1-decomposition can be used for deriving such properties.

4.1 Operator Decomposition and the Reconstruction Conjecture

4.1.1 Reconstruction Conjecture

As was stated in Chapter 1, the unique factorization property makes operator decomposition especially useful for problems associated with graph isomorphism. In this chapter, we will consider one of the most famous open problems of this kind—the *reconstruction conjecture*. This conjecture was formulated by Kelly

P. Skums & V. Zverovich, *Further Applications of Operator Decomposition*. In: *Methods of Graph Decompositions*. Edited by: Vadim Zverovich and Pavel Skums, Oxford University Press. © Vadim Zverovich & Pavel Skums (2024). DOI: 10.1093/oso/9780198882091.003.0004

[24] and Ulam [48] as early as 1942, and it is related to one of the ultimate general mathematical problems about the relation between the structure of an object and the structures of its subobjects. When restricted to graphs, the question can be formulated as follows: is it true that graphs are determined up to isomorphism by their induced subgraphs? Notice here that it is enough to consider only maximal (proper) induced subgraphs. It is obvious that not all graphs are determined up to isomorphism by their induced subgraphs because the two-vertex graphs K_2 and O_2 have the same sets of induced subgraphs (see Figure 4.1). However, there are reasons to believe that these two graphs are the only exceptions.

Formally, let us consider a simple graph G. The collection

$$D(G) = (G_v)_{v \in V(G)}$$

of vertex-deleted subgraphs of graph G is called the *deck* of G, and the elements of the deck are called the *cards* (see Figure 4.1). A graph H with the deck $D(H) = (H_u)_{u \in V(H)}$ is called a *reconstruction* of G if there exists a bijection $f : V(G) \to V(H)$ such that $G_v \simeq H_{f(v)}$. In this case, we say that the decks $D(G)$ and $D(H)$ are *equal*. The graph G is *reconstructible* if it is isomorphic to any of its reconstructions. The mapping f is sometimes called a *hypomorphism* [26], and the corresponding graphs *hypomorphic*. With a slight abuse of notation, we can consider f as both a mapping between vertex sets and a mapping between decks.

Kelly–Ulam Reconstruction Conjecture [24, 48] *Every graph with at least three vertices is reconstructible.*

In other words, the reconstruction conjecture states that all hypomorphic graphs on at least three vertices are isomorphic. Despite the simplicity of its formulation, this conjecture has remained open for 80 years. Furthermore, the lack of significant breakthroughs has produced a certain scepticism regarding the possibility of its proof or disproof in the foreseeable future. As early as 1969, Harary in his classical graph theory textbook urged its readers 'not to try to settle this conjecture since it appears to be rather difficult' [21]. More recent survey [34] discusses the conjecture as follows: 'even though it is one of the foremost unsolved problems in graph

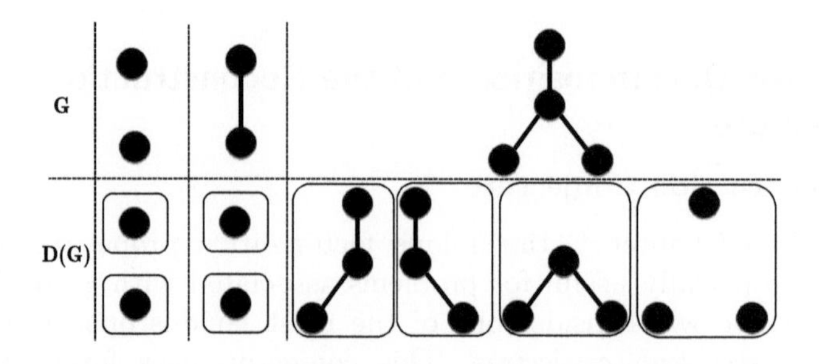

Figure 4.1 Examples of graphs and their decks.

theory, work on it has slowed down, may be due to the general feeling that existing techniques are not likely to lead to a complete solution'.

If the conjecture cannot be solved in general, it is natural to attempt to prove it for specific graph classes. A class of graphs is called *reconstructible* if all graphs from this class are reconstructible. Some known examples of reconstructible classes include disconnected graphs, complements of disconnected graphs, trees, regular graphs, unit interval graphs and outerplanar graphs [9]. Furthermore, it has been shown that almost all graphs (in the probabilistic sense) are reconstructible [32]. The reconstruction conjecture was also verified for graphs on up to 11 vertices using a computer [30]. Note that the analogues of the reconstruction conjecture do not hold for directed graphs [39, 40] and hypergraphs [25]. For more information, we refer our readers to old but still highly relevant survey papers by Nash-Williams [33] and Bondy [9]. There is also a recent survey by Kostochka and West [27], which is devoted to a more general conjecture posed by Kelly in 1957.

One of the possible ways to prove that the class of graphs is reconstructible is to consider a decomposition such that for every graph from the class under consideration the following properties hold:

(1) The graph is defined by this decomposition up to isomorphism.
(2) The parts of this decomposition and connections between them could be reconstructed from the deck of the graph.

As we have seen before, some variants of the operator decomposition may satisfy the first condition. In what follows, we will demonstrate when and how such decompositions satisfy the second condition. Based on these results, we will prove the reconstruction conjecture for several substantial classes of graphs.

We conclude this section with several basic definitions and facts that are usually used when dealing with graph reconstructibility. A class of graphs $\mathcal{R}$ is called *recognizable* if for any graph $G \in \mathcal{R}$ all its reconstructions also belong to $\mathcal{R}$. A class $\mathcal{R}$ is *weakly reconstructible* if for any $G \in \mathcal{R}$ every reconstruction of G that belongs to $\mathcal{R}$ is isomorphic to G. Clearly, $\mathcal{R}$ is reconstructible if and only if it is recognizable and weakly reconstructible. It is also obvious that a graph is reconstructible if and only if its complement is reconstructible.

Similar to graph classes, we say that a graph parameter is *recognizable* if its value for any pair of hypomorphic graphs is the same. For example, the numbers of vertices and edges of a graph are obviously recognizable: $|V(G)| = |D(G)|$ and

$$|E(G)| = \frac{\sum_{G_v \in D(G)} |E(G_v)|}{|D(G)| - 2}.$$

The latter is true because every edge of G belongs to every card except for two cards corresponding to its endpoints. More importantly for us here, the degree sequence d of a graph is recognizable; indeed, this follows from the facts that $d = (|E(G)| - |E(G_v)|)_{G_v \in D(G)}$ and $|E(G)|$ is recognizable.

Similar to the deck of a graph, we can naturally define the deck of a triad. Specifically, let $\sigma = (G, A, B)$ be a triad and $a \in A$, $b \in B$. By *vertex-deleted triads*, we mean the triads

$$\sigma_a = (G_a, A - \{a\}, B) \quad \text{and} \quad \sigma_b = (G_b, A, B - \{b\})$$

distinguished up to isomorphism of triads. Thus, the deck of a triad σ is the family of triads $(\sigma_v : v \in A \cup B)$, the order $|\sigma|$ of the triad σ is the order of the underlying graph G. We will also say that the graph G is *anticonnected* if $\overline{G}$ is connected.

4.1.2 Reconstruction of (P, Q)-decomposable Graphs

In this section, we will demonstrate that (P, Q)-decomposable graphs that are non-(P, Q)-split are reconstructible for any hereditary pair of classes (P, Q). This result is interesting for two reasons. First, it establishes reconstructibility not only for a particular class of graphs but for an infinite family of classes defined by rather general conditions. Second, the result demonstrates the methodology for approaching the reconstruction conjecture using the operator decomposition. The latter will be used in consecutive sections for some graph classes.

Let G be a (P, Q)-decomposable non-(P, Q)-split graph for some closed hereditary pair (P, Q), and let

$$G = \sigma_1 \circ \sigma_2 \circ \ldots \circ \sigma_k \circ H, \qquad \sigma_i = (T_i, A_i, B_i) \in (P, Q) \sum \tag{4.1}$$

be its (P, Q)-decomposition into the product of indecomposable factors. To start with, we can make the following simple observations:

- It is clear from the definition of a closed hereditary pair that all split graphs are (P, Q)-split. Furthermore, all graphs of orders at most 3 are split. Hence, if $|H| \leq 3$ in (4.1), then H is split and so G is (P, Q)-split.
- If $|T_1| \leq 3$, then either G or $\overline{G}$ is disconnected. In this case, G is reconstructible.

In what follows, $\mathcal{R}(P, Q)$ is the class of (P, Q)-decomposable non-(P, Q)-split graphs G with (P, Q)-decomposition (4.1) such that G is both connected and anticonnected. Given the above observations, it is true that $|H| \geq 4$, $|T_1| \geq 4$ and H is not (P, Q)-split. To prove that the class of (P, Q)-decomposable non-(P, Q)-split graphs is reconstructible, it is sufficient to show that $\mathcal{R}(P, Q)$ is reconstructible. In the rest of this section, we will establish this fact.

Now, let $G \in \mathcal{R}(P, Q)$ and (4.1) be a (P, Q)-decomposition of G. Let further

$$\sigma = (T, A, B) = \sigma_1 \circ \sigma_2 \circ \cdots \circ \sigma_k \tag{4.2}$$

be the operator part of decomposition (4.1) and $C = V(H)$. Since $|H| \geq 3$ and $|\sigma| \geq 3$, the deck $D(G)$ has the partition into disjoint subsets

$$D(G) = D_\sigma \cup D_H, \tag{4.3}$$

where

$$D_\sigma = \{\sigma \circ H_v : v \in V(H)\} \quad \text{and} \quad D_H = \{\sigma_u \circ H : u \in V(T)\}.$$

Here, T is a (P, Q)-split graph, while H is not (P, Q)-split and (P, Q)-indecomposable (see Figure 4.2).

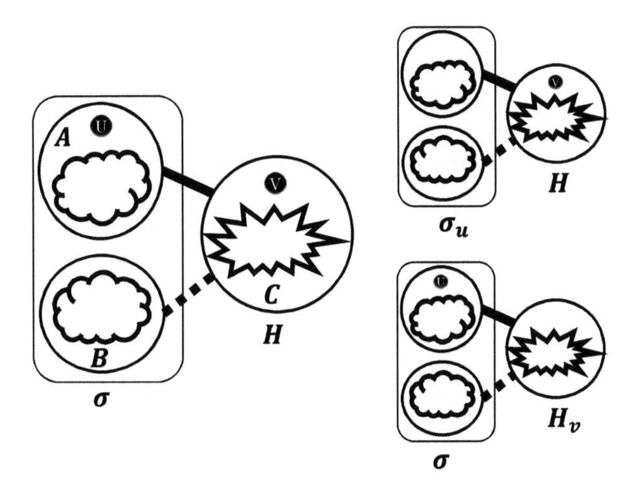

Figure 4.2 Schematic depiction of a graph $G = \sigma \circ H$ and its (P, Q)-decomposition. Left: the graph G. The thick solid line represents all possible edges between the part A of the triad σ and the subgraph H, while the thick dashed line represents the absence of any edges between B and H. Right: two cards $G_u = \sigma_u \circ H \in D_H$ and $G_v = \sigma \circ H_v \in D_\sigma$.

Lemma 4.1 *The class* $\mathcal{R}(P, Q)$ *is recognizable.*

Proof: Let G be a graph. We will prove that if $D(G)$ has partition (4.3), then $G \in \mathcal{R}(P, Q)$. Let $v \in V(G)$ and $G_v \in D_H$, i.e. $G_v \simeq \sigma_v \circ H$. Then, there exists a partition

$$V(G) - \{v\} = A_v \cup B_v \cup C_v \tag{4.4}$$

into disjoint subsets A_v, B_v, C_v such that $(G(A_v \cup B_v), A_v, B_v) \simeq T_v$ and $G(C_v) \simeq H$. Since P and Q are hereditary classes, it is true that $\sigma_v \in (P, Q)\sum$.

To complete the proof of the lemma, we need the following proposition.

Proposition 4.1 *Let* $u \in V(G)$. *Then,* $u \notin C_v$ *if and only if* $G_u \in D_H$.

Proof: Let us first prove the necessity. Suppose that $u \notin C_v$ and $G_u \in D_\sigma$. Then, $G_u \simeq \sigma \circ H_u$; that is,

$$V(G) - \{u\} = A_u \cup B_u \cup C_u, \tag{4.5}$$

where $(G(A_u \cup B_u), A_u, B_u) \simeq \sigma$ and $G(C_u) \simeq H_u$.

Since $|C_v| = |H| > |H_u| = |C_u|$, we have $C_v \nsubseteq C_u$. Thus, there are two possibilities to consider:

(1) $C_v \subseteq A_u \cup B_u$. In this case, due to the heredity of the classes P and Q, the induced subgraph $G(C_v \cap A_u)$ belongs to P, and the induced subgraph $G(C_v \cap B_u)$ belongs to Q. Therefore, $H \simeq G(C_v)$ is a (P, Q)-split graph. Thus, the contradiction with (4.3) is obtained.

(2) $C_v \cap (A_u \cup B_u) \neq \emptyset$ and $C_v \cap C_u \neq \emptyset$. In this case, as in (1), we have $G(C_v \cap A_u) \in P$ and $G(C_v \cap B_u) \in Q$. Thus, we can construct a (P, Q)-decomposition of H as follows:

$$H \simeq G(C_v) \simeq \Big(G(C_v \cap (A_u \cup B_u)), C_v \cap A_u, C_v \cap B_u\Big) \circ G(C_v \cap C_u).$$

This contradicts the assumption that the graph H is (P, Q)-indecomposable.

Thus, it is proved that $G_u \in D_H$ if $u \notin C_v$. On the other hand, the number of vertices u such that $u \notin C_v$ is equal to

$$|G| - |C_v| = |G| - |H| = |D_H|.$$

Therefore, the set D_H consists entirely of cards G_u such that $u \notin C_v$. The proposition is proved. □

Now, let us consider two vertex-deleted subgraphs $G_u, G_v \in D_H$. As above, cases (1) and (2) from the proof of Proposition 4.1 are impossible. Thus, $C_v \subseteq C_u$ and, by symmetry, $C_u \subseteq C_v$. Hence, we have

$$C_u = C_v = C \text{ for all } u, v \in V(G) \text{ such that } G_u, G_v \in D_H. \tag{4.6}$$

It is not difficult to see that C is a homogeneous set of G. Indeed, for the subgraph G_v we have $A_v \sim C$ and $B_v \not\sim C$. So, it remains to clarify the relations between v and C. By Proposition 4.1, $v \notin C$ and, therefore, $v \in A_u \cup B_u$. If $v \in A_u$, then $v \sim C$; if, on the other hand, $v \in B_u$, then $v \not\sim C$.

Now, without loss of generality, let us select a vertex $v \in A_u$, and set $A = A_v \cup \{v\}$ and $B = B_v$. Then, the previous considerations imply that the graph G can be factorized as follows:

$$G = (G(A \cup B), A, B) \circ G(C), \tag{4.7}$$

where $H \simeq G(C)$ is not (P, Q)-split and $G(B) \in Q$.

It remains to show that $G(A) \in P$. To demonstrate this, note first that by the definition of the class $\mathcal{R}(P, Q)$, the graph $\overline{G}$ is connected and, therefore, $B \neq \emptyset$. Let us consider a vertex $u \in B$. By Proposition 4.1, $G_u \in D_H$. Hence,

$$G_u \simeq (G(A_u \cup B_u), A_u, B_u) \circ G(C) \tag{4.8}$$

and $A \subset A_u \cup B_u$. Since $A \sim C$, we have $A \subseteq A_u$. Now, because

$$(G(A_u \cup B_u), A_u, B_u) \simeq \sigma_u \in (P, Q) \sum,$$

we obtain $G(A_u) \in P$ and, consequently, the assumption that the class P is hereditary implies $G(A) \in P$. This concludes the proof of Lemma 4.1. □

Lemma 4.2 *The class $\mathcal{R}(P, Q)$ is weakly reconstructible.*

Proof: Let G^1 and G^2 be two graphs from $\mathcal{R}(P,Q)$ with equal decks $D(G^1)$ and $D(G^2)$, with the corresponding hypomorphism f and with the following (P, Q)-decompositions:

$$G_1 = \sigma_1^1 \circ \sigma_2^1 \circ \cdots \circ \sigma_k^1 \circ H_0^1, \qquad G_2 = \sigma_1^2 \circ \sigma_2^2 \circ \cdots \circ \sigma_l^2 \circ H_0^2. \tag{4.9}$$

In the proof of Lemma 4.1, it was convenient to collapse decomposition (4.1) in such a way as to represent the graph G as the product of an operator part and an indecomposable subgraph. Here, in contrast, it will be more handy to consider the graph G_1 (G_2) as the product of an indecomposable triad and a (possibly decomposable) subgraph. Namely, let

$$\sigma^1 = \sigma_1^1, \quad H^1 = \sigma_2^1 \circ \sigma_3^1 \circ \cdots \circ \sigma_k^1 \circ H_0^1,$$

and

$$\sigma^2 = \sigma_1^2, \quad H^2 = \sigma_2^2 \circ \sigma_3^2 \circ \cdots \circ \sigma_l^2 \circ H_0^2.$$

So, we have

$$G^1 = \sigma^1 \circ H^1, \quad G^2 = \sigma^2 \circ H^2. \tag{4.10}$$

Here, $\sigma^1, \sigma^2 \in (P,Q)\sum$ are indecomposable triads and, as implied by the definition of the class $\mathcal{R}(P,Q)$, the graphs H^1 and H^2 are not (P, Q)-split graphs. So, the vertex sets of the graphs G^1 and G^2 can be partitioned as follows:

$$V(G^i) = A^i \cup B^i \cup C^i, \tag{4.11}$$

where $(T^i = G^i(A^i \cup B^i), A^i, B^i) \simeq \sigma^i$ and $G^i(C^i) \simeq H^i$, $i = 1, 2$.

Furthermore, the triads σ^1 and σ^2 are neither A-parts nor B-parts: this straightforwardly follows from the assumption that both G^1 and G^2 are connected and anticonnected. Therefore, Theorem 1.2 implies that $G^1 \simeq G^2$ if and only if

$$\sigma^1 \simeq \sigma^2 \quad \text{and} \quad H^1 \simeq H^2. \tag{4.12}$$

Just as in the proof of Lemma 4.1, the facts that $|T^i| \geq 3$ and $|H^i| \geq 3$, $i = 1, 2$, imply that the decks of G^1 and G^2 have partitions

$$D(G^i) = D_{\sigma^i} \cup D_{H^i}, \tag{4.13}$$

where $D_{\sigma^i} = \{\sigma^i \circ H_v^i : v \in C^i\}$ and $D_{H^i} = \{\sigma_v^i \circ H^i : v \in A^i \cup B^i\}$, $i = 1, 2$. All graphs from D_{H^i}, $i = 1, 2$, are (P, Q)-decomposable non-(P, Q)-split.

Our next step is to demonstrate that the Unique Factorization Theorem (Theorem 1.2) is applicable to all cards from the decks $D(G^1)$ and $D(G^2)$. It is certainly true for the cards from the subdecks D_{H^i}. So, it is sufficient to show that every graph $G_v^i \in D_{\sigma^i}$, $i = 1, 2$, is (P, Q)-heterogeneous. Consider, for example, the card $G_v^1 \simeq \sigma^1 \circ H_v^1$. It is easy to see that $H_v^1 \notin P \cup Q$. Indeed, if $H_v^1 \in P$ then, given that $O_1 \in Q$, the graph H^1 would be a (P, Q)-split graph with bipartition $(C^1 - \{v\}, \{v\})$. The case when $H_v^1 \in Q$ is analogous. Hence, it is established that (P, Q)-decompositions of all cards are unique.

The key to the proof of Lemma 4.2 is the next result:

Proposition 4.2 *The following statements are true:*

(1) The hypomorphism f maps the subdeck D_{σ^1} into the subdeck D_{σ^2} and the subdeck D_{H^1} into the subdeck D_{H^2}. Thus, for every $v \in C^1$, we have

$$G_v^1 = \sigma^1 \circ H_v^1 \simeq \sigma^2 \circ H_{f(u)}^2 = G_{f(u)}^2, \tag{4.14}$$

and for every $v \in A^1 \cup B^1$, we have

$$G_v^1 = \sigma_v^1 \circ H^1 \simeq \sigma_{f(v)}^2 \circ H^2 = G_{f(v)}^2. \tag{4.15}$$

(2) For the triads σ^1 and σ^2,

$$\sigma^1 \simeq \sigma^2. \tag{4.16}$$

(3) For all $v \in V(H^1)$,

$$H_v^1 \simeq H_{f(v)}^2. \tag{4.17}$$

Proof: We will start by proving a weaker result: there exists a card $G_v^1 \in D_{\sigma^1}$ such that $G_{f(v)}^2 \in D_{\sigma^2}$. Suppose the contrary. Then, there exist cards

$$G_{v_1}^1 \in D_{\sigma^1}, \quad G_{v_2}^1 \in D_{H^1}, \quad G_{u_1}^2 \in D_{\sigma^2}, \quad G_{u_2}^2 \in D_{H^2}$$

such that

$$\sigma^1 \circ H_{v_1}^1 = G_{v_1}^1 \simeq G_{u_2}^2 = \sigma_{u_2}^2 \circ H^2, \tag{4.18}$$

$$\sigma^2 \circ H_{u_1}^2 = G_{u_1}^2 \simeq G_{v_2}^1 = \sigma_{v_2}^1 \circ H^1. \tag{4.19}$$

The graphs at the right-hand parts of these expressions are not (P, Q)-split. This obviously implies that the same is true for the graphs at the left-hand parts. Let

$$\sigma_{u_2}^2 = \alpha_{u_2}^1 \circ \alpha_{u_2}^2 \circ \cdots \circ \alpha_{u_2}^r$$

be the canonical decomposition of the triad $\sigma_{u_2}^2$. From (4.18), by Corollary 1.2, we conclude that $\sigma^1 \simeq \alpha_{u_2}^1$ and, therefore,

$$|\sigma^1| \leq |\sigma_{u_2}^2| < |\sigma^2|. \tag{4.20}$$

Analogously, from (4.19), we obtain

$$|\sigma^2| \leq |\sigma_{v_2}^1| < |\sigma^1|. \tag{4.21}$$

This contradiction proves our claim.

Now, let us consider a vertex $v \in V(G^1)$ that satisfies (4.14). By the unique factorization property, (4.16) is true. This, in turn, implies that the hypomorphism f maps the subdeck D_{H^1} into the subdeck D_{H^2}. Thus, if $G_v^1 \in D_{H^1}$ and $G_v^1 \simeq G_u^2$, then $G_u^2 \in D_{H^2}$. Indeed, if there exist cards from D_{H^1} and D_{σ^2} such that (4.19) holds, then (4.21) is true. This contradicts (4.16). By symmetry, we conclude that $f(D_{H^1}) = D_{H^2}$ and, consequently, $f(D_{\sigma^1}) = D_{\sigma^2}$, which proves relations (4.18) and (4.19) for all $v \in V(G^1)$.

Finally, (4.17) follows from the unique factorization property applied to decompositions (4.15).

The proof of Lemma 4.2 can now be completed by induction on the numbers of operator factors k and l in (P, Q)-decompositions of G^1 and G^2. Let us choose some vertex $v \in A^1 \cup B^1$. According to Proposition 4.2, the cards corresponding to v and $f(v)$ satisfy relation (4.15). We already know that $\sigma^1 \simeq \sigma^2$ (relation (4.16)). Hence, it remains to prove that $H^1 \simeq H^2$.

(1) Suppose that $k=1$, i.e. H^1 is (P, Q)-indecomposable. Proposition 4.2 implies that $|H^1| = |H^2|$. Furthermore, it is easy to see that H^2 is also (P, Q)-indecomposable (i.e. $l=1$). Indeed, if

$$H^2 = \alpha_1^2 \circ \alpha_2^2 \circ \cdots \circ \alpha_r^2 \circ H_0^2$$

is the canonical (P, Q)-decomposition of H^2, then $H^1 \simeq H_0^2$ by the unique factorization property of operator decompositions (4.15) and, therefore, $|H_1| < |H_2|$. Hence, both H^1 and H^2 are (P, Q)-indecomposable and, consequently, decompositions (4.15) are canonical. Now, the isomorphism of H^1 and H^2 follows from the unique factorization property. By symmetry, the same is true when $l=1$.

(2) Suppose that $k>1$ and $l>1$; that is, the graphs H^1 and H^2 are (P, Q)-decomposable. In this case, H^1 and H^2 have operator decompositions with $k-1$ and $l-1$ operator factors. Moreover, by (4.17), these graphs have equal decks with $f|_{H^1}$ being the corresponding hypomorphism. If H^1 or $\overline{H^1}$ is disconnected, then we obtain $H^1 \simeq H^2$. Otherwise, H^1 and H^2 belong to $\mathcal{R}(P, Q)$ and the same conclusion follows from the inductive assumption. □

Together, Lemmas 4.1 and 4.2 imply the following result:

Theorem 4.1 [36] *(P, Q)-decomposable non-(P, Q)-split graphs are reconstructible for any closed hereditary pair of classes (P, Q).*

This theorem could be written in another way.

Theorem 4.2 [36] *Let P and Q be two hereditary graph classes such that P is closed with respect to the operation of join and Q is closed with respect to the operation of disjoint union. Let M be a module of graph G with associated partition (A, B, M) such that $G(A) \in P$, $G(B) \in Q$ and $G(M)$ is not (P, Q)-split. Then, the graph G is reconstructible.*

Notice that for 1-decomposition, which is the minimum element in the lattice of (P, Q)-decompositions, the situation is even better than for a general (P, Q)-decomposition. Recall that 1-decomposition has the unique factorization property for all graphs. Consequently, it is possible to prove that a stronger result is true in this case.

Theorem 4.3 [47] *All 1-decomposable graphs are reconstructible.*

We are not presenting the proof of this theorem here—it is basically a simplified version of the proof of Theorem 4.1. Moreover, in the next section, we will prove a more general result.

4.1.3 Reconstruction of P_4-disconnected Graphs

A graph G is called *P_4-connected* (or *p-connected*) if for every partition of $V(G)$ into two disjoint sets V_1 and V_2 there exists an induced P_4 (called *crossing* P_4) which contains vertices from both V_1 and V_2. Equivalently, a graph G is P_4-connected whenever any two vertices are connected by a P_4-chain; that is, a sequence of vertices such that every four consecutive vertices form an induced P_4 in G. Otherwise, G is called *P_4-disconnected* (or *p-disconnected*). P_4-disconnected graphs were introduced by Jamison and Olariu in [23]. The *p-connected component* of G is a maximal induced p-connected subgraph of G. It is obvious that every disconnected graph is p-disconnected, but the inverse inclusion is not true (see Figure 4.3). For example, complements of disconnected graphs are p-disconnected, and so are 1-decomposable graphs—it can easily be checked that if $G = (T, A, B) \circ G(C)$, then there is no induced P_4 that spans both $A \cup B$ and C.

As a matter of fact, P_4-connected and P_4-disconnected graphs are just two examples out of several substantial and interesting classes of graphs. Those classes are determined by the structural properties of sets of induced subgraphs P_4, which have been extensively studied in the literature. They include, in particular, P_4-free graphs (or *cographs*), P_4-extensible graphs, P_4-lite graphs, P_4-reducible graphs and P_4-sparse graphs (e.g. see [10] or [17]). The most general of them is the class of P_4-tidy graphs [17]. Let A be a subset of vertices in G such that $G(A) \simeq P_4$. A *partner* of A in G is a vertex $v \in G - A$ such that $G(A \cup v)$ contains at least two induced P_4. A graph G is *P_4-tidy* if any P_4 has at most one partner. The class of P_4-tidy graphs contains the aforementioned classes of P_4-extensible, P_4-lite, P_4-reducible, P_4-sparse and P_4-free graphs [17].

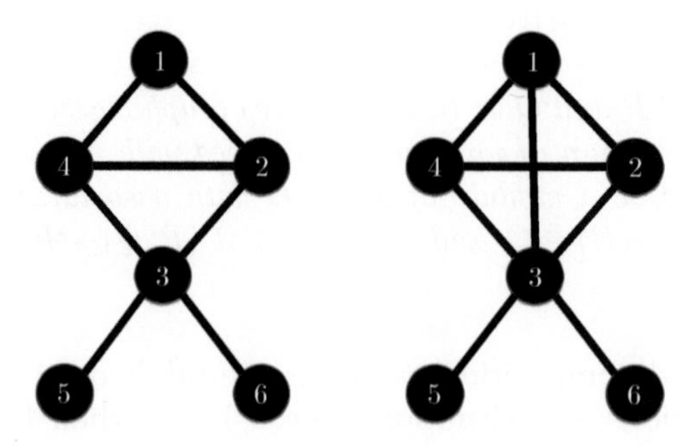

Figure 4.3 Left: an example of P_4-connected graph G. Right: an example of P_4-disconnected graph H, which is 1-decomposable because it can be represented in the form $H = (H(\{3,5,6\}), \{3\}, \{5,6\}) \circ H(\{1,2,4\})$.

We will use the previously introduced methodology to prove that P_4-disconnected graphs are reconstructible. This generalizes the results about reconstructibility of disconnected graphs, their complements and 1-decomposable graphs. In addition, we will show that the reconstructibility of P_4-disconnected graphs implies the reconstructibility of P_4-tidy graphs. Therefore, all aforementioned classes are also reconstructible. In particular, this fact generalizes the reconstructibility of P_4-reducible graphs established by Thatte [42].

Let us introduce several additional definitions and facts. A p-connected graph S is called *separable* [23] if there exists a disjoint partition of its vertex set $V(S) = A \cup B$ such that every crossing P_4 has its midpoints in A and its endpoints in B. In this case, a triad $\sigma = (S, A, B)$ is called a *separable p-connected triad.*

Lemma 4.3 [23] *Every separable p-connected graph induces a unique separable p-connected triad.*

A triad (G, A, B) is called *generalized split triad* if every connected component of $\overline{G(A)}$ and $G(B)$ is a module in G. For example, if all connected components of $\overline{G(A)}$ and $G(B)$ consist of one vertex, then G is a split graph.

Lemma 4.4 [23] *Let $\sigma = (G, A, B)$ be a separable p-connected triad. Then, σ is a generalized split triad. Moreover, the graphs $\overline{G(A)}$ and $G(B)$ are disconnected.*

Note that a separable p-connected triad contains at least four vertices. Further, a split graph G with bipartition (A, B) is called a *spider* (Figure 4.4) if there exists a bijection $f : B \to A$ such that one of the following conditions holds:

(1) $N(b) = \{f(b)\}$ for every vertex $b \in B$ (*thin spider*).
(2) $N(b) = A - \{f(b)\}$ for every vertex $b \in B$ (*thick spider*).

Lemma 4.5 *Spiders are reconstructible.*

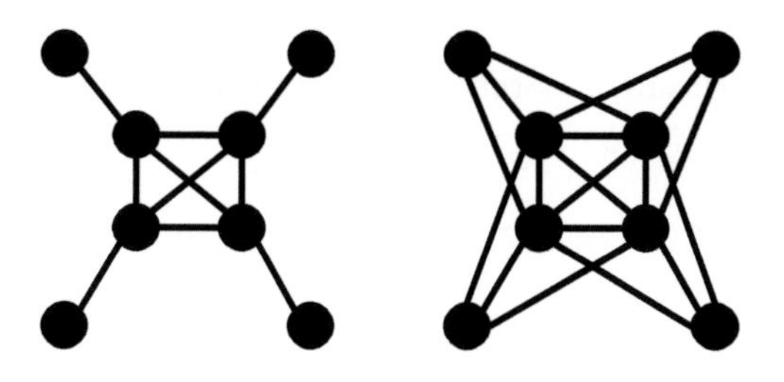

Figure 4.4 Left: example of a thin spider. Right: example of a thick spider.

Proof: Since thick spiders are complements of thin spiders, it is sufficient to prove that thin spiders are reconstructible.

Let G be a graph with $V(G) = \{v_1, v_2, \ldots, v_n\}$ and

$$\deg(v_1) \geq \deg(v_2) \geq \ldots \geq \deg(v_n).$$

By Theorem 2.1, G is a split graph if and only if the following equality holds:

$$\sum_{i=1}^{m} \deg(v_i) = m(m-1) + \sum_{i=m+1}^{n} \deg(v_i), \tag{4.22}$$

where

$$m = \max\{i : \deg(v_i) \geq i - 1\}.$$

Moreover, if (4.22) holds, then $A = \{v_1, v_2, \ldots, v_m\}$ is a maximal clique and $B = \{v_{m+1}, v_{m+2}, \ldots, v_n\}$ is an independent set.

Taking into account the above characterization, it is evident that G is a thin spider if and only if the following conditions hold:

(1) Equality (4.22) is satisfied.
(2) $\deg(v_i) = m(G)$ for every $i = 1, 2, \ldots, m(G)$.
(3) $\deg(v_i) = 1$ for every $i = m(G)+1, m(G)+2, \ldots, n$.

In other words, the property 'being a spider' is entirely determined by the degree sequence of a graph. Since the degree sequence is recognizable, thin spiders are reconstructible. □

A vertex v in a p-connected graph G is called a *p-articulation vertex* if G_v is p-disconnected. If every vertex of G is a p-articulation vertex, then G is called *minimally p-connected.*

Theorem 4.4 [2, 3] *A graph G is minimally p-connected if and only if G is a spider.*

Theorem 4.5 [2] *A p-connected graph that is not minimally p-connected contains at least two vertices which are not p-articulation vertices.*

The following structure theorem was proved in [23]. In our terms, it could be written in the following way:

Theorem 4.6 [23] *For an arbitrary graph G, exactly one of the following statements is true:*

(1) G is disconnected.
(2) $\overline{G}$ is disconnected.
(3) There is a unique separable component S of G with the corresponding partition $V(S) = A \cup B$ such that $G = (S, A, B) \circ H$.
(4) G is p-connected.

For example, all connected and anticonnected 1-decomposable graphs satisfy (3).

Now, let $\mathcal{R}$ be the class of graphs G such that

(a) G is p-disconnected;
(b) G is both connected and anticonnected;
(c) G is 1-indecomposable.

By Theorem 4.6, to prove that p-disconnected graphs are reconstructible it is sufficient to show that the class $\mathcal{R}$ is reconstructible.

Lemma 4.6. *Let T be a generalized split triad and let H be an arbitrary graph. Then, $G = T \circ H$ is p-disconnected.*

Proof: Let $V(G) = A \cup B \cup C$ such that $(G(A \cup B), A, B) \simeq T$, $G(C) \simeq H$ and $G = (G(A \cup B), A, B) \circ G(C)$. It is easy to see that for the partition

$$(A \cup B, C) \tag{4.23}$$

there is no crossing P_4. Indeed, let vertices x, y, z, t induce crossing P_4 for partition (4.23) with midpoints y, z and endpoints x, t such that $y \sim x$ and $z \sim t$. The only possibility is that $x \in C$, $y \in A$, $z, t \in B$. Then, the vertices z and t belong to the same connected component U of $S(B)$. However, because U is a homogeneous set and $y \sim z$, we have $y \sim t$, a contradiction. □

The previously noted fact that 1-decomposable graphs are p-disconnected is a special case of Lemma 4.6.

Lemma 4.7 *A graph is p-disconnected if and only if it is not a spider and at most one of its cards is p-connected.*

Proof: Assume that G is a p-disconnected graph. By Theorem 4.4, G is not a spider. Let us show that at most one card of G is p-connected.

If G $(\overline{G})$ is disconnected, then clearly at most one card of G is connected (anticonnected) and, therefore, our statement is true. Let $G = T \circ H$, where $T = (S, A, B)$ is a separable p-connected triad. If $|H| > 1$, then all cards of G have the form $T_v \circ H$ or $T \circ H_v$. Thus, by Lemma 4.6, all cards of G are p-disconnected. If $|H| = 1$, then $D(G) = \{T_v \circ H\} \cup \{S\}$. Therefore, by Lemma 4.6, there exists a unique p-connected card of G isomorphic to S.

Conversely, assume that G is not a spider and at most one of its card is p-connected. Suppose that G is p-connected. Then, by Theorem 4.5, there exist at least two p-connected cards of G, a contradiction. □

Since spiders are reconstructible, the following result is true.

Corollary 4.1 *p-Disconnected graphs are recognizable.*

Because disconnected graphs, their complements and 1-decomposable graphs are reconstructible, we obtain:

Corollary 4.2 *The class $\mathcal{R}$ is recognizable.*

For the remaining part of this section, we need the following technical lemma.

Lemma 4.8 *Let $G = (G(A \cup B), A, B) \circ G(C)$, where $(G(A \cup B), A, B)$ is a generalized split triad, and let D be a p-connected component of G. Then, $D \subseteq A \cup B$ or $D \subseteq C$.*

Proof: Suppose that $D \cap (A \cup B) \neq \emptyset$ and $D \cap C \neq \emptyset$. As was shown in Lemma 4.6, for the partition $(A \cup B, C)$ there is no crossing P_4 in G. Therefore, for the partition

$$(D \cap (A \cup B), D \cap C), \tag{4.24}$$

there is no crossing P_4 in the graph $G(D)$. This contradicts the fact that D is a p-connected component of G. □

Lemma 4.9 *The class $\mathcal{R}$ is weakly reconstructible.*

Proof: Let $G^1 = \sigma^1 \circ H^1$ and $G^2 = \sigma^2 \circ H^2$ be two graphs from $\mathcal{R}$ with equal decks $D(G^1)$ and $D(G^2)$, with the corresponding hypomorphism $f : V(G^1) \to V(G^2)$ and with

$$\sigma^1 = (S^1, A^1, B^1), \quad \sigma^2 = (S^2, A^2, B^2) \tag{4.25}$$

being separable p-connected triads from the definition of the class $\mathcal{R}$. By Theorem 4.6, $G^1 \simeq G^2$ if and only if $T^1 \simeq T^2$ and $H^1 \simeq H^2$.

Suppose first that $|H^1| = 1$. Then, $D(G^1) = \{\sigma^1_v \circ H^1\} \cup \{S^1\}$. It is evident that all vertex-deleted triads σ^1_v, σ^2_u are generalized split triads. Therefore, by Lemma 4.6, there exists a unique p-connected card of G^1, and this card is isomorphic to S^1.

Now, we necessarily have $|H^2| = 1$. Indeed, if $|H^2| > 1$ then $D(G^2) = \{\sigma^2_v \circ H^2\} \cup \{\sigma^2 \circ H^2_u\}$. Hence, by Lemma 4.6, all cards of G^2 are p-disconnected. This contradicts the equality of decks $D(G^1)$ and $D(G^2)$. Because $|H^2| = 1$, just as for G^1 there exists a unique p-connected card of G^2 isomorphic to S^2. Thus, we have $S^1 \simeq S^2$. By Lemma 4.3, $\sigma^1 \simeq \sigma^2$ and, consequently, $G^1 \simeq G^2$.

Let us suppose from now on that $|H^1| \geq 2$ and $|H^2| \geq 2$. The vertex sets of G^1 and G^2 are naturally 3-partitioned as follows:

$$V(G^i) = A^i \cup B^i \cup C^i, \tag{4.26}$$

where

$$(G(A^i \cup B^i), A^i, B^i) \simeq \sigma^i, \quad G^i(C^i) \simeq H^i$$

and

$$G^i \simeq (G(A^i \cup B^i), A^i, B^i) \circ G^i(C^i), \quad i = 1, 2.$$

Then, similar to the proofs of Lemmas 4.1 and 4.2, the decks of G^1 and G^2 can also be partitioned as follows:

$$D(G^i) = D_{\sigma^i} \cup D_{H^i}, \tag{4.27}$$

where for $i = 1, 2$,

$$D_{\sigma^i} = \{\sigma^i \circ H_v^i \ : \ v \in C^i\}, \quad D_{H^i} = \{\sigma_v^i \circ H^i \ : \ v \in A^i \cup B^i\}. \tag{4.28}$$

By Lemma 4.6, all cards from $D(G^1)$ and $D(G^2)$ are p-disconnected. Furthermore, all cards from D_{σ^1} and D_{σ^2} are both connected and anticonnected p-disconnected graphs (since so are G^1 and G^2).

Proposition 4.3 *Let $v \in C^1$ be a vertex of G^1 such that $f(v) \in A^2 \cup B^2$. Then, $|\sigma^1| < |\sigma^2|$.*

Proof: Suppose without loss of generality that $f(v) \in A^2$. We have,

$$G_v^1 = \sigma^1 \circ H_v^1 \in D_{\sigma^1} \quad \text{and} \quad G_{f(v)}^2 = \sigma_{f(v)}^2 \circ H^2 \in D_{H^2}.$$

Let $C_v^1 = C^1 - \{v\}$, $A_u^2 = A^2 - \{f(v)\}$, and let

$$\varphi : V(G^1) - \{v\} \to V(G^2) - \{f(v)\} \tag{4.29}$$

be an isomorphism of graphs G_v^1 and $G_{f(v)}^2$. If $\varphi(A^1 \cup B^1) \subseteq C^2$, then $\varphi(C_v^1) \supseteq A_u^2 \cup B^2$. In this case, $B^2 \sim C^2 \cap \varphi(A^1)$, which is impossible.

Therefore, by Lemma 4.8, it is true that $\varphi(A^1 \cup B^1) \subseteq (A_u^2 \cup B^2)$. Thus, $|\sigma^1| \leq |\sigma_u^2| < |\sigma^2|$. □

Proposition 4.3 immediately implies that there exists $v \in C^1$ such that $f(v) \in C^2$. Indeed, if it is not so, then there exist $G_{v_1}^1 \in D_{\sigma^1}$, $G_{v_2}^1 \in D_{H^1}$, $G_{u_1}^2 \in D_{\sigma^2}$ and $G_{u_2}^2 \in D_{H^2}$ such that

$$\sigma^1 \circ H_{v_1}^1 = G_{v_1}^1 \simeq G_{u_2}^2 = \sigma_{u_2}^2 \circ H^2 \tag{4.30}$$

and

$$\sigma^2 \circ H_{u_1}^2 = G_{u_1}^2 \simeq G_{v_2}^1 = \sigma_{v_2}^1 \circ H^1. \tag{4.31}$$

Further, by Proposition 4.3,

$$|\sigma^1| < |\sigma^2| \tag{4.32}$$

and

$$|\sigma^2| < |\sigma^1|, \tag{4.33}$$

a contradiction.

Now, let us consider $v \in C^1$ such that $f(v) \in C^2$. We have,

$$\sigma^1 \circ H_v^1 \simeq \sigma^2 \circ H_{f(v)}^2.$$

By Theorem 4.6,

$$\sigma^1 \simeq \sigma^2. \tag{4.34}$$

In addition, (4.34) implies that $f(v) \in A^2 \cup B^2$ for all $v \in A^1 \cup B^1$; otherwise, there exist cards from D_{σ^1} and D_{H^2} such that (4.30) holds. This results in inequality (4.32) contradicting (4.34).

Hence, it remains to prove that $H^1 \simeq H^2$, which can be done as follows. Since we assumed that G is 1-indecomposable, S^1 is not a split graph. Thus, there exists a connected component X of $\overline{G^1(A^1)}$ or $G^1(B^1)$ with at least three vertices. It is easy to see that for any $v \in X$ the triad σ_v^1 is a separable p-connected triad and the card G_v^1 is both a connected and anticonnected p-disconnected graph.

Let $v \in X$ and ψ be an isomorphism of the graphs G_v^1 and $G_{f(v)}^2$. As was demonstrated previously, $f(v)$ necessarily belongs to $A^2 \cup B^2$, i.e.

$$\sigma_v^1 \circ H^1 = G_v^1 \simeq G_{f(v)}^2 = \sigma_{f(v)}^2 \circ H^2.$$

By the same reasonings as in the proof of Proposition 4.3, we have

$$\psi((A^1 \cup B^1) - \{v\}) \subseteq (A^2 \cup B^2) - \{u\}.$$

Since $|\sigma^1| = |\sigma^2|$, it is true that

$$\psi((A^1 \cup B^1) - \{v\}) = (A^2 \cup B^2) - \{f(v)\}.$$

Therefore, $\psi(C^1) = C^2$ and $H^1 \simeq H^2$. □

Corollary 4.2 and Lemma 4.9 imply the following result:

Theorem 4.7 [38] *p-Disconnected graphs are reconstructible.*

A *quasi-starfish* (respectively, a *quasi-urchin*) [17] is a graph obtained from a thick spider (respectively, a thin spider) by replacing at most one vertex by K_2 or O_2.

Theorem 4.8 [17] *A graph G is P_4-tidy if and only if every p-component of G is isomorphic to either P_5, or $\overline{P_5}$, or C_5, or a quasi-starfish or a quasi-urchin.*

Corollary 4.3 *P_4-tidy graphs are reconstructible.*

Proof: Let G be a P_4-tidy graph. If G is p-disconnected, then G is reconstructible by Theorem 4.7. Suppose that G is p-connected. Then, G is isomorphic to either P_5, or $\overline{P_5}$, or C_5, or a quasi-starfish or a quasi-urchin. It is obvious that P_5, $\overline{P_5}$, C_5 are reconstructible. Moreover, quasi-starfishes are complements of quasi-urchins and, by Lemma 4.5, spiders are reconstructible. Thus, it is sufficient to consider the case when G is obtained from the thin spider H with bipartition (A, B) and with at least six vertices by replacing a vertex $v \in V(H)$ by K_2 or O_2. Consider the following cases:

(1) Suppose that $v \in A$ is replaced by K_2. From the complete description of the structure of 1-indecomposable split unigraphs presented in Chapter 1, one can see that G is a split unigraph. Therefore, G is reconstructible.
(2) Suppose that $v \in B$ is replaced by O_2. Again, G is a split unigraph and, therefore, it is reconstructible.

(3) Suppose that $v \in B$ is replaced by K_2. It is easy to see that a graph F is isomorphic to G if and only if $|V(F)| = |V(G)|$, there exist exactly two vertices $x, y \in V(F)$ with $\deg(x) = \deg(y) = 2$ and $F_x \simeq F_y$ is a thin spider. Therefore, G is reconstructible.

(4) Suppose that $v \in A$ is replaced by O_2. Then, it is not difficult to see that a graph F is isomorphic to G if and only if

$$|V(F)| = |V(G)| = 2k + 1, \quad k \geq 3,$$

and there exist two vertices $x, y \in V(F)$ such that $\deg(x) = \deg(y) = k$ and the cards F_x, F_y are thin spiders. Thus, in this case, G is also reconstructible.

□

4.1.4 Reconstruction of Bi-1-decomposable Graphs

In Section 4.1.2, we proved the reconstruction conjecture for a particular family of graph classes using the properties of (P, Q)-decomposition that is defined for a particular binary operation $\circ$ and for a particular closed hereditary pair of classes P and Q. As a matter of fact, different pairs of classes and different operations may just as well be applicable to the reconstruction conjecture, as long as the unique factorization property holds or perhaps some other less general properties are true. We will consider such decompositions in the next section. Here, we will demonstrate just one specific example for which the reconstruction conjecture was proved.

Basically, we can construct an analogue of the operator decomposition based on properties of bipartite graphs rather than split graphs. We define the *multiplication of triads* as follows (see Figure 4.5):

$$(G_1, A_1, B_1) \bullet (G_2, A_2, B_2) = (G_3, A_1 \cup A_2, B_1 \cup B_2), \tag{4.35}$$

where $V(G_3) = V(G_1) \cup V(G_2)$ and

$$E(G_3) = E(G_1) \cup E(G_2) \cup \{xy \,:\, x \in A_1, y \in B_2\}.$$

As before, the set $\sum$ is a semigroup with respect to the operation $\bullet$. Also, the definitions of the following terms remain the same: decomposable and indecomposable triads, undivided A-parts and B-parts, operator and canonical decompositions (in this case called *bipartite operator and canonical decompositions* to avoid confusion).

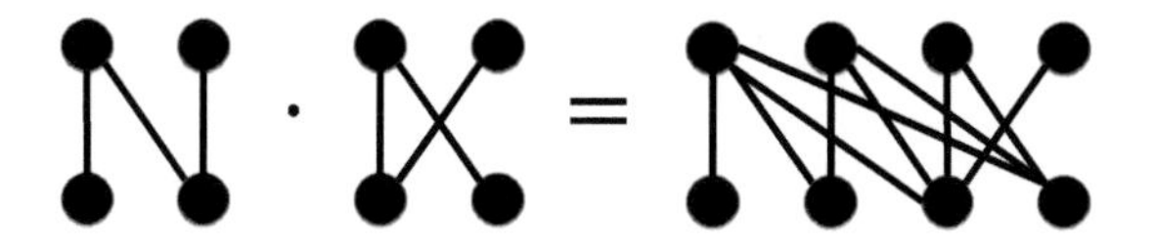

Figure 4.5 Example of the product $\bullet$ of two bipartite triads.

Theorem 4.9 [37] *Let T and S be two triads with canonical bipartite operator decompositions $T = C_1 \bullet C_2 \bullet \ldots \bullet C_k$ and $S = R_1 \bullet R_2 \bullet \ldots \bullet R_l$. Then, $T \simeq S$ if and only if $k = l$ and $C_i \simeq R_i$ for all $i = 1, 2, \ldots, k$.*

We define the product of a triad σ and a graph G as follows:

$$\sigma \bullet G = F, \tag{4.36}$$

where

$$(F, C, D) = \sigma \bullet (G, \emptyset, V(G)). \tag{4.37}$$

Note that unlike the 'classical' operator decomposition, this is not an action of the semigroup $\sum$ on the set of all graphs because in general

$$(\sigma_1 \bullet \sigma_2) \bullet G \neq \sigma_1 \bullet (\sigma_2 \bullet G).$$

The subset of triads (G, A, B), where G is a bipartite graph and (A, B) is the corresponding bipartition into independent sets, forms a subsemigroup of $\sum$. If we restrict operations (4.35) and (4.36) to this subsemigroup, then we arrive at *bipartite 1-decomposition*—a straightforward analogue of 1-decomposition defined for split graphs. Graphs and triads decomposable in terms of bipartite 1-decomposition will be called *bi-1-decomposable.* In fact, bi-1-decomposable triads are not something unheard of outside of the context of this chapter—they were introduced by Fouquet, Giakoumakis and Vanherpe in [16] under the name '$K + S$ bipartite graphs'.

The most important fact about the bipartite 1-decomposition is that it enjoys a very close analogue of the unique factorization property for 1-decomposition. Let G be a bipartite graph with bipartition (A, B). We say that G has a *dominating edge* uv if every vertex in $V(G) - \{u, v\}$ is adjacent to u or v.

Theorem 4.10 [37] *Let G and H be connected bi-1-decomposable graphs, which are not bipartite graphs with dominating edge. Let*

$$G = T_1 \bullet T_2 \bullet \ldots \bullet T_k \bullet G_0 \quad \text{and} \quad H = S_1 \bullet S_2 \bullet \ldots \bullet S_l \bullet H_0$$

be bipartite 1-decompositions of G and H. Then, $G \simeq H$ if and only if the following conditions hold:

(1) $k = l$;
(2) $T_i \simeq S_i$ *for* $i = 1, 2, \ldots, k$;
(3) $G_0 \simeq H_0$.

Using this theorem, it is possible to prove the next result:

Theorem 4.11 [37] *Bi-1-decomposable graphs are reconstructible.*

The proofs of these theorems are quite similar to the proofs of the corresponding theorems for 1-decomposition. Therefore, we are skipping them.

4.2 Generalizations of Operator Decomposition

4.2.1 Preliminaries

We have already considered the algebraic-type decompositions of graphs in the way they were originally introduced in [44] and further extended, studied and utilized in consecutive papers (e.g. [6, 36, 43, 46]). It should be noted that in all decomposition examples (with the exception of the bipartite operator decomposition briefly discussed in Section 4.1.4), the algebraic operation on the set of 2-partitioned graphs or triads was the same. Also, the variation of decompositions was associated with different properties of the subgraphs induced by A-parts and B-parts of partitions.

Further development of the theory of algebraic-type graph decomposition requires a further generalization. The natural next step is to consider arbitrary algebraic operations and turn the set of graphs into semigroups with respect to those operations. In this section, we study a general framework that defines and describes all such operations.

Let us consider a graph G together with some arbitrary partition of its vertex set into n disjoint subsets:

$$V(G) = A_1 \cup A_2 \cup \ldots \cup A_n.$$

Some of the sets A_i may be empty. We call this object an *n-partitioned graph* and denote it by $\tau = (G, A_1, \ldots, A_n)$; conversely, G is called *the basic graph* of τ. Denote the set of vertices and the set of edges of τ by $V(\tau)$ and $E(\tau)$, respectively.

An *isomorphism* of n-partitioned graphs can be naturally defined as an isomorphism of the corresponding graphs that preserves the partitions. Formally, an isomorphism f of τ and $\sigma = (F, B_1, \ldots, B_n)$ is an isomorphism of G and F such that $f(A_i) = B_i$ for every $i \in [n]$. Let Σ_n be the set of all isomorphism classes of n-partitioned graphs.

On the set Σ_n, we define a binary algebraic operation $\circ_H$ (*H-product* of graphs) determined by a fixed digraph H with $V(H) = \{1, 2, \ldots, n\}$. In what follows, we assume that the digraph H can have loops, but no multiple edges. Given $\tau = (G, A_1, \ldots, A_n)$ and $\sigma = (F, B_1, \ldots, B_n)$ with $V(G) \cap V(F) = \emptyset$, the product $\tau \circ_H \sigma$ is the following n-partitioned graph:

$$\tau \circ_H \sigma = (R, A_1 \cup B_1, \ldots, A_n \cup B_n), \tag{4.38}$$

where R is formed from the disjoint union $G \cup F$ by

- adding all edges joining A_i and B_j whenever $(i, j) \in A(H)$;
- adding no edges joining A_i and B_j if $(i, j) \notin A(H)$.

An n-partitioned graph $\tau \in \Sigma_n$ is *H-decomposable* if $\tau = \tau_1 \circ_H \tau_2$ where $\tau_1, \tau_2 \in \Sigma_n$; and *H-prime* otherwise. It is clear that every n-partitioned graph $\tau \in \Sigma_n$ can be represented as the product

$$\tau = \tau_1 \circ_H \tau_2 \circ_H \ldots \circ_H \tau_k$$

of H-prime factors, $k \geq 1$. Such a representation is an *H-decomposition* of τ. It is easy to see that the standard operator decomposition discussed in the previous chapters is H_0-decomposition, where the digraph H_0 is shown in Figure 4.6.

Note that those kinds of graph operations were considered within the studies of other types of graph decompositions and associated parameters such as clique-width [14] and NLC-width [49]. However, in these studies the operations were introduced with different purposes and with different applications in mind, and hence algebraic properties of the operations were not essential. In contrast, we will consider this idea here from a different perspective—as a study of binary algebraic operations. The main question for us is again the existence of the unique factorization property.

In the second part of this section, we will define and study *H-threshold graphs* and *threshold-width* of graphs [35]—the family of graph classes and the graph-theoretic parameter that naturally arise from our general definition of the algebraic-type decomposition. The ideas of H-threshold graphs and threshold-width came both from the well-known notion of a threshold graph [13] and from the study of clique-width [14] and NLC-width [49]. As was discussed in Chapter 3, threshold graphs form an important class with many interesting properties and applications. In particular, threshold graphs are the graphs with the simplest operator decomposition (Theorem 3.3): all factors of such decompositions are single-vertex triads. Thus, when we move to the more general decomposition, we can define H-threshold graphs as the graphs that can be represented as products of one-vertex factors under $\circ_H$. For instance, standard threshold graphs are H_0-threshold graphs, where the digraph H_0 is depicted in Figure 4.6. It turns out that every graph is an H-threshold graph for some digraph H. Thus, it is natural to look for such a representation with the smallest H. We define *threshold-width* $\theta(G)$ of a graph G to be the minimum order of a digraph H such that G is H-threshold. The analogues of threshold-width in the theories of NLC-width and clique-width are linear NLC-width and linear clique-width, respectively.

The class of *difference graphs* [19] (or *bipartite chain graphs* [50]) is another class of graphs related to threshold graphs. This class is described by the following theorem, which in some aspects mirrors the aforementioned result for threshold graphs:

Theorem 4.12 [19] *For a graph G, the following properties are equivalent and they determine membership in the class of difference graphs.*

- *(a) There exist real weights $(\beta_v : v \in V(G))$ and a threshold s such that $|\beta_v| \leq s$ for all $v \in V(G)$, and $uv \in E(G)$ if and only if $|\beta_u - \beta_v| \geq s$.*
- *(b) G is bipartite with 2-partition (A, B), and the sets $\{N_A(b) : b \in B\}$ and $\{N_B(a) : a \in A\}$ are ordered by inclusion.*

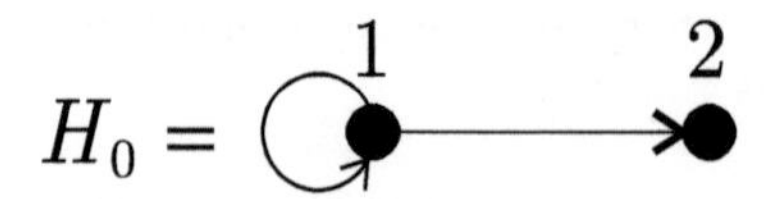

Figure 4.6 The digraph H_0.

In general, the properties of difference graphs and threshold graphs are similar [19]. We will see that this is no accident, since difference graphs are H-threshold graphs for a particular H. Notice that graphs with bounded threshold-width generalize threshold and difference graphs, and we will show that they extend other properties of those classes. In particular, it is possible to characterize graphs with fixed threshold-width in terms of vertex partitions into cliques and independent sets and orderings of vertex neighbourhoods (analogously to Theorem 4.12 and the property of total vicinal preorder for threshold graphs discussed in Chapter 3), even though the characterization becomes more complicated.

4.2.2 H-product of Graphs and Its Properties

Suppose that the digraph H with the vertex set $V(H) = [n]$ and an arc set $A(H)$ is fixed, as well as the corresponding operation $\circ_H$. For convenience, we will sometimes denote the operation $\circ_H$ simply by $\circ$ if it is clear which digraph H is meant. It is easy to check that for every digraph H, the operation $\circ_H$ is associative. Thus, the set Σ_n with the operation $\circ_H$ is a semigroup. A digraph H is *symmetric* if $(i, j) \in A(H)$ whenever $(j, i) \in A(H)$. It is clear that the operation $\circ_H$ is commutative if and only if H is symmetric.

The major question for us here, however, is the existence of the unique factorization property. As a starting point, we can use the fact that the unique factorization property of the operation $\circ_H$ can be established for 2-vertex digraphs H. Indeed, we already proved this fact for the digraph shown in Figure 4.6 (Theorem 1.1), and for its loopless analogue (Theorem 4.10). The unique factorization property for other 2-vertex digraphs follows from the results of [28]. It is possible to establish a much more general fact:

Theorem 4.13 (Unique Factorization Theorem for Operation $\circ_H$) [35] *For every n-vertex digraph H, every n-partitioned graph $\tau \in \Sigma_n$ has a unique H-decomposition up to exchange of commuting factors.*

Proof: It is clear that if two n-partitioned graphs have H-decompositions that differ only by some exchange of commuting factors, then they are isomorphic. So, we will need to prove only the converse statement. We will use the induction on the number of factors in the H-decomposition of an n-partitioned graph. The fact that we are trying to prove is evident for prime n-partitioned graphs.

Let $\upsilon = (G, X_1, \ldots, X_n)$ and $\omega = (F, Y_1, \ldots, Y_n)$ be isomorphic n-partitioned graphs with the H-decompositions

$$\upsilon = \tau_1 \circ \tau_2 \circ \ldots \circ \tau_k, \quad \omega = \rho_1 \circ \rho_2 \circ \ldots \circ \rho_l, \quad k, l \geq 2. \tag{4.39}$$

We may assume that $X_i \cup Y_i \neq \emptyset$ for all $i = 1, 2, \ldots, n$.

Let $f : V(\upsilon) \to V(\omega)$ be an isomorphism of υ and ω. We will use the following notation. For a set $X \subseteq V(\upsilon)$, let $f(X) = \{f(x) : x \in X\}$; for a subgraph H of G, let $f(H) = F(f(V(H)))$; and let $f(\tau) = (f(H), f(A_1), \ldots, f(A_n))$ for an n-partitioned graph $\tau = (H, A_1, \ldots, A_n)$, where H is a subgraph of G.

Let $\tau_1 = (G_1, A_1, \dots, A_n)$ and $\rho_1 = (F_1, B_1, \dots, B_n)$. By setting

$$\tau' = \tau_2 \circ \dots \circ \tau_k = (G', S_1, \dots, S_n) \quad \text{and} \quad \rho' = \rho_2 \circ \dots \circ \rho_l = (F', Q_1, \dots, Q_n),$$

we have

$$\upsilon = \tau_1 \circ \tau' \quad \text{and} \quad \omega' = \rho_1 \circ \rho'. \tag{4.40}$$

By the definition of isomorphism, we obtain $f(A_i \cup S_i) = B_i \cup Q_i$.

Suppose that there exists $i \in [n]$ such that $f(A_i) \cap B_i \neq \emptyset$ and $f(A_i) \cap Q_i \neq \emptyset$. In this case, we have

$$f(\tau_1) = \tau'' \circ \tau''',$$

where

$$\tau'' = \Big(F\big(f(V(\tau_1)) \cap V(\rho_1)\big), f(A_1) \cap B_1, \dots, f(A_n) \cap B_n\Big),$$

$$\tau''' = \Big(F\big(f(V(\tau_1)) \cap V(\rho')\big), f(A_1) \cap Q_1, \dots, f(A_n) \cap Q_n\Big).$$

Here, our assumption implies $V(\tau'') \neq \emptyset$ and $V(\tau''') \neq \emptyset$. So, an isomorphic image of τ_1 is decomposable, and thus we arrive at a contradiction to the fact that τ_1 is prime. The same contradiction is obtained in a similar way when we assume that there exist $i, j \in [n]$, $i \neq j$, such that $f(A_i) \subseteq B_i$ and $f(A_j) \subseteq Q_j$. Finally, by symmetry, the assumption that $f^{-1}(B_i) \cap A_i \neq \emptyset$ and $f^{-1}(B_i) \cap S_i \neq \emptyset$ for some $i \in [n]$ also leads to a contradiction.

We are left with two possibilities: $f(A_i) \subseteq B_i$ for all $i \in [n]$ and $f(A_i) \subseteq Q_i$ for all $i \in [n]$. Let us consider them one by one.

(1) Suppose that $f(A_i) \subseteq B_i$ for every $i \in [n]$. Then, the facts proved above imply that $f(A_i) = B_i$ and $f(S_i) = Q_i$ for all $i \in [n]$. Thus,

$$\tau_1 \simeq \rho_1, \quad \tau' \simeq \rho'.$$

After applying the induction hypothesis to τ' and ρ', we obtain $k = l$ and, under a respective ordering and without loss of generality,

$$\tau_2 \simeq \rho_2, \quad \tau_3 \simeq \rho_3, \quad \dots \quad \tau_k \simeq \rho_k.$$

Hence, the statement of the theorem in this case is true.

(2) Suppose that $f(A_i) \subseteq Q_i$ for every $i \in [n]$. This condition implies $B_i \subseteq f(S_i)$. In this case, there are two subcases to consider.

(2.1) Let us assume first that $f(S_i) \cap Q_i = \emptyset$ for all $i \in [n]$. In this subcase, we have $f(S_i) = B_i$ and $f(A_i) = Q_i$ for every $i \in [n]$. In other words, this means that $\tau' \simeq \rho_1$ and $\tau_1 \simeq \rho'$. Thus,

$$\upsilon \simeq \tau_1 \circ \rho_1 \simeq \omega \simeq \rho_1 \circ \tau_1.$$

Therefore, the statement of Theorem 4.13 is true.

(2.2) In the second subcase, suppose that there exists $i \in [n]$ such that $f(S_i) \cap Q_i \neq \emptyset$. Let

$$\zeta = \Big(F\big(f(V(\tau')) \cap V(\rho')\big), f(S_1) \cap Q_1, \ldots, f(S_n) \cap Q_n\Big).$$

By our assumption, $V(\zeta) \neq \emptyset$, which implies $f(\tau') = \rho_1 \circ \zeta$, $\rho' = f(\tau_1) \circ \zeta$ and, thus,

$$\tau' \simeq \rho_1 \circ \zeta, \quad \rho' \simeq \tau_1 \circ \zeta.$$

This means that τ_1 is the first factor in some H-decomposition of the n-partitioned graph ρ'. By applying the induction hypothesis to ρ', we may assume without loss of generality that $\tau_1 \simeq \rho_2$ and $\zeta \simeq \rho_3 \circ \rho_4 \circ \ldots \circ \rho_l$. Thus,

$$\tau_2 \circ \tau_3 \circ \ldots \circ \tau_k \simeq \tau' \simeq \rho_1 \circ \rho_3 \circ \rho_4 \circ \ldots \circ \rho_l.$$

As before, by applying the induction hypothesis to the n-partitioned graph τ', we obtain $k = l$ and $\rho_1 \simeq \tau_2$, $\tau_3 \simeq \rho_3$, $\ldots$,$\tau_k \simeq \rho_k$ under an appropriate ordering.

To complete the proof, it remains to show that τ_1 and ρ_1 commute. For this purpose, it is sufficient to prove that for every pair $i, j \in [n]$, $i \neq j$, such that $(i,j) \in A(H)$ and $(j,i) \notin A(H)$, one of the following four conditions hold: $A_i \cup A_j = \emptyset$, $A_j \cup B_j = \emptyset$, $A_i \cup B_i = \emptyset$ or $B_i \cup B_j = \emptyset$.

We have, $f(A_i) \sim f(S_j)$ and $f(A_j) \not\sim f(S_i)$ because $A_i \sim S_j$, $A_j \not\sim S_i$ and f is an isomorphism. Since $f(A_i) \subseteq Q_i$, $f(A_j) \subseteq Q_j$, $B_i \subseteq f(S_i)$, $B_j \subseteq f(S_j)$, it follows that $f(A_i) \not\sim f(S_j)$ and $f(A_j) \sim f(S_i)$. These two facts imply that one of the following is true:

(a) $A_i \cup A_j = \emptyset$;
(b) $A_i \cup S_i = \emptyset$, which implies $B_i = \emptyset$;
(c) $A_j \cup S_j = \emptyset$, which implies $B_j = \emptyset$;
(d) $S_i \cup S_j = \emptyset$, which implies $B_i = \emptyset$ and $B_j = \emptyset$.

This concludes the proof. □

4.2.3 *H*-threshold Graphs and Threshold-width of Graphs

In this section, we study analogues of threshold graphs—the graphs with the simplest H-decompositions. To formalize the definition, we will denote by K_i^k the k-partitioned graph $(K_1, \emptyset, \ldots, \emptyset, \{v\}, \emptyset, \ldots, \emptyset)$ (the only non-empty set of the partition is the i-th set).

Let H be a digraph on k vertices. A graph G is an *H-threshold graph* if it is a basic graph for a k-partitioned graph of the form

$$K_{i_1}^k \circ_H K_{i_2}^k \circ_H \ldots \circ_H K_{i_n}^k. \tag{4.41}$$

In this case, for simplicity, we will write $G = K_{i_1}^k \circ_H K_{i_2}^k \circ_H \ldots \circ_H K_{i_n}^k$ (though strictly speaking the left part of this equality is a graph and the right part is a

k-partitioned graph). The representation of a graph G in form (4.41) is a *threshold representation* of G.

Figure 4.7 illustrates the notion of an H-threshold graph by depicting the threshold representations of the graphs P_4 and C_4 for different 2-vertex digraphs H (2-partitioned factors K_i^2 are represented by ovals). Note that these graphs are not threshold graphs in the standard sense.

Proposition 4.4 *Every graph G is an H-threshold graph for some digraph H.*

Proof: Let $V(G) = [n]$. Define the digraph H as follows: $V(H) = [n]$, $(i, j) \in A(H)$ if and only if $ij \in E(G)$ and $i < j$; that is, H is obtained from G by assigning an orientation to every edge of G. It is easy to see that $G = K_1^n \circ_H K_2^n \circ_H \dots \circ_H K_n^n$. □

The digraph H constructed in the proof of Proposition 4.4 has $|V(G)|$ vertices. However, threshold graphs are H_0-threshold graphs for the digraph H_0 with only 2 vertices. Hence, it is natural to consider the minimum order of a digraph H such that a given graph G is an H-threshold graph. Thus, the corresponding graph parameter can be defined as follows. The *threshold-width* of a graph G is the parameter

$$\theta(G) = \min\{|H| : G \text{ is an } H\text{-threshold graph}\}. \tag{4.42}$$

By Proposition 4.4, every graph G has a threshold-width and, moreover, $\theta(G) \le n$. The next proposition establishes another elementary property of threshold-width.

Proposition 4.5 *For every graph G, it is true that $\theta(G) = \theta(\overline{G})$.*

Proof: Suppose that G is an H-threshold graph for a digraph H with the vertex set $V(H) = [k]$, i.e. $G = K_{i_1}^k \circ_H K_{i_2}^k \circ_H \dots \circ_H K_{i_n}^k$. Let $\{v_j\} = V(K_{i_j}^k)$, where

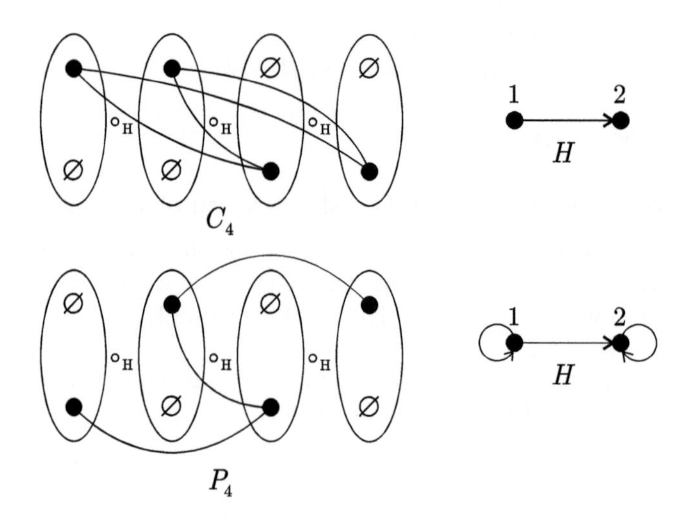

Figure 4.7 C_4 and P_4 as H-threshold graphs.

$j = 1, 2, \ldots, n$. Consider vertices v_p and v_q such that $p < q$. We have, $v_p \sim v_q$ if and only if one of the following conditions hold:

(1) $i_p = i_q$ and $(i_p, i_p) \in A(H)$;
(2) $i_p \neq i_q$ and $(i_p, i_q) \in A(H)$.

Let $\overline{H}$ be the complement of H; that is, the digraph with the same vertex set and with the arc set $A(\overline{H}) = \{(i, j) : (i, j) \notin A(H)\}$. Taking into account (1) and (2), it is easy to see that $\overline{G} = K^k_{i_1} \circ_{\overline{H}} K^k_{i_2} \circ_{\overline{H}} \ldots \circ_{\overline{H}} K^k_{i_n}$. □

Our next goal is to describe graphs G with $\theta(G) \leq k$. As we will see, similar to the standard threshold graphs, we can provide a description in terms of relations between vertex neighbourhoods. Firstly, we need to introduce some auxiliary definitions and lemmas.

For a digraph H and $v \in V(H)$, the *in-neighbourhood* and the *out-neighbourhood* of v, denoted $N_{\text{in}}(v)$ and $N_{\text{out}}(v)$, are the sets

$$\{u \in V(H) - \{v\} : (u, v) \in A(H)\}$$

and

$$\{w \in V(H) - \{v\} : (v, w) \in A(H)\},$$

respectively. A digraph H is an *oriented graph* if, for every $u, v \in V(H)$, it contains at most one arc from the set $\{(u, v), (v, u)\}$.

Let H be a digraph and let $(v_1, v_2, \ldots, v_n)$ be an ordering of its vertices. This ordering is an *acyclic ordering* or *topological sort* if all arcs in H have the form (v_i, v_j), where $i < j$. A digraph is *acyclic* if it does not contain directed cycles. The following property of acyclic digraphs is well known.

Proposition 4.6 [5] *A digraph is acyclic if and only if there exists an acyclic ordering of its vertices.*

Let

$$S = \Big(\{X^1_1, X^1_2\}, \{X^2_1, X^2_2\}, \ldots, \{X^k_1, X^k_2\}\Big)$$

be the sequence of set pairs, where $X^i_j \subseteq \{1, 2, \ldots, k\} - \{i\}$, $i \in [k]$, $j \in [2]$ (some sets X^i_j may be empty). The sequence S is an *orgraphical sequence of pairs of sets* or simply *orgraphical sequence* if there exists an oriented graph D on a vertex set $V(D) = [k]$ such that

$$S = \Big(\{N_{\text{in}}(1), N_{\text{out}}(1)\}, \{N_{\text{in}}(2), N_{\text{out}}(2)\}, \ldots, \{N_{\text{in}}(k), N_{\text{out}}(k)\}\Big).$$

In this case, D is called a *realization* of S.

The evident necessary conditions for S to be orgraphical are as follows:

(i) $X^i_1 \cap X^i_2 = \emptyset$;
(ii) $i \in X^j_1 \cup X^j_2$ whenever $j \in X^i_1 \cup X^i_2$.

When S has these properties, we say that S is a *proper sequence*.

We will now establish necessary and sufficient conditions for a proper sequence to be orgraphical. These conditions will be instrumental in characterizing the graphs with fixed threshold-widths. Suppose that S is a proper sequence and define the graph $R(S)$ as follows:

- $V(R(S)) = \{X_j^i : i \in [k], j \in [2]\}$;
- For each choice of $i, j \in [k]$ and $q, p \in [2]$, $X_q^i \sim X_p^j$ if and only if either $i = j$, $q \neq p$ or $i \in X_p^j$, $j \in X_q^i$.

The reason behind this definition becomes apparent via the following lemma:

Lemma 4.10 *A proper sequence S is orgraphical if and only if the graph $R(S)$ is bipartite.*

Proof: Suppose that D is a realization of S. Let

$$l(X_q^i) = \begin{cases} 1 & \text{if } X_q^i = N_{\text{out}}(i); \\ 2 & \text{if } X_q^i = N_{\text{in}}(i). \end{cases}$$

It can be seen that l is a proper 2-colouring of $R(S)$. Indeed, suppose that $X_q^i \sim X_p^j$. If $i = j$ then, by definition, we have $l(X_1^i) \neq l(X_2^i)$. Suppose that $i \neq j$ and, without loss of generality, $X_q^i = N_{\text{in}}(i)$. Thus, by definition, $j \in N_{\text{in}}(i)$. Together with the facts that D is an oriented graph and $i \in X_p^j$, this implies $X_p^j = N_{\text{out}}(j)$. Thus, $l(X_q^i) \neq l(X_p^j)$. This proves that any two adjacent vertices receive different colours.

Conversely, let l be a proper 2-colouring of $R(S)$. Let us define a digraph D on the vertex set $[k]$ as follows: $(i, j) \in A(D)$ if and only if $i \in X_p^j$, $j \in X_q^i$, $l(X_q^i) = 1$ and $l(X_p^j) = 2$. Since l is a proper 2-colouring, this definition correctly determines the oriented graph; and for every $i \in [k]$ if, for example, $l(X_1^i) = 1$, $l(X_2^i) = 2$, then $X_1^i = N_{\text{out}}(i)$, $X_2^i = N_{\text{in}}(i)$. □

Corollary 4.4 *If D_1 and D_2 are two different realizations of S, then D_1 can be obtained from D_2 by reversal of all arcs of some of its connected components.*

For a sequence $\pi = (a_1, a_2, \ldots, a_n)$, the sequence $(a_n, a_{n-1}, \ldots, a_1)$ is denoted by $\text{inv}(\pi)$. Let

$$V(G) = V_1 \cup V_2 \cup \ldots \cup V_k \tag{4.43}$$

be a partition of the vertex set of a graph G, where each V_i is a clique or an independent set. We will say that partition (4.43) satisfies the *neighbourhood's ordering property* if, for every $i \in [k]$, there exists an ordering $\psi(i) = (u_1^i, u_2^i, \ldots, u_{r_i}^i)$ of the set V_i such that for every $j \in [k] - \{i\}$ the set $\{N_{V_j}(u) : u \in V_i\}$ is ordered by inclusion, and this ordering either coincides with $\psi(i)$ or with $\text{inv}(\psi(i))$. In other words, for every $j \in [k] - \{i\}$, either

$$N_{V_j}(u_1^i) \supseteq N_{V_j}(u_2^i) \supseteq \ldots \supseteq N_{V_j}(u_{r_i}^i) \tag{4.44}$$

or

$$N_{V_j}(u_1^i) \subseteq N_{V_j}(u_2^i) \subseteq \ldots \subseteq N_{V_j}(u_{r_i}^i). \tag{4.45}$$

Assume that the orderings $\psi(i)$ are fixed. For every $i \in [k]$, the set $[k] - \{i\}$ is naturally partitioned into two classes. For convenience, let us denote those classes by Y_1^i (if it contains j satisfying (4.44)) and by $Y_2^i = ([k] - \{i\}) - Y_1^i$ (if it contains j satisfying (4.45)).

Let

$$X_r^i = Y_r^i - \{j \,:\, V_i \sim V_j \text{ or } V_i \not\sim V_j\},\ r = 1, 2,$$

and

$$S = S(V_1, V_2, \ldots, V_k) = \Big(\{X_1^1, X_2^1\}, \{X_1^2, X_2^2\}, \ldots, \{X_1^k, X_2^k\}\Big).$$

Suppose that S is an orgraphical sequence. By Lemma 4.10, $R(S)$ is a bipartite graph, which further will also be denoted by $R(V_1, V_2, \ldots, V_k)$. Let D be a realization of S. Assume without loss of generality that $N_{\text{out}}(i) = X_1^i$; if this is not the case, replace $\psi(i)$ by $\text{inv}(\psi(i))$. Recall that this means that D is an oriented graph.

Using the digraph D, let us define the digraph

$$F = F(V_1, V_2, \ldots, V_k) = F_D(V_1, V_2, \ldots, V_k)$$

as follows:

$$V(F) = V(G), \quad A(F) = A_1 \cup A_2 \cup A_3^1 \cup \ldots \cup A_3^k,$$

where

$$A_1 = \Big\{(u, v) \,:\, u \in V_i, v \in V_j, uv \in E(G), (i, j) \in A(D);\ i, j \in [k]\Big\};$$

$$A_2 = \Big\{(v, u) \,:\, u \in V_i, v \in V_j, uv \notin E(G), (i, j) \in A(D);\ i, j \in [k]\Big\};$$

$$A_3^i = \begin{cases} \{(u_1^i, u_2^i), \ldots, (u_{r_i-1}^i, u_{r_i}^i)\} \text{ if } X_1^i = N_{\text{out}}(i) \text{ in } D; \\ \{(u_{r_i}^i, u_{r_{i-1}}^i), \ldots, (u_2^i, u_1^i)\} \text{ if } X_2^i = N_{\text{out}}(i) \text{ in } D; \end{cases} \quad i \in [k].$$

In other words, the digraph F is constructed in the following way. First, consider every pair V_i, V_j such that neither $V_i \sim V_j$ nor $V_i \not\sim V_j$. Without loss of generality, suppose that $(i, j) \in A(D)$. Now, consider the set $E_{i,j}$ of edges of the complete bipartite graph with the parts V_i and V_j. If an edge $uv \in E_{i,j}$ belongs to $E(G)$, then orient it in the direction from V_i to V_j; otherwise, orient it in the direction from V_j to V_i. Next, turn every set V_i into an oriented path. The order of vertices in that path is determined either by $\psi(i)$ or by $\text{inv}(\psi(i))$, depending on which of the sets X_1^i or X_2^i is the out-neighbourhood of i in D.

Now, we are ready to formulate a characterization of graphs with threshold-width $\theta(G) \leq k$.

Theorem 4.14 [35] *For a graph G, $\theta(G) \leq k$ if and only if there exists partition* (4.43) *such that*

(1) it satisfies the neighbourhood's ordering property;
(2) the sequence $S = S(V_1, V_2, \ldots, V_k)$ is orgraphical; that is, the graph $R(S) = R(V_1, V_2, \ldots, V_k)$ is bipartite;
(3) the digraph $F = F(V_1, V_2, \ldots, V_k)$ is acyclic.

Proof: Let us prove the sufficiency first. Suppose that D is a realization of S that determines F. Let us expand D by adding the set of arcs

$$\{(i,i) \,:\, V_i \text{ is a clique}\} \cup \{(i,j),(j,i) \,:\, V_i \sim V_j\}.$$

Denote the obtained graph by H.

Let $(v_1, v_2, \ldots, v_n)$ be an acyclic ordering of the digraph F. We will show that $G = K^k_{i_1} \circ_H K^k_{i_2} \circ_H \ldots \circ_H K^k_{i_n}$, where $V(K^k_{i_j}) = \{v_j\}$, $v_j \in V_{i_j}$. Denote $K^k_{i_1} \circ_H K^k_{i_2} \circ_H \ldots \circ_H K^k_{i_n} = Z$. Consider an edge $ab \in E(G)$. Then, it can be shown that $ab \in E(Z)$. Indeed, if V_i is a clique in G, then $(i,i) \in A(H)$, which implies that V_i is a clique in Z. Analogously, if $V_i \sim_G V_j$ then (i,j), $(j,i) \in A(H)$ and so $V_i \sim_Z V_j$ by the definition of the operation $\circ_H$.

It remains to consider the case when $a \in V_i$, $b \in V_j$, $i \neq j$ and neither $V_i \sim V_j$ nor $V_i \not\sim V_j$. In this case, i and j are connected by an arc in D. Let us assume without loss of generality that $(i,j) \in A(D)$. In this case, by the definition of F, we have $(a,b) \in A(F)$. Therefore, in the acyclic ordering, a goes before b, i.e. $a = V(K^k_{i_r})$, $b = V(K^k_{i_s})$, $r < s$. Together with the fact that $(i,j) \in A(H)$, this implies $ab \in E(Z)$.

Conversely, suppose that $ab \in E(Z)$. Let $a = v_r$, $b = v_s$, $r < s$ (i.e. a precedes b in the acyclic ordering), $a \in V_i$, $b \in V_j$. So, we know that the relation $V_i \not\sim V_j$ does not hold. By the definition of the operation $\circ_H$, we have $(i,j) \in A(H)$. If $i = j$ then V_i is a clique, and so $ab \in V(G)$. Further, let $i \neq j$ and suppose that $V_i \sim V_j$ is not true. This implies that $(i,j) \in A(D)$ and so a and b are adjacent in F. Since a precedes b in the acyclic ordering, $(a,b) \in A(F)$. Thus, the arc (a,b) is directed from V_i to V_j, which implies $ab \in E(G)$.

Now, we will prove the necessity. Assume that

$$G = K^k_{i_1} \circ_H K^k_{i_2} \circ_H \ldots \circ_H K^k_{i_n},$$

where $\{v_j\} = V(K^k_{i_j})$. In this case,

$$V(G) = V_1 \cup V_2 \cup \ldots \cup V_k, \tag{4.46}$$

where $V_i = \{v \,:\, \{v\} = V(K^k_i)\}$, $i \in [k]$. If $(i,i) \in A(H)$ then V_i is a clique; otherwise, it is an independent set.

Suppose that $V_i = \{v_{l_1}, v_{l_2}, \ldots, v_{l_i}\}$, $l_1 < l_2 < \ldots < l_i$. If $(i,j) \in A(H)$ then

$$N_{V_j}(v_{l_1}) \supseteq N_{V_j}(v_{l_2}) \supseteq \ldots \supseteq N_{V_j}(v_{l_i}); \tag{4.47}$$

otherwise,

$$N_{V_j}(v_{l_i}) \supseteq \ldots \supseteq N_{V_j}(v_{l_2}) \supseteq N_{V_j}(v_{l_1}). \tag{4.48}$$

Therefore, partition (4.46) satisfies the neighbourhood's ordering property.

Let D be a digraph obtained from H by deleting loops and arcs of the set $\{(i,j) : V_i \sim V_j\}$. For the digraph D, we have

$$N_{\text{out}}(i) = \Big\{j : \text{(4.47) holds and neither } V_i \sim V_j \text{ nor } V_i \not\sim V_j\Big\}$$

and

$$N_{\text{in}}(i) = \Big\{j : \text{(4.48) holds and neither } V_i \sim V_j \text{ nor } V_i \not\sim V_j\Big\}.$$

Thus, D is a realization of $S(V_1, V_2, \ldots, V_k)$. It remains to show that $(v_1, v_2, \ldots, v_n)$ is an acyclic ordering of $F = F(V_1, V_2, \ldots, V_k)$. All arcs with both ends in V_l, $l \in [k]$, have the form (v_i, v_{i+1}). So, let us consider $v_i \in V_l$, $v_j \in V_s$, $l \neq s$, such that v_i and v_j are adjacent in F. By the definition of F, neither $V_l \sim V_s$ nor $V_l \not\sim V_s$. Therefore, l and s are adjacent in H. Let $(l, s) \in A(H)$. If $(v_i, v_j) \in A(F)$ then $v_i v_j \in E(G)$, which can be true only for $i < j$. If $(v_j, v_i) \in A(F)$ then $v_i v_j \notin E(G)$, which can be true only for $j < i$. Theorem 4.14 is proved. □

Proposition 4.7 *If partition (4.43) is given, it can be tested in polynomial time whether it satisfies the conditions of Theorem 4.14. In case of an affirmative answer, the proofs of Lemma 4.10 and Theorem 4.14 provide an algorithm for reconstruction of the graph H such that G is an H-threshold graph.*

The definition of a digraph $F(V_1, V_2, \ldots, V_k)$ depends on a realization D of the sequence $S(V_1, V_2, \ldots, V_k)$, which may not be unique. The next proposition shows that from the viewpoint of Theorem 4.14 all realizations are equivalent.

Proposition 4.8 *Let D_1, D_2 be two realizations of $S(V_1, V_2, \ldots, V_k)$ for partition (4.43). If $F_{D_1}(V_1, V_2, \ldots, V_k)$ is acyclic, then $F_{D_2}(V_1, V_2, \ldots, V_k)$ is also acyclic.*

Proof: Suppose that $F_{D_1}(V_1, V_2, \ldots, V_k)$ is acyclic. By Corollary 4.4, D_1 and D_2 have the same sets of connected components. It follows from the definition that $\{i_1, i_2, \ldots, i_j\}$ is a connected component of D_l if and only if $V_{i_1} \cup V_{i_2} \cup \ldots \cup V_{i_j}$ is a connected component of $F_{D_l}(V_1, V_2, \ldots, V_k)$, $l = 1, 2$. Thus, the definition of F and Corollary 4.4 imply that $F_{D_2}(V_1, V_2, \ldots, V_k)$ can be obtained from $F_{D_1}(V_1, V_2, \ldots, V_k)$ by reversal of all arcs of some of its connected components. Therefore, $F_{D_2}(V_1, V_2, \ldots, V_k)$ is acyclic. □

4.2.4 Graphs with Small Threshold-width

By definition, the only graphs with threshold-width 1 are complete and empty graphs. Threshold graphs have threshold-width at most 2, but there are non-threshold graphs with this property. For example, in Figure 4.7, we can see that the non-threshold graphs C_4 and P_4 have threshold-width equal to 2. The next proposition demonstrates that threshold graphs and difference graphs can still be considered as 'basic' graphs with threshold-width at most 2:

Theorem 4.15 [35] *For a graph G, $\theta(G) \leq 2$ if and only if G or $\overline{G}$ is a threshold graph or a difference graph.*

Proof: It is well known that a graph is threshold if it can be obtained from K_1 by adding isolated or dominating vertices [13]. This fact and Theorems 4.12, 4.14 imply the necessity. Let us prove the sufficiency. We will use Theorem 4.14. By definition, there exists a partition $V(G) = V_1 \cup V_2$ such that each of V_1, V_2 is a clique or an independent set. This partition satisfies the neighbourhood's ordering property. It is clear that a realization of the family $S(V_1, V_2)$ is either an empty digraph (if $V_1 \sim V_2$ or $V_1 \not\sim V_2$) or a digraph D with $A(D) = \{(1,2)\}$.

Let us prove that $F = F(V_1, V_2)$ is acyclic. If $V_1 \sim V_2$ or $V_1 \not\sim V_2$, then F is empty. Otherwise, suppose that $A(D) = \{(1,2)\}$. Let $V_1 = \{u_1, u_2, \ldots, u_r\}$ and $V_2 = \{v_1, v_2, \ldots, v_s\}$, where

$$N_{V_2}(u_1) \supseteq N_{V_2}(u_2) \supseteq \ldots \supseteq N_{V_2}(u_r)$$

and

$$N_{V_1}(v_s) \supseteq N_{V_1}(v_{s-1}) \supseteq \ldots \supseteq N_{V_1}(v_1).$$

All arcs of F with both ends in V_1 (respectively, V_2) have the form (u_i, u_{i+1}), $i = 1, 2, \ldots, r-1$ (respectively, (v_i, v_{i+1}), $i = 1, 2, \ldots, s-1$). Therefore, if there exists a directed cycle in F, it should contain the arcs (u_j, v_l), (v_p, u_i), $i \leq j$, $l \leq p$ (since F contains no loops, we may assume without loss of generality that $i \neq j$). By the definition of F, this means that $u_j v_l \in E(G)$, $u_i v_p \notin E(G)$. Since $N_{V_2}(u_i) \supseteq N_{V_2}(u_j)$, we have $u_i v_l \in E(G)$. If $l = p$, then we obtain a contradiction. If $l \neq p$ then, as $N_{V_1}(v_p) \supseteq N_{V_1}(v_l)$, we again have $u_i v_p \in E(G)$. This contradiction completes the proof. □

It is interesting to compare the characterization described by Theorem 4.15 with the characterization of graphs with small linear NLC-width [18]: a graph G has linear NLC-width 1 if and only if G is a threshold graph.

Corollary 4.5 *The class of difference graphs coincides with the class of H'-threshold graphs, where H' consists of two vertices and one arc joining them.*

Proof: Assume without loss of generality that $V(H') = \{1, 2\}$ and $A(H') = \{(1,2)\}$. All H'-threshold graphs are difference graphs by the definition of $\circ_{H'}$ and by Theorem 4.12. Let us show that all difference graphs are H'-threshold graphs. Consider a difference graph G. If G is disconnected, then it is a disjoint union of a connected difference graph F and r isolated vertices. If T is a threshold representation of F with respect to $\circ_{H'}$, then

$$G = (K_2^2 \circ_{H'} K_2^2 \circ_{H'} \ldots \circ_{H'} K_2^2) \circ_{H'} T,$$

where there are r multipliers in the parentheses. So, it is sufficient to consider a connected difference graph G with 2-partition (A, B). If G is a complete bipartite graph, then

$$G = K_{m,n} = (K_1^2 \circ_{H'} K_1^2 \circ_{H'} \dots \circ_{H'} K_1^2) \circ_{H'} (K_2^2 \circ_{H'} K_2^2 \circ_{H'} \dots \circ_{H'} K_2^2),$$

where there are m and n multipliers in the first and the second pair of parentheses, respectively. Further, suppose that G is not a complete bipartite graph, which implies $|A| \geq 2$ and $|B| \geq 2$. By Proposition 4.15, G is an H-threshold graph for some digraph H with two vertices. It is evident that H has no loops. If $A(H) = \emptyset$ then G is an empty graph, and if $A(H) = \{(1,2),(2,1)\}$ then $G = K_{m,n}$. Thus, $H = H'$. □

We can also provide a characterization of graphs with $\theta(G) \leq 2$ in terms of a finite list of forbidden induced subgraphs.

Theorem 4.16 [35] *Let G be a graph. Then, $\theta(G) \leq 2$ if and only if neither G nor $\overline{G}$ contains one of the graphs from the set $L = \{C_5, P_5, P_3 \cup P_2, W_4, Bull, X, Y, Z\}$ (see Figure 4.8) as an induced subgraph.*

Proof: It is straightforward to check that every graph from the set L does not satisfy Proposition 4.15. We will prove that every graph that does not contain induced subgraphs from the set L is either a threshold graph or a difference graph or the complement of a difference graph.

Let us prove first that G is either a split graph or a bipartite graph or the complement of a bipartite graph. After that, we will prove that for each of its parts, the neighbourhoods of its vertices in the other part are ordered by inclusion.

Suppose that neither G nor $\overline{G}$ is bipartite. We will show that G is split. To do that, let us consider a maximum clique A of G such that the subgraph induced by the set $B = V(G) - A$ has the smallest possible number of edges. We will prove that B is an independent set. Suppose to the contrary that B is not an independent set. Since B is not a clique, there exist $x, y, z \in B$ such that $x \sim y$ and $z \nsim x$ or $z \nsim y$.

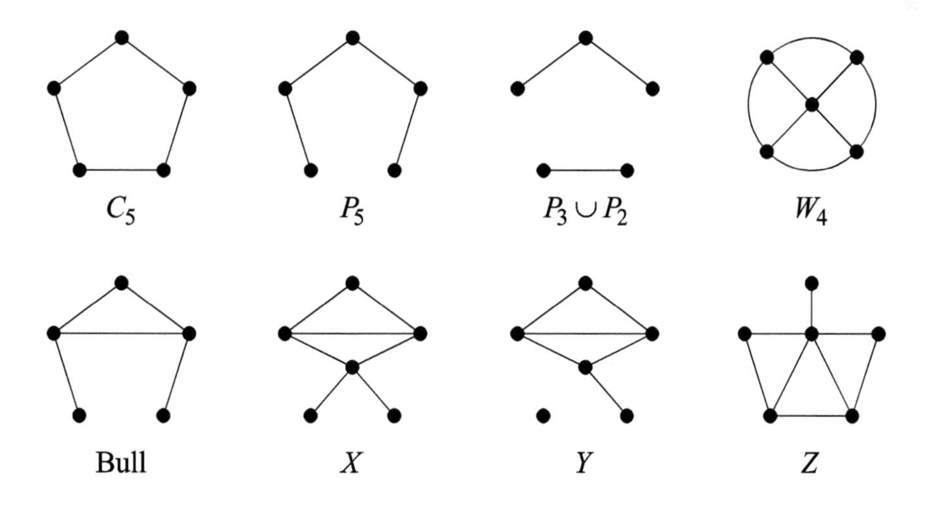

Figure 4.8 The set L.

Since A is a maximum clique, there exists at least one vertex of A that is not adjacent to x (respectively, y). If all vertices of A, except for one vertex u, are adjacent to both x and y, then $A - \{u\} \cup \{x, y\}$ is a clique, which contradicts the maximality of A. Hence, there exist $u, v \in A$ such that $u \not\sim x$ and $v \not\sim y$.

It is easy to see that $|A| \geq 3$. Indeed, if $|A| = 2$ then G is triangle-free. Together with the fact that G is $\{C_5, P_5\}$-free, this implies that G does not contain odd cycles, which contradicts the fact that G is not a bipartite graph.

An induced cycle C is a *bad* C_4 if $C \simeq C_4$ and there exists a vertex $a \in V(G) - C$ such that $G(C \cup \{a\})$ contains at least one triangle. By the assumption of the theorem, G does not contain a bad C_4.

If $u \sim y$ and $v \sim x$, then G contains a bad C_4. Therefore, the following cases are possible: (1) $u \not\sim y$, $v \not\sim x$ and (2) $u \sim y$, $v \not\sim x$. Consider these cases one by one.

Case (1): $u \not\sim y$, $v \not\sim x$. Suppose that there exists $t \in B - \{x, y\}$ such that $t \sim x$ and $t \not\sim y$. In this case, without loss of generality, $t \sim v$ because $G(u, v, y, x, t) \not\simeq P_3 \cup P_2$. Since $G(y, x, t, v, u) \not\simeq P_5$, we have $t \sim u$. Thus, the maximality of A implies the existence of $w \in A - \{u, v\}$ such that $w \not\sim t$. As $G(y, x, t, v, w) \not\simeq P_5$ and C_5, it is true that $w \sim x$. This implies that $\{w, v, t, x\}$ forms a bad C_4.

So, it is shown that $t \sim \{x, y\}$ or $t \not\sim \{x, y\}$ for every $t \in B - \{x, y\}$. Let $T_1 = \{t \in B - \{x, y\} : t \sim \{x, y\}\}$, $T_2 = \{t \in B - \{x, y\} : t \not\sim \{x, y\}\}$. We know from the above considerations that $T_2 \neq \emptyset$.

Let $t \in T_2$. Because $G(u, v, y, x, t) \not\simeq \overline{W_4}$, without loss of generality $t \sim v$. Since $G(t, v, u, y, x) \not\simeq P_3 \cup P_2$, it is true that $t \sim u$. Therefore, we have $T_2 \sim \{u, v\}$.

We proceed by proving a series of auxiliary statements.

Lemma 4.11 *For every* $q \in A - \{u, v\}$, *we have* $q \sim T_2$ *or* $q \sim \{x, y\}$. *Moreover,* T_2 *is a clique.*

Proof: Suppose that there exists $t \in T_2$ such that $q \not\sim t$. Now, $q \sim \{x, y\}$ because $G(t, v, q, y, x) \not\simeq P_3 \cup P_2$ and P_5. If there exist $t_1, t_2 \in T_2$ such that $t_1 \not\sim t_2$, then $G(t_1, v, t_2, y, x) \simeq P_3 \cup P_2$. □

Let $Q_1 = \{q \in A - \{u, v\} : q \sim T_2\}$ and $Q_2 = (A - \{u, v\}) - Q_1$. By Lemma 4.11, $Q_2 \sim \{x, y\}$. Moreover, $Q_2 \neq \emptyset$ because A is a maximal clique.

Lemma 4.12 $Q_2 \sim T_1$. *Moreover,* T_1 *is a clique.*

Proof: Suppose that there exist $t_1, t_2 \in T_1$ such that $t_1 \not\sim t_2$. Since $G(u, v, y, t_1, t_2) \not\simeq P_3 \cup P_2$, without loss of generality we have $t_2 \sim v$. This implies $t_2 \sim u$ or $t_1 \sim v$ because $G(u, v, t_2, y, t_1) \not\simeq P_5$ and C_5. We have, $t_1 \not\sim v$ because otherwise v, t_2, x, t_1 form a bad C_4. Thus, $t_1 \not\sim v$, $t_2 \sim u$. Analogously, $t_1 \not\sim u$.

By the maximality of the clique A, there exists $q \in A - \{u, v\}$ such that $q \not\sim t_2$. Because $G(q, v, t_2, x, t_1) \not\simeq P_5$ and C_5, we have $q \sim x$. Therefore, $G(q, v, t_2, x)$ is a bad C_4. So, it is proved that T_1 is a clique.

Let us show now that $T_1 \sim Q_2$. Suppose the contrary; that is, let there exist $t \in T_1$ and $q \in Q_2$ such that $t \not\sim q$. By the definition of Q_2, there exists $z \in T_2$ such that $q \not\sim z$. Since $G(z, u, q, x, t) \not\simeq P_5$ and C_5, it follows that $t \sim u$. In this case, $G(u, q, x, t)$ is a bad C_4. □

By Lemmas 4.11 and 4.12, $V_1 = Q_2 \cup T_1 \cup \{x, y\}$ and $V_2 = Q_1 \cup T_2 \cup \{u, v\}$ are cliques, and $V_1 \cup V_2 = V(G)$. This is a contradiction to the fact that $\overline{G}$ is not bipartite. Case (1) is complete.

Case 2: $u \sim y$, $v \not\sim x$.

Lemma 4.13 *For every $z \in B - \{x, y\}$, it is true that $z \sim \{x, y\}$ or $z \not\sim \{x, y\}$.*

Proof: Assume to the contrary that there exists $z \in B - \{x, y\}$ such that the lemma is not satisfied for it. Let $z \sim x$, $z \not\sim y$. Since $G(v, u, y, x, z) \not\simeq P_5$ and C_5, we have $z \sim u$. Consider $w \in A - \{u, v\}$. As $G(y, x, z, v, w) \not\simeq P_3 \cup P_2$, there are edges between $\{v, w\}$ and $\{x, y, z\}$. If there exists at least one edge from the set $\{wy, wz, vz\}$, then $G(u, y, x, z)$ is a bad C_4. Otherwise, $w \sim x$ and $G(w, u, y, x)$ is a bad C_4.

Thus, $z \sim y$, $z \not\sim x$. Suppose that $z \sim v$. In this case, $z \sim u$ because $G(z, y, u, v)$ is not a bad C_4. Therefore, by the maximality of A, there exists $w \in A$ such that $w \not\sim z$. For this vertex, we have $w \sim y$ because $G(x, y, z, v, w) \not\simeq P_5$ and C_5. This implies that $G(w, v, z, y)$ is a bad C_4. Hence, $z \not\sim v$. We obtain, $z \not\sim u$, otherwise $G(z, y, u, v, x) \simeq Bull$. Since $G(x, y, z, w, v) \not\simeq P_3 \cup P_2$, there exists at least one edge from the set $\{wx, wy, wz\}$.

Suppose that $w \sim x$. This implies $w \sim y$, since otherwise $G(w, x, y, u)$ is a bad C_4. Hence, $w \sim z$ because $G(w, y, x, z, v) \not\simeq Bull$. We obtain $G(x, y, z, u, v, w) = \overline{Y}$. Thus, $w \not\sim x$. Now, if $w \sim z$ then $w \sim y$, since $G(w, u, y, z)$ is not a bad C_4. This implies $G(w, z, y, v, x) \simeq Bull$. Therefore, $w \not\sim z$ and, thus, $w \sim y$ and $G(w, u, v, y, z, x) \simeq X$. This contradiction completes the proof of Lemma 4.13. □

Let $B - \{x, y\} = S_1 \cup S_2$, where

$$S_1 = \{z \in B : z \sim \{x, y\}\} \quad \text{and} \quad S_2 = \{z \in B : z \not\sim \{x, y\}\}.$$

Since $\overline{G}$ is not a bipartite graph, we have $S_2 \neq \emptyset$.

Lemma 4.14 *For every $r \in A - \{u, v\}$, we have $r \sim \{x, y\}$ or $r \sim S_2$.*

Proof: Assume that there exists $z \in S_2$ such that $r \not\sim z$. Let $z \sim v$. Since $G(z, v, r, y, x) \not\simeq P_3 \cup P_2$, it follows that $r \sim y$ or $r \sim x$. The situation when $r \sim x$ and $r \not\sim y$ is impossible because otherwise $G(r, u, y, x)$ would be a bad C_4. If $r \sim y$ then $r \sim x$, since $G(x, y, r, v, z) \not\simeq P_5$.

It remains to consider the case when $z \not\sim v$. In this case, $r \sim y$ or $r \sim x$, since $G(r, v, y, x, z) \not\simeq \overline{W_4}$. As above, the case when $r \sim x$ and $r \not\sim y$ is impossible. Thus, $r \sim y$. Because $G(v, u, r, y, x, z) \not\simeq Y$, it is true that $z \sim u$ or $r \sim x$. The situation when $z \sim u$ and $r \not\sim x$ contradicts the fact that $G(r, u, y, x, z) \not\simeq Bull$. Hence, $r \sim x$. □

Let $A - \{u, v\} = R_1 \cup R_2$, where

$$R_1 = \{r \in A - \{u, v\} : r \sim S_2\} \quad \text{and} \quad R_2 = (A - \{u, v\}) - R_1.$$

By Lemma 4.14, $R_2 \sim \{x, y\}$.

Lemma 4.15 $S_2 \sim \{u, v\}$. *Moreover, S_2 is a clique.*

Proof: We start with the proof of the first statement of the lemma. Let $z \in S_2$ and assume that $z \not\sim v$. We will show that this is impossible.

Suppose that there exists $r \in A - \{u, v\}$ such that $r \not\sim y$. By Lemma 4.14, we have $r \sim z$. It follows that $r \sim x$, since $G(z, r, v, y, x) \not\simeq P_3 \cup P_2$. Then, $G(r, u, y, x)$ is a bad C_4. Hence, it is proved that $y \sim A - \{v\}$. Therefore, there exists $s \in B - \{x, y, z\}$ such that $s \sim v$ and $s \not\sim y$. Indeed if, on the contrary, $N_B(v) \subseteq N_B(y)$ then $A' = (A - \{v\}) \cup \{y\}$ is a maximum clique, and for the subgraph induced by the set $B' = V(G) - A'$, we have $|E(G(B'))| < |E(G(B))|$. This contradicts the definition of the clique A.

Since $G(x, y, u, v, s) \not\simeq P_5$ and C_5, we have $s \sim u$. Moreover, $s \not\sim x$ by Lemma 4.13, and $s \sim z$ because otherwise $G(v, s, y, x, z) \simeq \overline{W_4}$. We obtain, $G(z, s, v, y, x) \simeq P_3 \cup P_2$. Hence, $z \sim v$. Since $G(z, v, u, x, y) \not\simeq P_5$, we have $z \sim u$. Now, it is easy to see that S_2 is a clique. Indeed, if there exist $s_1, s_2 \in S_2$ such that $s_1 \not\sim s_2$, then $G(s_1, v, s_2, y, x) \simeq P_3 \cup P_2$. □

In particular, Lemma 4.15 and the maximality of A imply $R_2 \neq \emptyset$.

Lemma 4.16 $R_2 \sim S_1$. *Moreover, S_1 is a clique.*

Proof: Let there exist $r \in R_2$ and $s \in S_1$ such that $r \not\sim s$. By Lemma 4.14, $r \sim \{x, y\}$. By definition, there exists $z \in S_2$ such that $z \not\sim r$. Lemma 4.15 implies $z \sim \{u, v\}$. Since $G(z, v, r, x, s) \not\simeq P_5$ and C_5, we have $s \sim v$, which implies that $G(v, r, x, s)$ is a bad C_4. Thus, it is proved that $R_2 \sim S_1$.

Let us show now that S_1 is a clique. Suppose that there exist $z_1, z_2 \in S_1$ such that $z_1 \not\sim z_2$. Because $G(z_1, x, z_2, u, v) \not\simeq P_3 \cup P_2$, there exists at least one edge between $\{z_1, z_2\}$ and $\{u, v\}$. At the same time, if $z_1 \sim v$ and $z_1 \not\sim u$, then $G(z_1, v, u, y)$ is a bad C_4. Thus, without loss of generality, $z_1 \sim u$. This implies $z_2 \not\sim u$ because otherwise $G(z_1, x, z_2, u)$ is a bad C_4. Since $G(v, u, z_1, x, z_2) \not\simeq P_5$ and C_5, we have $z_1 \sim v$. This implies $z_2 \not\sim v$, otherwise $G(v, z_1, x, z_2)$ is a bad C_4. The maximality of A implies the existence of $w \in A$ such that $w \not\sim z_1$. It follows that $w \not\sim x$ because $G(w, v, z_1, x)$ is not a bad C_4. But in this case, we have $G(w, v, z_1, x, z_2) \simeq P_5$ or C_5. □

By Lemmas 4.15 and 4.16, $V_1 = R_2 \cup S_1 \cup \{x, y\}$ and $V_2 = R_1 \cup S_2 \cup \{u, v\}$ are cliques, and $V_1 \cup V_2 = V(G)$. This is a contradiction to the fact that $\overline{G}$ is not bipartite. Case (2) is complete.

Thus, it is proved that G or $\overline{G}$ is a split graph or a bipartite graph. Let (A, B) be a 2-partition of G.

Lemma 4.17 *The neighbourhoods of vertices in A (respectively, B) are ordered by inclusion.*

Proof: Assume to the contrary that there exist $u, v \in A$ and $x, y \in B$ such that $u \sim x$, $v \not\sim x$, $u \not\sim y$, $v \sim y$. Now, suppose that G is a bipartite graph. If $|V(G)| = 4$, then the statement of the theorem obviously holds. Let there exist $z \in B - \{x, y\}$. Since $G(u, v, x, y, z) \not\simeq \overline{W_4}$ and $P_3 \cup P_2$, it is true that $z \sim \{u, v\}$, which implies $G(u, v, x, y, z) \simeq P_5$. This contradiction proves the lemma for bipartite graphs. Therefore, the statement of the lemma also holds for the complements of bipartite graphs. Thus, it remains to consider the case when G is a split graph and neither a bipartite graph nor the complement of a bipartite graph.

The following statements hold:

(a) $N(x) \cup N(y) = A$ (since G does not contain the graph *Bull*).
(b) For every $z \in B - \{x, y\}$, we have $|N(z) \cap \{u, v\}| \leq 1$ (by the same reason as in (a)).
(c) $|A| \geq 3$, $|B| \geq 3$ (otherwise G or $\overline{G}$ is a bipartite graph).

Let $z \in B - \{x, y\}$, $w \in A - \{u, v\}$, $w \sim x$. As $G(u, v, x, y, w, z) \not\simeq Y$ and $\overline{Z}$, the edge set $E(G)$ contains at least one of the edges zu, zv, zw. If there exists exactly one of these edges, then $G(u, v, w, z, y) \simeq Bull$, $G(u, v, w, x, y, z) \simeq X$ and $G(u, v, w, z, y) \simeq Bull$, respectively. Therefore, taking into account (b), one of the following two conditions holds: $zw, zv \in E(G)$, $zu \notin E(G)$ or $zw, zu \in E(G)$, $zv \notin E(G)$.

Let $F = G(u, v, x, y, w, z)$. In the first case, since $G(w, v, z, y, x) \not\simeq Bull$, we have $w \sim y$, which implies $F \simeq \overline{Y}$. In the second case, $w \sim y$ because $F \not\simeq \overline{X}$, which implies $F \simeq \overline{Y}$. This contradiction completes the proof of Lemma 4.17 □

Theorem 4.16 is proved. □

4.3 Realization Graphs of Degree Sequences

As we saw earlier, the canonical decomposition was used to characterize unigraphs and unigraphical sequences. The bounds on the number of unigraphs cited in Section 1.2.2 indicate that the majority of the graphs are not unigraphs, even though the number of n-vertex unigraphs is exponential. Consequently, most degree sequences have more than one realization. Thus, it is interesting to study properties of such realizations and their relations between one another. In particular, these relations can be described by so-called realization graphs [7].

For a given degree sequence d, the *realization graph* $G(d)$ has the vertices corresponding to different (possibly isomorphic) labelled realizations of d, and two vertices being adjacent whenever the corresponding realizations can be obtained from each other by a single 2-switch. The examples illustrating this concept are shown in Figure 4.9. In particular, the degree sequence $d = (1, 1, 1, 1)$ has exactly three labelled realizations, all of which can be obtained from one another by a single 2-switch (Figure 4.9, top). Thus, its realization graph $G(d)$ is the complete graph K_3.

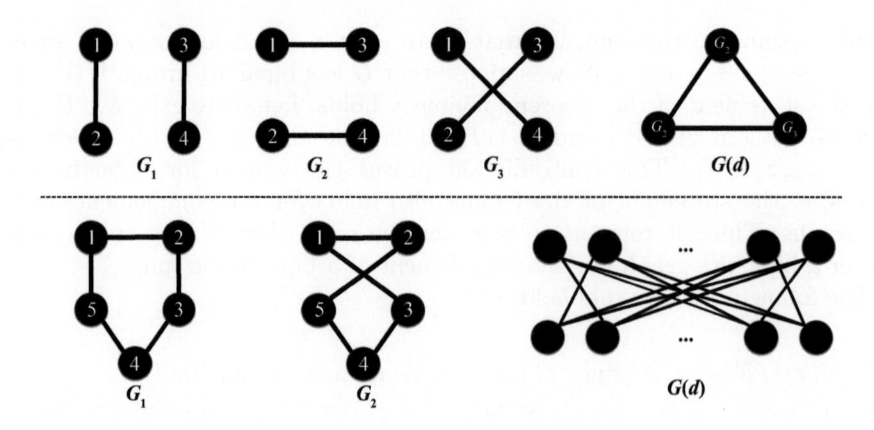

Figure 4.9 Realizations and realization graphs of the sequences $(1,1,1,1)$ and $(2,2,2,2,2)$.

The sequence $d = (2,2,2,2,2)$ provides a more complicated example. All realizations of d are isomorphic to a 5-cycle C_5. Each realization can be described by a permutation of the set $\{1,2,3,4,5\}$, yielding 12 permutations for all labelled realizations. With a single 2-switch, exactly five distinct realizations can be derived from a given realization in the following manner: choose an edge uv in a cycle, then create a new cycle by traversing the vertices, excluding u and v, in their original order, while reversing the traversal order for u and v (see Figure 4.9, bottom). Thus, a realization represented by permutation $\sigma = (i_1, i_2, i_3, i_4, i_5)$ changes to a realization described by permutation $\sigma' = (i_j, i_{j+1})\sigma$, where (i_j, i_{j+1}) is a transposition and $j+1$ is computed modulo 5. As a result, the realization graph $G(d)$ is 5-regular. Additionally, applying a 2-switch to a realization alters the parity of the corresponding permutation indicating that the graph $G(d)$ is bipartite with parts corresponding to realizations represented by even and odd permutations. Taking into account the symmetry of the realizations, $G(d)$ is isomorphic to the complete bipartite graph $K_{6,6}$ minus a perfect matching.

Properties of realization graphs have been studied in a number of papers. In particular, the fact that every two realizations of a graphical sequence can be obtained from each other by a sequence of 2-switches (see Chapter 2) is obviously equivalent to the claim that the graphs $G(d)$ are connected for all degree sequences d. Several studies investigated Hamiltonian properties of $G(d)$ [1, 11].

It turns out that 1-decomposition naturally leads to decomposition of realization graphs that in certain settings allows for their characterization. Recall that the *Cartesian product* $G \times H$ of graphs G and H is a graph such that:

- $V(G \times H)$ is the set-theoretical Cartesian product $V(G) \times V(H)$;
- two vertices (u_1, u_2) and (v_1, v_2) are adjacent in $G \times H$ whenever $u_1 = v_1$ and $u_2 v_2 \in E(H)$, or $u_2 = v_2$ and $u_1 v_1 \in E(G)$.

Suppose that

$$d = \alpha_1 \circ \alpha_2 \circ \ldots \circ \alpha_k \circ \beta \tag{4.49}$$

is 1-decomposition of the graphical sequence d, where $\alpha_1, \alpha_2, \ldots, \alpha_k$ are indecomposable split sequences and β is an indecomposable graphical sequence. Such a decomposition exists by Corollary 2.9. The following theorem describes the realization graph $G(d)$ in terms of the graphs $G(\alpha_i)$ and $G(\beta)$.

Theorem 4.17 [7]

$$G(d) \simeq G(\alpha_1) \times G(\alpha_2) \times \ldots \times G(\alpha_k) \times G(\beta). \tag{4.50}$$

Proof: We outline the idea of the proof given in [7]. It is sufficient to prove (4.50) for $k=1$, as the validity of the theorem for an arbitrary k immediately follows by induction. Hence, we assume that $d = \alpha \circ \beta$ where $\alpha = \alpha_1$. Note that if $|\alpha| = 1$, then it is easy to check that (4.50) holds. Thus, we may also assume that α is non-trivial.

As we know from the previous chapters, splitness and 1-decomposability are both determined by the degree sequence of a graph. Thus, all realizations of a split degree sequence α are split graphs; similarly, all realizations of a decomposable sequence d are decomposable. Furthermore, Lemma 2.1 states that any split realization of α has exactly one splitting. Thus, as implied by definitions, if (H, A, B) is a labelled splitted realization of α and F is a labelled realization of β, then $(H, A, B) \circ F$ is a labelled realization of d. Given these facts, one can define a bijective mapping

$$\varphi : V(G(\alpha) \times G(\beta)) \to V(G(d))$$

by setting

$$\varphi((H, F)) = (H, A, B) \circ F. \tag{4.51}$$

To prove Theorem 4.17, it is sufficient to show that φ is a graph isomorphism. The key observation here is as follows: if $U \subseteq V((H, A, B) \circ F)$ is a set of four vertices involved in a 2-switch, then $U \subseteq V(H)$ or $U \subseteq V(F)$. The validity of this observation straightforwardly follows from the definition of 1-decomposition.

Now, suppose that (H_1, F_1) and (H_2, F_2) are adjacent in $G(\alpha) \times G(\beta)$. This means that, for instance, $H_1 = H_2$, whereas F_1 and F_2 differ by a single 2-switch, which also transforms the graph $(H_1, A_1, B_1) \circ F_1$ into the graph $(H_2, A_2, B_2) \circ F_2$. Note that both these graphs are realizations of d. This demonstrates that $\varphi((H_1, F_1))$ and $\varphi((H_2, F_2))$ are adjacent in $G(d)$.

Conversely, assume that

$$\varphi((H_1, F_1)) = (H_1, A_1, B_1) \circ F_1$$

and

$$\varphi((H_2, F_2)) = (H_2, A_2, B_2) \circ F_2$$

are adjacent in $G(d)$. According to the key observation, the 2-switch transforming the first realization into the second involves vertices that are entirely contained in $V(H_1)$ or entirely contained in $V(F_1)$. In the former case, H_1 and H_2 are adjacent in $G(\alpha)$, and $F_1 = F_2$. In the latter case, $H_1 = H_2$, whereas F_1 and F_2 are adjacent in $G(\beta)$. In either case, (H_1, F_1) and (H_2, F_2) are adjacent in $G(\alpha) \times G(\beta)$. $\square$

Theorem 4.17 can be used to characterize realization graphs of graphs from particular classes. Recall that $(2K_2, C_4)$-free graphs are called *pseudo-split.*

Theorem 4.18 [7] *Let d be a degree sequence. The following statements are equivalent:*

(1) G(d) is bipartite;

(2) G(d) is triangle-free;

(3) d is the degree sequence of a pseudo-split matrogenic graph.

A graph is P_4-reducible if every vertex belongs to at most one induced path P_4. It was also proved in [7] that $G(d)$ is isomorphic to a hypercube if and only if d is the degree sequence of a split P_4-reducible graph.

The next theorem gives a characterization of the degree sequences whose realization graphs are complete graphs:

Theorem 4.19 [8] *Let d be a degree sequence. The following statements are equivalent:*

(1) $G(d)$ *is isomorphic to the complete graph* K_n *with* $n \geq 4$*;*

(2) $d = \tau \circ \alpha \circ \tau'$, *where each of* τ, τ' *is either empty (i.e. omitted) or the degree sequence of a threshold graph, whereas* $\alpha = (n, 2, 1^n)$ *or* $\alpha = (n^n, n-1, 1)$.

Acknowledgements

This chapter is based on the following publications: (A) Reprinted by permission from Elsevier: Discrete Mathematics, 310, P. V. Skums, S. V. Suzdal and R. I. Tyshkevich, Operator decomposition of graphs and the reconstruction conjecture, (3), 423–429, ©2010. (B) Reprinted by permission from Elsevier: Electronic Notes in Discrete Mathematics, 29, P. V. Skums and R. I. Tyshkevich, Bipartite operator decomposition of graphs and the reconstruction conjecture, 201–205, ©2007. (C) Reprinted by permission from Elsevier: Discrete Mathematics, 313, P. V. Skums, H-product of graphs, H-threshold graphs and threshold-width of graphs, (21), 2390–2400, ©2013.

4.4 References

[1] S. R. Arikati and U. N. Peled, The realization graph of a degree sequence with majorization gap 1 is Hamiltonian, *Linear Algebra and Its Applications*, **290** (1–3)(1999), 213–235.

[2] L. Babel, Tree-like P_4-connected graphs, *Discrete Mathematics*, **191** (1–3) (1998), 13–23.

[3] L. Babel and S. Olariu, On the structure of graphs with few P_4s, *Discrete Applied Mathematics*, **84** (1–3)(1998), 1–13.

[4] L. Babel and S. Olariu, On the p-connectedness of graphs—a survey, *Discrete Applied Mathematics*, **95** (1–3)(1999), 11–33.

[5] J. Bang-Jensen and G. Gutin, *Digraphs: Theory, Algorithms and Applications*, London: Springer Science & Business Media, 2008.

[6] M. D. Barrus, Antimagic labeling and canonical decomposition of graphs, *Information Processing Letters*, **110** (2010), 261–263.

[7] M. D. Barrus, On realization graphs of degree sequences, *Discrete Mathematics*, **339** (8)(2016), 2146–2152.

[8] M. D. Barrus and N. Haronian, Cliques in realization graphs, *Discrete Mathematics*, **346** (1)(2023), 113184.

[9] A. Bondy, A graph reconstructor's manual, *Surveys in Combinatorics*, **166** (1991), 221–252.

[10] A. Brandstädt, V. B. Le and J. Spinrad, *Graph Classes: a Survey*, Philadelphia, PA: Society for Industrial and Applied Mathematics, 1999.

[11] R. A. Brualdi, Matrices of zeros and ones with fixed row and column sum vectors, *Linear Algebra and Its Applications*, **33** (1980), 159–231.

[12] V. Chvatal, Star-cutsets and perfect graphs, *Journal of Combinatorial Theory, Ser. B*, **39** (3)(1985), 189–199.

[13] V. Chvatal and P. L. Hammer, Aggregation of inequalities in integer programming, *Annals of Discrete Mathematics*, **1** (1977), 145–162.

[14] B. Courcelle and S. Olariu, Upper bounds to the clique width of graphs, *Discrete Applied Mathematics*, **101** (2000), 77–114.

[15] S. Földes and P. L. Hammer, Split graphs, *Congressus Numerantium*, **19** (1977), 311–315.

[16] J.-L. Fouquet, V. Giakoumakis and J.-M. Vanherpe, Bipartite graphs totally decomposable by canonical decomposition, *International Journal of Foundations of Computer Science*, **10** (04)(1999), 513–533.

[17] V. Giakoumakis, F. Roussel and H. Thuillier, On P_4-tidy graphs, *Discrete Mathematics and Theoretical Computer Science*, **1** (1997), 17–41.

[18] F. Gurski, Characterizations for co-graphs defined by restricted NLC-width or clique-width operations, *Discrete Mathematics*, **306** (2) (2006), 271–277.

[19] P. L. Hammer, U. Peled and X. Sun, Difference graphs, *Discrete Applied Mathematics*, **28** (17) (1990), 35–44.

[20] P. L. Hammer and B. Simeone, The splittence of a graph, *Combinatorica*, **3** (1)(1981), 275–284.

[21] F. Harary, *Graph Theory*, Reading, MA: Addison–Wesley Publishing Company, 1969.

[22] C. T. Hoáng and V. B. Le, P_4-free colorings and P_4-bipartite graphs, *Discrete Mathematics and Theoretical Computer Science*, **4** (2001), 109–122.

[23] B. Jamison and S. Olariu, p-Components and the homogeneous decomposition of graphs, *SIAM Journal of Discrete Mathematics*, **8** (3) (1995), 448–463.

[24] P. J. Kelly, *On Isometric Transformations*, PhD thesis, University of Wisconsin, 1942.

[25] W. L. Kocay, A family of nonreconstructible hypergraphs, *Journal of Combinatorial Theory, Ser. B*, **42** (1)(1987), 46–63.
[26] W. L. Kocay, Hypomorphisms, orbits, and reconstruction, *Journal of Combinatorial Theory, Ser. B*, **44** (2)(1988), 187–200.
[27] A. V. Kostochka and D. B. West, On reconstruction of graphs from the multiset of subgraphs obtained by deleting ℓ vertices, *IEEE Transactions on Information Theory*, **67** (6)(2020), 3278–3286.
[28] V. Limouzy, F. de Montgolfier and M. Rao, NLC-2 graph recognition and isomorphism, *Lecture Notes in Computer Science*, **4769** (2007), 86–98.
[29] N. V. R. Mahadev and U. N. Peled, *Threshold Graphs and Related Topics*, Annals of Discrete Mathematics, Vol. 56, Amsterdam: Elsevier, 1995.
[30] B. D. McKay, Small graphs are reconstructible, *Australasian Journal of Combinatorics*, **15** (1997), 123–126.
[31] R. Merris, Split graphs, *European Journal of Combinatorics*, **24** (4) (2003), 413–430.
[32] V. Muller, Probabilistic reconstruction from subgraphs, *Commentationes Mathematicae Universitatis Carolinae*, **17** (4)(1976), 709–719.
[33] C. S. J. N. Nash-Williams, The reconstruction problem, In: L. W. Beineke, and R. J. Wilson (eds.), *Selected Topics in Graph Theory*, **1** (1978), 205–236.
[34] S. Ramachandran and S. Arumugam, Graph reconstruction—some new developments, *AKCE International Journal of Graphs and Combinatorics*, **1** (1)(2004), 51–61.
[35] P. Skums, H-product of graphs, H-threshold graphs and threshold-width of graphs, *Discrete Mathematics*, **313** (21)(2013), 2390–2400.
[36] P. V. Skums, S. V. Suzdal and R. I. Tyshkevich, Operator decomposition of graphs and the reconstruction conjecture, *Discrete Mathematics*, **310** (3) (2010), 423–429.
[37] P. V. Skums and R. I. Tyshkevich, Bipartite operator decomposition of graphs and the reconstruction conjecture, *Electronic Notes in Discrete Mathematics*, **29** (2007), 201–205.
[38] P. V. Skums and R. I. Tyshkevich, Reconstruction conjecture for graphs with restrictions on 4-vertex paths, *Diskretnyi Analiz i Issledovanie Operatsiy*, **16** (4)(2009), 87–96.
[39] P. K. Stockmeyer, A census of non-reconstructable digraphs, I: Six related families, *Journal of Combinatorial Theory, Ser. B*, **31** (2)(1981), 232–239.
[40] P. K. Stockmeyer, The falsity of the reconstruction conjecture for tournaments, *Journal of Graph Theory*, **1** (1)(1977), 19–25.
[41] S. V. Suzdal and R. I. Tyshkevich, (P,Q)-decomposition of graphs, Rostock University, Rostock, Germany, 1999, 1–16. (Preprint aus dem Fachbereich Informatik, No 5-1999).
[42] B. D. Thatte, Some results on the reconstruction problem: p-claw-free, chordal and P_4-reducible graphs, *Journal of Graph Theory*, **19** (4)(1995), 549–561.
[43] R. I. Tyshkevich, Decomposition of graphical sequences and unigraphs, *Discrete Mathematics*, **220** (2000), 201–238.

[44] R. I. Tyshkevich and A. A. Chernyak, Canonical decomposition of a unigraph, *Izvestiya Akademii Nauk BSSR*, Ser. fiz.-mat. nauk, (5)(1979), 14–26.

[45] R. I. Tyshkevich and A. A. Chernyak, Decomposition of graphs, *Kibernetika*, **21** (2)(1985), 231–242.

[46] R. I. Tyshkevich and S. V. Suzdal, Decomposition of graphs, *Izbrannye Trudy Belorusskogo Gosudarstvennogo Universiteta (Selected Scientific Proceedings of Belarus State University)*, **6** (2001), 482–504.

[47] V. Tyurin, Reconstructibility of decomposable graphs, *Izvestiya Akademii Nauk BSSR*, Ser. fiz.-mat. nauk, (3)(1987), 16–20.

[48] S. M. Ulam, *A Collection of Mathematical Problems*, Vol. 29, New York: Wiley (Interscience), 1960.

[49] E. Wanke, k-NLC graphs and polynomial algorithms, *Discrete Applied Mathematics*, **54** (1994), 251–266.

[50] M. Yannakakis, The complexity of the partial order dimension problem, *SIAM Journal on Algebraic Discrete Methods*, **3** (3)(1982), 351–358.

5

Line Graphs and Hypergraphs

V. Zverovich, Yu. Metelsky & P. Skums

This chapter is devoted to line graphs and line hypergraphs. The key role here is played by decompositions of a graph into cliques. Such decompositions must possess special properties and they will be called Krausz coverings. In the first section, we discuss a multivalued function $\mathcal{L}$ called the line hypergraph. The function $\mathcal{L}$ generalizes two classical concepts at once, namely, of the line graph and the dual hypergraph. In terms of this function, proofs of some fundamental theorems of Berge, Krausz and Whitney on line graphs can be unified and their more general versions can be obtained. The next two sections explore the class L_3^ℓ of line graphs of linear 3-uniform hypergraphs. It is known that this class cannot be characterized by a finite list of forbidden induced subgraphs. However, under some restrictions on vertex degrees, such a characterization can be provided. Also, we will study the recognition problem for the class L_3^ℓ, in particular if it is polynomially solvable. Finally, in the last section, a natural generalization of the notion of domino is considered. A graph is called an *r-mino* if each of its vertices belongs to at most r maximal cliques; thus, a domino is a 2-mino. We will show that the class of r-minoes coincides with the class of line graphs of Helly hypergraphs with rank at most r. Also, we will explore whether this class can be characterized by a finite list of forbidden induced subgraphs.

5.1 Line Hypergraphs

We will explore a multivalued function $\mathcal{L}$ called the line hypergraph, which simultaneously generalizes the classical concepts of the line graph and the dual hypergraph. In terms of this function, proofs of some fundamental theorems on line graphs can be unified and their more general versions can be obtained. Three such theorems will be considered: the Berge theorem describing all hypergraphs for a given line graph G in terms of clique coverings of G [2] (p. 400), the Krausz global characterization of line graphs for simple graphs [18] and the Whitney theorem on isomorphisms of line graphs [35].

V. Zverovich, Yu. Metelsky & P. Skums, *Line Graphs and Hypergraphs*. In: *Methods of Graph Decompositions*. Edited by: Vadim Zverovich and Pavel Skums, Oxford University Press. DOI: 10.1093/oso/9780198882091.003.0005

The necessary terminology and the definition of the function 'line hypergraph' are given in Sections 5.1.1 and 5.1.2, respectively. All values of this multivalued function $\mathcal{L}(H)$ are called *line hypergraphs* of a hypergraph H. Section 5.1.3 contains a technical lemma on isomorphisms. In particular, the lemma transfers the statement of Lovász about trivial isomorphisms of line graphs [21] (p. 506) onto line hypergraphs. Theorem 5.2 from Section 5.1.4 describes the inverse image $\mathcal{L}^{-1}(G)$ in terms of clique coverings of a hypergraph G. For a simple graph G, Theorem 5.2 yields Proposition 1 from [2] (p. 400) together with a criterion of isomorphism of elements in $\mathcal{L}^{-1}(G)$.

A characterization of values of the function 'line hypergraph' for hypergraphs with a prescribed property P is given in Section 5.1.5. The characterization generalizes the Krausz theorem about line graphs of simple graphs [18] and is formulated in terms of clique coverings, similar to the Krausz theorem. So, we call this characterization and the corresponding coverings the *Krausz-type characterization* and *Krausz coverings*. In Section 5.1.5, we consider an important concept of the P-large clique for some given properties P. It enables one to prove the uniqueness of Krausz coverings in some cases.

Section 5.1.6 deals with the Whitney theorem [35] formulated in Section 1.3.3. It is known that this theorem has no simple analogue for line graphs of hypergraphs. However, its variants are possible in a more general situation. One of them is related to Harary's conjecture on reconstruction of a graph from the list of its subgraphs obtained by deleting one edge. There are a number of papers [4–6, 8, 10–12, 14, 30] devoted to this version where a strong isomorphism of hypergraphs (denoted by $\cong$) is mainly considered, that is, a tougher relationship than isomorphism. The closest results to the Whitney theorem are the theorems of Berge and Rado [6] and Gardner [14]. In [6], for every integer $p \geq 2$, a pair of hypergraphs

$$\mathcal{B}_p = (V, (B_1, B_2, \ldots, B_p)) \quad \text{and} \quad \mathcal{D}_p = (W, (D_1, D_2, \ldots, D_p))$$

is constructed in a certain way. For $p = 3$, this pair coincides with the pair excluded in the Whitney theorem: $\mathcal{B}_3 = K_{1,3}$ and $\mathcal{D}_3 = K_3$. Now, let us consider two hypergraphs:

$$H_k = (V_k, (E_1^k, E_2^k, \ldots, E_m^k)), \quad k = 1, 2, \quad m \geq p. \tag{5.1}$$

If there exist $X_k \subseteq V_k$ $(k = 1, 2)$ and $I \subseteq \{1, 2, \ldots, m\}$ with $|I| = p$ such that one of the partial subhypergraphs

$$H_k(X_k, I_k) = (X_k \cap V_k, (E_i^k : i \in I))$$

is strongly isomorphic to $\mathcal{B}_p$ and the other to $\mathcal{D}_p$, then the pair H_1, H_2 is said to *contain the $\mathcal{B}_p$–$\mathcal{D}_p$ pair.*

Berge–Rado Theorem [6] *The following statements are true:*

(i) Let (5.1) be two hypergraphs whose partial hypergraphs $H_1(I)$ and $H_2(I)$ are strongly isomorphic for all $I \subseteq \{1, 2, \ldots, m\}$ with $|I| = p - 1$. Suppose that the pair H_1, H_2 does not contain the $\mathcal{B}_p$–$\mathcal{D}_p$ pair. Then $H_1 \cong H_2$.

(ii) $\mathcal{B}_p \ncong \mathcal{D}_p$, whereas $\mathcal{B}_p - B_k \cong \mathcal{D}_p - D_k$ for every $k = 1, 2, \ldots, p$.

Based on the concept of a $\mathcal{B}_p$–$\mathcal{D}_p$ pair and the above theorem, Gardner [14] obtained in close terms a criterion of strong isomorphism for hypergraphs with the same line graph.

In this section, another version of the Whitney theorem will be discussed. The Whitney theorem may be formulated as the following uniqueness theorem: *if G is a simple connected graph and P is the class of simple graphs whose orders are greater than 3, then*

$$|\mathcal{L}^{-1}(G) \cap P| \leq 1. \tag{5.2}$$

In Section 5.1.6, the condition (5.2) for a hypergraph G is connected with the action of the automorphism group $\mathrm{Aut}(G)$. This enables one to prove inequality (5.2) for a number of hypergraph classes and some given properties P. In particular, a direct generalization of the Whitney theorem is obtained.

5.1.1 Basic Terminology

With minor adaptations, we adopt the terminology of Berge [2]. Let us remind ourselves of some definitions, and note differences from Berge's terminology.

A *hypergraph* is a pair $(V, \mathcal{E})$, where V is a finite non-empty set (the *vertex set*) and $\mathcal{E}$ is a finite family of non-empty subsets of V (the *edge family*). In contrast to the generally accepted definition, isolated vertices are permitted. This is necessary because the line graph of any hypergraph must be a hypergraph. The vertex set and the edge family of a hypergraph H are denoted by $V(H)$ and $\mathcal{E}(H)$, respectively. The number $|V(H)|$ is called the *order* of H and is denoted by $|H|$.

For a vertex $v \in V(H)$, the edge subfamily

$$(E \in \mathcal{E}(H) : v \in E) = \mathcal{E}_H(v) = \mathcal{E}(v)$$

is called the *star of the vertex v*. Note that $\mathcal{E}(v)$ is a family of edges but not a partial hypergraph $H(v)$ as in [3]. The family

$$\mathcal{S}(H) = (\mathcal{E}(v) : v \in V(H))$$

is called the *star family* of H. The number $|\mathcal{E}(v)| = \deg v$ is the *degree* of a vertex $v \in V(H)$, and $|E|$ is the *degree* of an edge $E \in \mathcal{E}(H)$. An edge of degree 1 is called a *loop*.

Further,

$$\operatorname{rank} H = \max_{E \in \mathcal{E}(H)} \deg E.$$

A hypergraph without multiple edges is called a *simple hypergraph*. The edge family of a simple hypergraph H can be considered as a set. In this case, we write $E(H)$ instead of $\mathcal{E}(H)$ as for simple graphs. The *union* $G \cup H$ of simple hypergraphs G and H is defined as follows:

$$V(G \cup H) = V(G) \cup V(H), \quad E(G \cup H) = E(G) \cup E(H).$$

A hypergraph is called an *antichain* if no edge is a subset of another edge. We say that a hypergraph is *l-linear* ($l \geq 1$) if any two different edges have at most

l common vertices. In particular, the class of 1-linear hypergraphs is exactly the well-known class of *linear hypergraphs.*

A hypergraph is called an *r-uniform (or uniform) hypergraph* if all its edges have the same degree r. The *complete r-uniform hypergraph K_n^r* is a simple hypergraph of order n whose edge set coincides with the set of r-subsets of the vertex set $(1 \le r \le n)$. A hypergraph G is called a *partial hypergraph* of a hypergraph H if

$$\mathcal{E}(G) \subseteq \mathcal{E}(H) \quad \text{and} \quad V(G) = \bigcup_{E \in \mathcal{E}(G)} E.$$

If a partial hypergraph of H is a connected complete uniform hypergraph, then it is called the *clique* of H. More precisely, if $r \ge 2$ then K_n^r is a clique of rank r; K_1^1 is a clique of rank 1; a single vertex is a clique of rank 0. A *maximal clique* is maximal with respect to inclusion.

A finite family

$$Q = (C_j : j \in J) \tag{5.3}$$

of cliques C_j is called a *clique covering* of a hypergraph H if H is the union of the cliques from Q. The cliques C_j are called the *clusters* of Q, and the family

$$\text{rank } Q = (r_j : j \in J), \quad r_j = \text{rank } C_j,$$

is called the *rank of* Q. A cluster of rank 0 is called *trivial.* The covering Q is called *irreducible* if no clique C_j is a partial hypergraph of other clique from Q. The minimal number of cliques taken over all clique coverings of H is denoted by $\text{cc}(H)$. It is evident that only simple hypergraphs have clique coverings.

We say that two clique coverings $(C_j : j \in J_1)$ and $(D_j : j \in J_2)$ are *equal* if there exists a bijection $\alpha : J_1 \to J_2$ such that $C_j = D_{\alpha(j)}$ for every $j \in J_1$. A *covering* and the *equality of coverings* for the vertex set $V(H)$ are defined analogously. The vertex set of a clique will be called a *clique* too if there is no confusion. If (5.3) is a clique covering of a hypergraph H, then the family

$$V(Q) = (V(C_j) : j \in J)$$

is a *clique covering of the vertex set* $V(H)$.

Let H_1 and H_2 be two hypergraphs and let

$$\mathcal{E}(H_k) = (E_i^k : i \in I_k), \quad k = 1, 2.$$

An *isomorphism* $(\alpha, \beta) : H_1 \to H_2$ is a pair of bijections

$$\alpha : V(H_1) \to V(H_2) \quad \text{and} \quad \beta : I_1 \to I_2$$

such that if $E_i^1 = \{v_j : 1 \le j \le r\}$ then

$$\alpha(E_i^1) = \{\alpha(v_j) : 1 \le j \le r\} = E_{\beta(i)}^2.$$

If there exists an isomorphism $H_1 \to H_2$, then we say that H_1 and H_2 are *isomorphic* and write $H_1 \simeq H_2$. For simple hypergraphs, their isomorphism is defined more easily. This is a bijection $\alpha : V(H_1) \to V(H_2)$ such that $\alpha(X) \in E(H_2)$ if and only if $X \in E(H_1)$ for any subset $X \subseteq V(H_1)$.

For a hypergraph H, the *line graph* $L(H)$ is defined as follows:

(i) $V(L(H)) = \mathcal{E}(H)$. In accordance with the definition of a hypergraph, $V(L(H))$ is a set and $\mathcal{E}(H)$ is a family. In this situation, the above equality means that if

$$\mathcal{E}(H) = (E_i : 1 \leq i \leq m),$$

then

$$V(L(H)) = \{E_1, E_2, \ldots, E_m\}$$

is an m-element set. In other words, multiple edges of H give rise to different vertices of $L(H)$.

(ii) Vertices E_i and E_j are adjacent in $L(H)$ if and only if $E_i \cap E_j \neq \emptyset$.

The *dual hypergraph* H^* of a hypergraph H without isolated vertices is the following object. The vertices of H^* are exactly the edges of H, and the edges of H^* are exactly the vertices of H. A vertex E and an edge v are incident in H^* if and only if the corresponding edge E and vertex v are incident in H. The 2-*section* $[H]_2$ of a hypergraph H is defined as the simple graph whose vertex set is $V(H)$ and two different vertices are adjacent if and only if they are adjacent in H. Notice that $[H]_2$ has no loops.

It is obvious that H^* can be realized in the following form:

$$V(H^*) = \mathcal{E}(H), \quad \mathcal{E}(H^*) = (E_v : v \in V(H)), \quad E_v = \mathcal{E}_H(v).$$

For the line graph $L(H)$, we have

$$L(H) = \bigcup_{v \in V(H)} F_v, \tag{5.4}$$

where F_v is the complete graph with the vertex set $\mathcal{E}_H(v)$. Therefore,

$$[H^*]_2 = L(H). \tag{5.5}$$

A hypergraph H is called *conformal* if the vertex set of any clique in the 2-section $[H]_2$ is contained in some edge of H. A hypergraph H without isolated vertices with $\mathcal{E}(H) = (E_i : i \in I)$ is said to *satisfy the Helly property* if the condition $J \subseteq I$ and $E_i \cap E_j \neq \emptyset$ for all $i, j \in J$ implies that

$$\bigcap_{j \in J} E_j \neq \emptyset.$$

5.1.2 Multivalued Function 'Line Hypergraph'

Let H be a hypergraph without isolated vertices. Let us compare the line graph $L(H)$ and the dual hypergraph H^*. According to equality (5.4), the line graph $L(H)$

is obtained from H^* by replacing each edge E by a complete graph (i.e. a clique of rank 2) with the vertex set E. Possible multiple edges are replaced by a single edge. If we begin to replace edges of H^* by cliques of different ranks, not only 2, then we obtain hypergraphs 'similar' to both $L(H)$ and H^*. In particular, for any such a hypergraph H', equality (5.5) holds:

$$[H']_2 = L(H).$$

Thus, the idea of a multivalued function $\mathcal{L}$ whose set of values is the set of all hypergraphs H' naturally appears.

To put it more precisely, let H be a hypergraph without isolated vertices, $V(H) = \{v_1, v_2, \ldots, v_n\}$ be its vertex set and

$$\mathbf{1}_H = (\deg\, v_i : 1 \leq i \leq n)$$

be the degree sequence of H. We set

$$\mathbf{0}_H = (0_{v_i} : 1 \leq i \leq n), \quad \text{where} \quad 0_{v_i} = \begin{cases} 0 \text{ if } \deg\, v_i = 1; \\ 2 \text{ if } \deg\, v_i > 1. \end{cases}$$

Furthermore, let $\mathbb{Z}_+^n$ be the lattice of integer-valued vectors

$$\boldsymbol{x} = (x_1, x_2, \ldots, x_n), \quad x_i \geq 0, \quad 1 \leq i \leq n,$$

with the following order:

$$\boldsymbol{x} \leq \boldsymbol{y} \;\Leftrightarrow\; x_i \leq y_i, \;\; 1 \leq i \leq n.$$

Now, let $\mathcal{D}_H = [\mathbf{0}_H, \mathbf{1}_H]$ be an interval in $\mathbb{Z}_+^n$, and

$$\mathcal{D} = (d_1, d_2, \ldots, d_n) \in \mathcal{D}_H.$$

For $v_i \in V(H)$, let F_{v_i} denote the clique of rank d_i with the vertex set $\mathcal{E}(v_i)$, and let us consider the following hypergraph:

$$\mathcal{L}_{\mathcal{D}}(H) = \bigcup_{i=1}^{n} F_{v_i}.$$

The hypergraph $\mathcal{L}_{\mathcal{D}}(H)$ is called the *line hypergraph of H with respect to the vector* $\mathcal{D}$. If we write $\mathcal{D}$ in the form $\mathcal{D} = (d_v : v \in V(H))$, where $d_v = d_i$ for $v = v_i$, then the previous definition takes the form

$$\mathcal{L}_{\mathcal{D}}(H) = \bigcup_{v \in V(H)} F_v.$$

Let us denote by $\mathcal{H}$ the set of hypergraphs without isolated vertices and define the multivalued function $\mathcal{L}$ on $\mathcal{H}$ as follows:

$$\mathcal{L}(H) = \{\mathcal{L}_{\mathcal{D}}(H) : \mathcal{D} \in \mathcal{D}_H\}, \quad \text{where} \quad H \in \mathcal{H}.$$

The function $\mathcal{L}$ is called the *line hypergraph*. Any element from the image $\mathcal{L}(H)$ is called a *line hypergraph of H*. Notice that the hypergraphs in $\mathcal{L}(H)$ are simple. Examples of line hypergraphs can be found in Section 1.3.2.

It is evident that $\mathcal{L}_{0_H}(H) = L(H)$, and $\mathcal{L}_{1_H}(H)$ is obtained from the dual hypergraph H^* if multiple edges are replaced by a single edge. We have $\mathcal{L}_{1_H}(H) = H^*$ if H does not contain similar vertices, that is, $\mathcal{E}(v_i) \neq \mathcal{E}(v_j)$ for $i \neq j$. It is also evident that $|\mathcal{L}(H)| = 1$ if and only if $\mathbf{1}_H = (2, 2, \ldots, 2)$.

It follows from (5.4) that the family of cliques $(F_v : v \in V(H))$ is a covering of the line hypergraph $\mathcal{L}_{\mathcal{D}}(H)$. Let us call this covering *standard* and denote it by $\mathrm{SC}(H, \mathcal{D})$.

5.1.3 On Isomorphisms of Line Hypergraphs

Let H_1 and H_2 be two hypergraphs without isolated vertices:

$$V(H_k) = V_k, \quad \mathcal{E}(H_k) = (E_i^k : i \in I_k), \quad k = 1, 2.$$

Further, let $\mathcal{D}_k$ be a vector from the interval $\mathcal{D}_{H_k}$, which was defined in the previous section, and

$$G_k = \mathcal{L}_{\mathcal{D}_k}(H_k).$$

Also, let

$$\alpha : V_1 \to V_2, \quad \beta : I_1 \to I_2, \quad \varphi = (\alpha, \beta) : H_1 \to H_2 \tag{5.6}$$

be an isomorphism of hypergraphs. The isomorphism (5.6) is called an *isomorphism of pairs* $(H_1, \mathcal{D}_1) \to (H_2, \mathcal{D}_2)$ if

$$d_v^1 = d_{\alpha(v)}^2$$

for every vertex $v \in V_1$, where $(d_v^k : v \in V_k) = \mathcal{D}_k$.

Define a mapping $\bar{\varphi}$ as follows:

$$\bar{\varphi} : V(G_1) \to V(G_2), \quad E_i^1 \to E_{\beta(i)}^2.$$

It is evident that $\bar{\varphi}$ is a bijection. We say that $\bar{\varphi}$ *is induced* by the isomorphism φ.

Let $\bar{\delta} : V(G_1) \to V(G_2)$ be a bijection. We shall say that $\bar{\delta}$ *preserves stars* if there exists a bijection $\gamma : V_1 \to V_2$ such that

$$\bar{\delta}(\mathcal{E}_{H_1}(v)) = \mathcal{E}_{H_2}(\gamma(v)) \tag{5.7}$$

for every vertex $v \in V_1$. The bijection $\bar{\delta}$ naturally acts on the star family $\mathcal{S}(H_1)$:

$$\bar{\delta}(\mathcal{S}(H_1)) = \Big(\bar{\delta}(\mathcal{E}_{H_1}(v)) : v \in V_1\Big).$$

The star family $\mathcal{S}(H_k)$ is a covering of the vertex set $V(G_k)$ and, by the definition of the equality of coverings (see Section 5.1.1), equalities (5.7) give

$$\bar{\delta}(\mathcal{S}(H_1)) = \mathcal{S}(H_2).$$

Now, let us consider the standard coverings, which were defined in the previous section:

$$\mathrm{SC}(H_k, \mathcal{D}_k) = (F_v^k : v \in V_k).$$

Suppose that in addition to (5.7), the following conditions hold:

$$d_v^1 = d_{\gamma(v)}^2, \quad v \in V_1. \tag{5.8}$$

Equalities (5.7) and (5.8) mean that $\bar{\delta}$ induces an isomorphism of cliques $F_v^1 \to F_{\gamma(v)}^2$, that is, $\bar{\delta}(F_v^1) = F_{\gamma(v)}^2$. By the definition of the equality of coverings, we have

$$\bar{\delta}(\mathrm{SC}(H_1, \mathcal{D}_1)) = \mathrm{SC}(H_2, \mathcal{D}_2).$$

In this case, we say that $\bar{\delta}$ *preserves the standard covering.*

It is evident that for any isomorphism φ in (5.6), the induced mapping $\bar{\varphi}$ preserves stars. And if φ is an isomorphism of pairs, then $\bar{\varphi}$ preserves the standard covering.

Lemma 5.1 *For every bijection $\bar{\delta} : V(G_1) \to V(G_2)$ preserving stars, there exists an isomorphism of hypergraphs $H_1 \to H_2$ that induces $\bar{\delta}$. If $\bar{\delta}$ preserves the standard covering, then it is an isomorphism of line hypergraphs $G_1 \to G_2$ such that there exists an isomorphism of pairs $(H_1, \mathcal{D}_1) \to (H_2, \mathcal{D}_2)$ that induces $\bar{\delta}$.*

Proof: Let $\bar{\delta} : V(G_1) \to V(G_2)$ be a bijection that preserves stars. Then, there exists a bijection γ satisfying (5.7). Let us define the bijection $\delta : I_1 \to I_2$ as follows:

$$\delta(i) = j \quad \Leftrightarrow \quad \bar{\delta}(E_i^1) = E_j^2 \quad \text{for} \quad i \in I_1. \tag{5.9}$$

We will show that the pair (γ, δ) is an isomorphism of hypergraphs $H_1 \to H_2$. If

$$E_i^1 \in \mathcal{E}(H_1), \quad \bar{\delta}(E_i^1) = E_j^2 \quad \text{and} \quad E_i^1 = \{v_1, v_2, \ldots, v_d\},$$

then

$$E_i^1 \in \mathcal{E}_{H_1}(v_1) \cap \mathcal{E}_{H_1}(v_2) \cap \ldots \cap \mathcal{E}_{H_1}(v_d).$$

It follows from (5.7) that

$$E_j^2 \in \mathcal{E}_{H_2}(\gamma(v_1)) \cap \mathcal{E}_{H_2}(\gamma(v_2)) \cap \ldots \cap \mathcal{E}_{H_2}(\gamma(v_d))$$

and

$$E_j^2 = \{\gamma(v_1), \gamma(v_2), \ldots, \gamma(v_d), \ldots\}.$$

If $\gamma(v_{d+1}) \in E_j^2$, then

$$E_j^2 \in \mathcal{E}_{H_2}(\gamma(v_{d+1})), \quad E_i^1 \in \mathcal{E}_{H_1}(v_{d+1}) \quad \text{and} \quad v_{d+1} \in E_i^1.$$

This is impossible because $|E_i^1| = d$. Consequently,

$$E_j^2 = \{\gamma(v_1), \gamma(v_2), \ldots, \gamma(v_d)\},$$

and thus

$$\gamma(E_i^1) = \{\gamma(v_1), \gamma(v_2), \ldots, \gamma(v_d)\} = E_j^2 = \bar{\delta}(E_i^1) = E_{\delta(i)}^2.$$

Hence, $\psi = (\gamma, \delta)$ is an isomorphism of hypergraphs $H_1 \to H_2$. It follows from (5.9) that ψ induces the bijection $\bar{\delta}$. By definition, ψ is an isomorphism of pairs $(H_1, \mathcal{D}_1) \to (H_2, \mathcal{D}_2)$ if (5.8) holds.

It remains to prove that $\bar{\delta} : G_1 \to G_2$ is an isomorphism of hypergraphs. Suppose that $\emptyset \neq X \subseteq V(G_1)$. If $X \in E(G_1)$ then, by (5.4), $X \subseteq \mathcal{E}_{H_1}(v)$ for some vertex $v \in V_1$ and $|X| = d^1_v$. Taking into account (5.7) and (5.8), we have

$$\bar{\delta}(X) \subseteq \mathcal{E}_{H_2}(\gamma(v)) \quad \text{and} \quad |\bar{\delta}(X)| = d^1_v = d^2_{\gamma(v)}.$$

Hence, $\bar{\delta}(X) \in E(G_2)$.

Using the inverse bijection $\bar{\delta}^{-1}$, we analogously obtain

$$\bar{\delta}(X) \in E(G_2) \Rightarrow X = \bar{\delta}^{-1}\bar{\delta}(X) \in E(G_1).$$

Thus,

$$\bar{\delta}(X) \in E(G_2) \Leftrightarrow X \in E(G_1).$$

Consequently, $\bar{\delta}$ is an isomorphism of hypergraphs G_1 and G_2. □

The next two corollaries are obvious.

Corollary 5.1 *Statements (i) and (ii) are true:*

(i) We have $H_1 \simeq H_2$ if and only if there exists a bijection $V(G_1) \to V(G_2)$ preserving stars.

(ii) We have $(H_1, \mathcal{D}_1) \simeq (H_2, \mathcal{D}_2)$ if and only if there exists a bijection $V(G_1) \to V(G_2)$ that preserves the standard covering. This bijection is an isomorphism of line hypergraphs $G_1 \to G_2$.

Following the terminology of Lovász [21], we say that an isomorphism of line hypergraphs $G_1 \to G_2$ is *trivial* if there exists an isomorphism $H_1 \to H_2$ that induces it. In the next corollary, the statement from [21] (p. 506) on trivial isomorphisms of line graphs is transferred onto line hypergraphs.

Corollary 5.2 *Statements (i) and (ii) are true:*

(i) If φ is an isomorphism of pairs $(H_1, \mathcal{D}_1) \to (H_2, \mathcal{D}_2)$, then the induced mapping $\bar{\varphi}$ is a trivial isomorphism of line hypergraphs.

(ii) An isomorphism $\bar{\delta} : G_1 \to G_2$ is trivial if and only if it preserves stars. If $\bar{\delta}$ is trivial, then the isomorphism $H_1 \to H_2$ that induces $\bar{\delta}$ can be chosen among isomorphisms of pairs $(H_1, \mathcal{D}_1) \to (H_2, \mathcal{D}_2)$.

5.1.4 Inverse Image $\mathcal{L}^{-1}(G)$

As pointed out above, all values of the function $\mathcal{L}(H)$ are simple hypergraphs for any hypergraph H. Let G be an arbitrary simple hypergraph. The set of hypergraphs

$$\{H : \mathcal{L}_{\mathcal{D}}(H) \simeq G \text{ for some } \mathcal{D} \in \mathcal{D}_H\}$$

is called the *inverse image* $\mathcal{L}^{-1}(G)$. Further considerations aim to investigate $\mathcal{L}^{-1}(G)$ for a given G, thus developing Berge's idea [2] (p. 400). For a simple graph G, the inverse image $\mathcal{L}^{-1}(G)$ is described in [2] in terms of clique coverings of G. A generalization of that result will be discussed here.

Let G be a simple hypergraph, $\mathcal{A}_G$ be the set of triads $(H, \mathcal{D}, \gamma)$, where

$$H \in \mathcal{L}^{-1}(G), \quad \mathcal{D} \in \mathcal{D}_H, \quad \mathcal{L}_\mathcal{D}(H) \simeq G$$

and

$$\gamma : \mathcal{L}_\mathcal{D}(H) \to G$$

is an isomorphism. Let $\mathcal{B}_G$ be the set of clique coverings of G. Obviously, $\mathcal{B}_G \neq \emptyset$.

The concept of a canonical hypergraph plays an important role here. Let

$$Q = (C_i : 1 \leq i \leq t) \in \mathcal{B}_G.$$

Define the hypergraph F_Q as follows:

$$V(F_Q) = V(G), \quad \mathcal{E}(F_Q) = (E_i : 1 \leq i \leq t), \quad E_i = V(C_i). \tag{5.10}$$

It is clear that F_Q does not contain isolated vertices. Hence, there exists the dual hypergraph $(F_Q)^*$. The hypergraph $(F_Q)^*$ is called *canonical* (*with respect to* Q) and is denoted by $C(Q)$. For graphs, the construction of a canonical hypergraph was used in [2]. The concept of a canonical hypergraph was introduced by Gardner [14] for the particular case when Q is a covering of edges.

Lemma 5.2 *We have* rank $Q \in \mathcal{D}_{C(Q)}$. *Consequently, there exists a line hypergraph* $\mathcal{L}_\mathcal{D}(C(Q))$ *with* $\mathcal{D} =$ rank Q.

Proof: By definition,

$$V(C(Q)) = \{E_i : 1 \leq i \leq t\}, \quad E_i = V(C_i). \tag{5.11}$$

If $V(G) = \{v_j : 1 \leq j \leq n\}$, then

$$\mathcal{E}(C(Q)) = \mathcal{S}(F_Q) = (V_j : 1 \leq j \leq n), \quad V_j = \mathcal{E}_{F_Q}(v_j).$$

For $E_i \in V(C(Q))$, we have

$$\mathcal{E}_{C(Q)}(E_i) = (V_j : E_i \in V_j) = (V_j : v_j \in E_i). \tag{5.12}$$

Consequently,

$$\deg_{C(Q)} E_i = |E_i| = |C_i|.$$

Therefore, rank $C_i \leq |E_i|$ and the lemma is proved. □

Let us set

$$\mathcal{D}_Q = \text{rank } Q = (d_i : 1 \le i \le t).$$

The pair $(C(Q), \mathcal{D}_Q)$ is called the *canonical pair* $\text{CP}(Q)$. Now, define the mapping

$$\gamma_Q : V(\mathcal{L}_{\mathcal{D}_Q}(C(Q))) \to V(G) \ \text{ such that } \ V_j \to v_j, \ \ 1 \le j \le n.$$

The triad $(C(Q), \mathcal{D}_Q, \gamma_Q)$ is called the *canonical triad* $\text{CT}(Q)$.

Lemma 5.3 *The mapping* γ_Q *is an isomorphism of hypergraphs*

$$\mathcal{L}_{\mathcal{D}_Q}(C(Q)) \to G,$$

and $\text{CT}(Q) \in \mathcal{A}_G$.

Proof: Let us denote $H' = C(Q)$ and consider an arbitrary subset

$$E = \{V_{i_1}, V_{i_2}, \ldots, V_{i_d}\} \subseteq V(\mathcal{L}_{\mathcal{D}_Q}(H')).$$

It is evident that $E \in E(\mathcal{L}_{\mathcal{D}_Q}(H'))$ if and only if there is a vertex $E_i \in V(H')$ such that

$$E \subseteq \mathcal{E}_{H'}(E_i) \quad \text{and} \quad d = d_i. \tag{5.13}$$

According to (5.12), the condition (5.13) is equivalent to the following:

$$\{v_{i_1}, v_{i_2}, \ldots, v_{i_d}\} \subseteq E_i, \quad d = d_i = \text{rank } C_i.$$

Hence,

$$\gamma_Q(E) = \{v_{i_1}, v_{i_2}, \ldots, v_{i_d}\} \in E(G).$$

Thus, $\gamma_Q : \mathcal{L}_{\mathcal{D}_Q}(H') \to G$ is an isomorphism. □

Let us define the mapping

$$\psi : \mathcal{A}_G \to \mathcal{B}_G, \quad (H, \mathcal{D}, \gamma) \to \gamma(\text{SC}(H, \mathcal{D})),$$

where $\text{SC}(H, \mathcal{D})$ is a standard covering.

Theorem 5.1 [34] *Under the previous notation, the following three statements hold:*

(i) The mapping ψ *is a surjection. If* Q *is an arbitrary clique covering of* G*, then*

$$\psi(\text{CT}(Q)) = Q.$$

(ii) Let

$$(H_k, \mathcal{D}_k, \gamma_k) \in \mathcal{A}_G, \quad k = 1, 2, \tag{5.14}$$

and denote $Q_k = \psi(H_k, \mathcal{D}_k, \gamma_k)$*. Also, let* $V(Q_k)$ *be the clique covering of the vertex set* $V(G)$ *corresponding to the clique covering* Q_k*. Then* $H_1 \simeq H_2$ *if and only if there exists a bijection* $\alpha : V(G) \to V(G)$ *such that*

$$\alpha(V(Q_1)) = V(Q_2). \tag{5.15}$$

(iii) *We have* $(H_1, \mathcal{D}_1) \simeq (H_2, \mathcal{D}_2)$ *for triads (5.14) if and only if there exists an automorphism* $\alpha \in \mathrm{Aut}(G)$ *such that* $\alpha(Q_1) = Q_2$.

Proof: (i) By Lemma 5.3, $\mathrm{CT}(Q) \in \mathcal{A}_G$. Let us find $\psi(\mathrm{CT}(Q))$. Since

$$\mathrm{CT}(Q) = (C(Q), \mathcal{D}_Q, \gamma_Q),$$

it follows that $\psi(\mathrm{CT}(Q))$ coincides with the image of the standard covering $\mathrm{SC}(C(Q), \mathcal{D}_Q)$ under the action of the isomorphism γ_Q. Let

$$Q = (C_i : 1 \leq i \leq t).$$

The above standard covering is the family of cliques F_v of the line hypergraph $\mathcal{L}_{\mathcal{D}_Q}(C(Q))$, where v is taken over all $V(C(Q))$ in accordance with (5.11). The vertex set of the clique F_{E_i} is the star $\mathcal{E}_{C(Q)}(E_i)$ and rank $F_{E_i} = d_i$, where $(d_i : 1 \leq i \leq t)$ $= \mathrm{rank} Q$. By (5.12),

$$V(F_{E_i}) = (V_j : v_i \in E_j),$$

where $\{v_j : 1 \leq j \leq n\} = V(G)$. Consequently, $\gamma_Q(F_{E_i})$ is a clique of rank d_i with the vertex set

$$\{v_j : v_j \in E_i\} = E_i = V(C_i).$$

Thus, $\psi(\mathrm{CT}(Q)) = Q$.

(ii) This follows from Corollary 5.1. Indeed, by this corollary, $H_1 \simeq H_2$ if and only if there exists a bijection

$$\bar{\delta} : V(\mathcal{L}_{\mathcal{D}_1}(H_1)) \to V(\mathcal{L}_{\mathcal{D}_2}(H_2))$$

preserving stars. Now, if there exists $\bar{\delta}$, then we set $\alpha = \gamma_2 \bar{\delta} \gamma_1^{-1}$. It is obvious that α is a bijection on the set $V(G)$. Further,

$$V(Q_k) = \gamma_k(\mathcal{S}(H_k)),$$

where $\mathcal{S}(H_k)$ is the star family of H_k. Hence,

$$\alpha(V(Q_1)) = \gamma_2 \bar{\delta} \gamma_1^{-1} \gamma_1(\mathcal{S}(H_1)) = \gamma_2 \bar{\delta}(\mathcal{S}(H_1)) = \gamma_2(\mathcal{S}(H_2)) = V(Q_2).$$

Conversely, if $\alpha : V(G) \to V(G)$ is a bijection satisfying (5.15), then we set $\bar{\delta} = \gamma_2^{-1} \alpha \gamma_1$. We have

$$\bar{\delta} : V(\mathcal{L}_{\mathcal{D}_1}(H_1)) \to V(\mathcal{L}_{\mathcal{D}_2}(H_2))$$

and

$$\bar{\delta}(\mathcal{S}(H_1)) = \gamma_2^{-1} \alpha \gamma_1 \gamma_1^{-1}(V(Q_1)) = \mathcal{S}(H_2).$$

(iii) This case is handled in the same way. The proof of the theorem is complete. □

Corollary 5.3 *If* $\psi(H, \mathcal{D}, \gamma) = Q$, *then* $(H, \mathcal{D}) \simeq \mathrm{CP}(Q)$.

The next theorem follows directly from Theorem 5.1.

Theorem 5.2 [34] *Statements (i)–(iv) are true.*

(i) *For any simple hypergraph* G, *the inverse image* $\mathcal{L}^{-1}(G)$ *coincides with the set of canonical hypergraphs* $C(Q)$, *where* Q *is taken over all clique coverings of* G.

(ii) *We have* $C(Q_1) \simeq C(Q_2)$ *if and only if there exists a bijection* $\bar{\delta} : V(G) \to V(G)$ *such that*

$$\bar{\delta}(V(Q_1)) = V(Q_2), \tag{5.16}$$

where $V(Q_k)$ *is a clique covering of* $V(G)$ *corresponding to* Q_k.

(iii) *We have* $\mathrm{CP}(Q_1) \simeq \mathrm{CP}(Q_2)$ *if and only if there exists an automorphism* $\bar{\delta} \in \mathrm{Aut}(G)$ *satisfying (5.16). The automorphism* $\bar{\delta}$ *satisfying (5.16) is trivial.*

(iv) *If a hypergraph* G *is uniform, then* $C(Q_1) \simeq C(Q_2)$ *if and only if* $\mathrm{CP}(Q_1) \simeq \mathrm{CP}(Q_2)$.

5.1.5 Krausz-type Characterizations and Coverings

The following characterization of line graphs is well known.

Krausz Theorem [18] *A graph* G *is the line graph of some simple graph if and only if there exists a clique covering of* G *satisfying the next two conditions:*

(i) *Every vertex of* G *belongs to exactly two clusters of the covering.*

(ii) *Any two clusters of the covering have at most one vertex in common.*

For line hypergraphs, the Krausz theorem can be generalized in the following way. Let P be a hypergraph-theoretic property. We can understand P as a class of hypergraphs closed with respect to isomorphism. Without loss of generality, we may assume that any hypergraph from P has no isolated vertices. We set

$$P^* = \{H^* : H \in P\}, \quad \mathcal{L}(P) = \bigcup_{H \in P} \mathcal{L}(H).$$

Furthermore, let G be a simple hypergraph, Q be a clique covering of G and F_Q be the hypergraph determined by (5.10). If $F_Q \in P$, then we say that Q *has the property* P or briefly write $Q \in P$.

Theorem 5.3 [34] *For a hypergraph* G, *the following statements are equivalent:*

(i) $G \in \mathcal{L}(P)$.

(ii) *There exists a clique covering of* G *with the property* P^*.

Proof: From the definition of the canonical hypergraph $C(Q)$, it follows that $C(Q) \in P$ if and only if $Q \in P^*$. The result now follows from Theorem 5.2 (i). □

Many known characterizations of line graphs of hypergraphs with a given property are direct corollaries of Theorem 5.3 and are transferred onto line hypergraphs. Let us consider some examples.

Let P_r be the class of r-uniform hypergraphs and P^l be the class of l-linear hypergraphs. Then, $(P_r)^*$ is the class of hypergraphs whose vertex degrees are equal to r, and $(P^l)^*$ is the class of hypergraphs for which any $l+1$ edges have at most one vertex in common. In particular, $(P^1)^* = P^1$.

Suppose that Q is a clique covering of a hypergraph G. By definition, Q is an *r-covering* if every vertex of G belongs to at most r clusters of Q; Q is a *strict r-covering* if every vertex of G belongs to exactly r clusters of Q; and Q is an *l-linear covering* if any $l+1$ clusters of Q have at most one vertex in common. A 1-linear covering is called a *linear covering.*

The following Corollaries 5.4–5.6 are true for any hypergraph G. For the case when G is a graph, these corollaries were proved in [3].

Corollary 5.4 *$\mathcal{L}^{-1}(G) \cap P_r \neq \emptyset$ if and only if there exists a strict r-covering of G.*

Corollary 5.5 *$\mathcal{L}^{-1}(G) \cap P^l \neq \emptyset$ if and only if there exists an l-linear covering of G.*

Let P_r^l denote the class of l-linear r-uniform hypergraphs. In particular, P_r^{r-1} is the class of simple r-uniform hypergraphs.

Corollary 5.6 *$\mathcal{L}^{-1}(G) \cap P_r^l \neq \emptyset$ if and only if there exists an l-linear strict r-covering of G.*

It is obvious that the word 'strict' in the statements of Corollaries 5.4 and 5.6 can be omitted. If $l=1$ and $r=2$, then Corollary 5.6 becomes the Krausz theorem. Hence, Theorem 5.3 and all its variants as well as the corresponding clique coverings are called *Krausz-type characterizations* and *Krausz P-coverings.*

Corollary 5.7 *Any linear hypergraph, in particular any simple graph, is a line hypergraph of some linear hypergraph.*

For graphs, Corollary 5.7 was proved in [3]. Furthermore, let $P = P_r$, P^l or P_r^1, let $G \in \mathcal{L}(P)$ and let C be a maximal clique of G. The clique C is called *P-large* if

- rank $C > r$ for $P = P_r$,
- rank $C > l+1$ for $P = P^l$,
- rank $C > 2$ or $|C| > r^2 - r + 1$ for $P = P_r^1$.

If rank $C \leq 1$, then C is called *P-large* as well. In Theorem 5.4, the word 'strict' is omitted from the definitions of Krausz P-coverings. This theorem will often be used in Sections 5.2 and 5.3 for analysing some line graphs.

Theorem 5.4 [34] *Each P-large clique of G is a cluster of every Krausz P-covering of G.*

Proof: Let Q denote a Krausz P-covering of G, A be a clique of G such that it does not belong to any cluster of Q, $E_1 \in E(A)$ and $s = \text{rank } A$. We construct inductively the sequences of edges $(E_i : 1 \leq i \leq s)$ and clusters $(C_i : 1 \leq i \leq s)$ in the following way:

$$E_1 = \{a_1, a_2, \ldots, a_s\} \in E(C_1),$$

$$E_2 = \{b_1, a_2, a_3, \ldots, a_s\} \in E(C_2),$$

$$\vdots$$

$$E_s = \{b_1, b_2, \ldots, b_{s-1}, a_s\} \in E(C_s),$$

where $b_i \in V(A) - V(C_i)$. It is obvious that all s cliques C_i are pairwise distinct. If $P = P_r$, then Q is an r-covering. Since $a_s \in V(C_i)$ for $1 \leq i \leq s$, it follows that $s \leq r$. If $P = P^l$, then Q is an l-linear covering. However, $\{a_{s-1}, a_s\} \subseteq V(C_i)$ for $1 \leq i \leq s-1$. Hence, $s - 1 \leq l$ or $s \leq l + 1$.

It remains to consider the case $P = P_r^1$ and $s = 2$. In this case, Q is a linear 2-covering. Suppose that $a \in V(A)$ and $(D_1, D_2, \ldots, D_t)$ is the list of all clusters of Q such that D_i contains the vertex a for $1 \leq i \leq t$. Then $t \leq r$. Let us denote

$$B_i = V(A) \cap V(D_i) = \{a, b_{i,1}, b_{i,2}, \ldots, b_{i,m_i}\}, \quad 1 \leq i \leq t.$$

We see that $V(A)$ is divided into pairwise disjoint subsets $\{a\}$ and $B_i - \{a\}$ for $1 \leq i \leq t$. If $b \in B_i$ and c is taken over all B_j with $j \neq i$, then all edges bc belong to different clusters of Q and $bc \notin E(B_i)$. Since Q is an r-covering, it follows that $|B_j| \leq r - 1$ and hence $|A| \leq r(r-1) + 1$.

Thus, the clique A is not P-large for any of the above properties P. This completes the proof. □

Corollary 5.8 *If $G \in \mathcal{L}(P)$ and each maximal clique of G is P-large, then the set of maximal cliques is a unique irreducible P-covering of G.*

Corollary 5.9 *Suppose that each maximal clique of G is P-large. Then $G \in \mathcal{L}(P)$ if and only if the following condition holds:*

(i) *For $P = P_r$, every $r+1$ maximal cliques of G have no vertex in common.*

(ii) *For $P = P^l$, every $l+1$ maximal cliques of G have at most one vertex in common.*

(iii) *For $P = P_r^l$, every $r+1$ maximal cliques of G have no vertex in common and every $l+1$ maximal cliques have at most one vertex in common.*

Denote by P_{Hl} the class of hypergraphs satisfying the Helly property, and let P_{cn} stand for conformal hypergraphs.

Corollary 5.10 *Suppose that $P = P_{\mathrm{Hl}} \cap P^1$ and $G \in \mathcal{L}(P)$. Then the set of maximal cliques of G is a unique irreducible P-covering.*

Proof: It is known that

$$(P_{\mathrm{Hl}})^* = P_{\mathrm{cn}}.$$

Therefore, by Theorem 5.3, there exists a clique covering Q of G such that $Q \in P_{\mathrm{cn}} \cap P^1$. Without loss of generality, we may assume that Q is irreducible. Let C be a maximal clique of G. By Theorem 5.4, C is a cluster of Q if rank $C \neq 2$. If rank $C = 2$, then C is a clique of the 2-section $[G]_2$. Hence, there exists a cluster $C' \in Q$ such that $V(C) \subseteq V(C')$. If $E \in E(C)$, then $E \in E(C'')$ for some cluster $C'' \in Q$. Since the covering Q is linear, we have

$$C'' = C', \quad \text{rank } C' = 2, \quad C = C', \quad C \in Q.$$

Thus, Q coincides with the set of maximal cliques because Q is irreducible. The corollary is proved. □

5.1.6 Walking Around the Whitney Theorem

The Whitney theorem has no simple analogue for line hypergraphs in contrast to the Berge theorem and the Krausz theorem. For the function 'line hypergraph', it is natural to consider the version of the Whitney theorem in the form of a list of two properties (P, P') for which the following implication holds:

$$G \in P' \Rightarrow |\mathcal{L}^{-1}(G) \cap P| \leq 1. \tag{5.17}$$

If P is the class of connected graphs of order not equal to 3 and P' is the class of all graphs, then the classical Whitney theorem is a uniqueness theorem in the above form.

The next corollary follows directly from Theorem 5.2.

Corollary 5.11 *If P is the class of uniform hypergraphs, then*

$$|\mathcal{L}^{-1}(G) \cap P| \leq 1$$

if and only if the action of $(\mathrm{Aut}(G), P^* \cap \mathcal{B}_G)$ is transitive, where $\mathcal{B}_G$ is the set of clique coverings of G.

The properties of P-coverings from Section 5.1.5 enable us to indicate a number of pairs (P, P') such that implication (5.17) holds. Corollaries 5.8 and 5.10 immediately imply the following:

Corollary 5.12 *If P'' is the class of hypergraphs dual to antichains and (P, P') is one of the pairs described in (i)–(iv), then implication (5.17) holds.*

(i) $P = P_r$; *P' is the class of hypergraphs from P'' whose edge degrees are greater than r.*

(ii) $P = P^l$; P' *is the class of hypergraphs from* P'' *whose edge degrees are greater than* $l+1$.

(iii) $P = P_r^1$; P' *is the class of hypergraphs from* P'' *such that orders of its maximal cliques of* rank 2 *are greater than* $r(r-1)+1$.

(iv) $P = P^1 \cap P_{\text{HI}}$; $P' = P''$.

It is easy to see that the condition $P' \subseteq P''$ is quite natural. The classical Whitney theorem states that (5.17) is true if $P = P_0$, where P_0 is the class of connected simple graphs of order not equal to 3, and P' is the class of all simple graphs.

Theorem 5.5 [34] *For any hypergraph* G, *implication (5.17) is true if* $P = P_0$.

To prove this theorem, we need two lemmas.

Lemma 5.4 *Let* C *be a clique of a connected hypergraph* $G \in \mathcal{L}(P_2^1)$ *and* rank $C \neq 0$. *Then, the clique* C *is a cluster of at most one linear strict 2-covering of* G.

Proof: Let Q be a linear strict 2-covering of G:

$$Q = (C_i : i = 1, 2, \ldots, k, k+1, \ldots)$$

and $C = C_1$. Further, let P be some subfamily of Q, for example

$$P = (C_i : 1 \leq i \leq k).$$

Now, let $V'(G)$ and $E'(G)$ denote the subsets of vertices and edges of G, respectively, belonging to the cliques of the family P. We set

$$E_1(G) = E(G) - E'(G).$$

If $E_1(G) = \emptyset$, then Q is obtained from P by addition of trivial clusters, the cluster with a vertex v being added only if v belongs to exactly one cluster of Q. Thus, we can reconstruct Q from P.

Suppose that $E_1(G) \neq \emptyset$. Since G is connected, there exist a vertex $v \in V'(G)$ and an edge $e \in E_1(G)$ such that $v \in e$. Denote by $E_2(G)$ the set of edges from $E_1(G)$ that contain the vertex v, and by W the union of edges from $E_2(G)$, so that W is a set of vertices. Since the vertex v can belong to at most two clusters of Q, we see that the degrees of all edges in $E_2(G)$ are equal to $\deg e = r$. Moreover, if H is a clique of rank r with the vertex set W, then H is a cluster of Q. Add this cluster to P. Repeating this construction, we obtain the covering Q. This yields the lemma. □

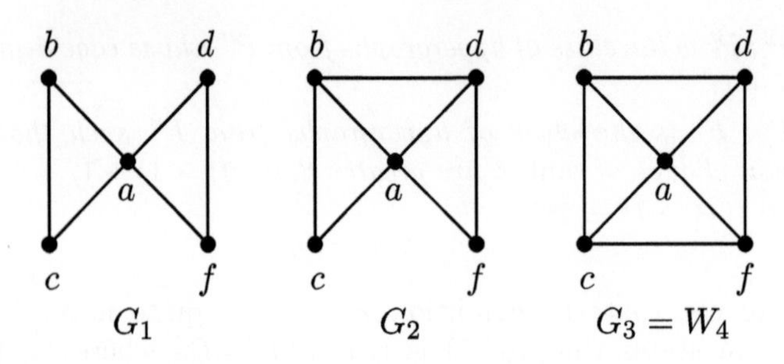

Figure 5.1 Three possibilities for the graph F.

The graph W_4 in the next lemma is the wheel shown in Figure 5.1 (the right picture).

Lemma 5.5 *If G is a connected hypergraph from $\mathcal{L}(P_2^1)$, then one of the following statements holds:*

(i) *There exists a unique linear strict 2-covering of G.*
(ii) *G is a graph isomorphic to $K_4 - e$, $\overline{3K_2}$ or W_4; and there exist exactly two linear strict 2-coverings of G, which are interchanged by an automorphism of G.*
(iii) *$G \simeq K_3$ or K_1^1.*

Proof: Let $G \not\simeq K_3$ and K_1^1. By Theorem 5.2, there exists a linear strict 2-covering Q of G. Suppose first that there is a maximal clique A in G such that rank $A \neq 2$ or $|A| > 3$; that is, A is a P_2^1-large clique. By Theorem 5.4, the clique A is a cluster of any linear 2-covering of G. In particular, A is a cluster of Q. Taking into account Lemma 5.4, we obtain that the covering Q is a unique linear strict 2-covering of G. Thus, it remains to consider the case where each maximal clique of G has rank 2 and order 3. Hence, the graph G is a union of triangles and does not contain K_4. Because Q is a strict 2-covering of G, the maximum degree $\Delta = 3$ or 4.

Suppose that $\Delta = 3$, $a \in V(G)$ and deg $a = 3$. Then, the induced subgraph

$$F = G(N(a) \cup \{a\})$$

coincides with $K_4 - cd$, where the vertex set of K_4 is $\{a, b, c, d\}$. An arbitrary linear strict 2-covering Q of the graph G contains one of two triangles

$$T_1 = (a, b, c) \text{ or } T_2 = (a, b, d).$$

Hence, by Lemma 5.4, there are at most two such coverings: $T_1 \in Q_1$ and $T_2 \in Q_2$. If $|G| = 4$, then there are exactly two such coverings:

$$Q_1 = \{T_1, da, db, \{c\}\} \text{ and } Q_2 = \{T_2, ca, cb, \{d\}\}.$$

The transposition (c,d) is an automorphism of G interchanging these coverings. If $|G| > 4$, then there is the fifth vertex f. Without loss of generality, we may assume that f is adjacent to c. The vertex f is not adjacent to a or b because deg a = deg $b = \Delta$. Therefore, $T_1 \in Q$ and $Q = Q_1$.

Now, suppose that $\Delta = 4$, $a \in V(G)$ and deg $a = 4$. Then, there are three possibilities for the graph F shown in Figure 5.1. For the graphs G_1 and G_2, both triangles (a, b, c) and (a, d, f) are clusters of any linear strict 2-covering. By Lemma 5.4, such a covering is unique. For the graph G_3, there are two possibilities: $(a, b, c) \in Q_1$ and $(a, b, d) \in Q_2$. If $|G| = 5$, then there exist exactly two such coverings:

$$Q_1 = \{(a, b, c), (a, d, f), bd, cf\} \quad \text{and} \quad Q_2 = \{(a, b, d), (a, c, f), bc, df\}.$$

Moreover,

$$t = (c, d) \in \text{Aut}(G)$$

and $t(Q_1) = Q_2$. Further, suppose that there is a vertex g adjacent to b. It is evident that g must be adjacent to c or d. Assume that g is adjacent to d. If g is not adjacent to c, then the triangle (b, c, a) is a cluster of any linear strict 2-covering and the result follows from Lemma 5.4. If g is adjacent to c and is not adjacent to f, then the same holds for (d, f, a). Thus, it remains to consider only the case when g is adjacent to b, c, d and f. Since G is connected and $\Delta = 4$, we have $G \simeq \overline{3K_2}$. This completes the proof of the lemma. □

Theorem 5.5 follows from Corollary 5.11 and Lemma 5.5.

5.2 Line Graphs of Linear 3-Uniform Hypergraphs

This section explores the class L_3^ℓ of line graphs of linear 3-uniform hypergraphs. It is known that this class cannot be characterized by a finite list of forbidden induced subgraphs [27]. However, such a characterization can be given, provided that vertex degrees are at least 19. We will also discuss the recognition problem '$G \in L_3^\ell$?' in the class of graphs with minimum vertex degree $\delta(G) \geq 19$.

5.2.1 Preliminaries

The following two characterizations of line graphs of simple graphs are well known. The Krausz theorem was already discussed in Section 5.1.5; however, we formulate it again for convenience.

Krausz Theorem [18] *A graph G is the line graph $L(H)$ of some simple graph H if and only if there exists a clique covering of G satisfying the following conditions:*

(i) Each vertex of G belongs to at most two cliques of the covering.
(ii) Any two cliques have at most one common vertex.

Beineke Theorem [1] *A graph G is the line graph of some simple graph if and only if none of the nine graphs shown in Figure 5.2 is an induced subgraph of G.*

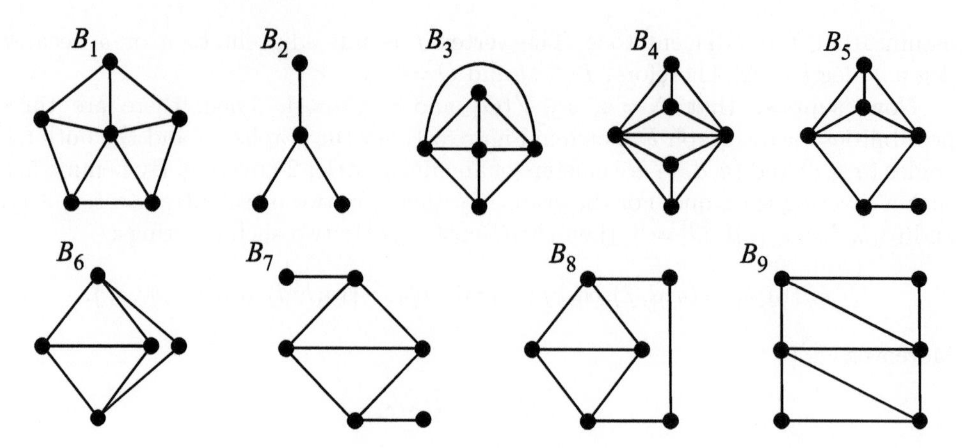

Figure 5.2 Beineke graphs.

Recall that the *line graph* $L(H)$ of a hypergraph H is the intersection graph of the edges of H. Thus, the vertices of $L(H)$ are in one-to-one correspondence with the edges of H, and two vertices are adjacent in $L(H)$ if and only if the corresponding edges are adjacent in H. A hypergraph is called *linear* if any two edges have at most one common vertex. A hypergraph whose edge degrees are all equal to the same number r is called *r-uniform*. Denote by L_r and L_r^ℓ the classes of line graphs of r-uniform hypergraphs and line graphs of linear r-uniform hypergraphs, respectively. Thus, L_2^ℓ is the class of line graphs of simple graphs.

Lovász [22] posed the problem of characterizing the class L_3 and noted that it cannot be characterized by a finite list of forbidden induced subgraphs (see also [13]). Such a characterization is also impossible for the class L_3^ℓ [27]. A global characterization of Krausz type for the class L_r for arbitrary r was given by Berge [3]. Similar to the previous section, such characterizations are called *Krausz-type characterizations*. A Krausz-type characterization of the class L_r^ℓ was obtained in [27]. It is pointed out in [3] that any graph is the line graph of some linear hypergraph.

For an arbitrary graph belonging to L_r^ℓ, the list of maximal cliques can be constructed in polynomial time [20], where the corresponding polynomial is in the vertex number and the power of the polynomial increases together with r. Therefore, the minimal value of r such that $G \in L_r^\ell$, which is called the *Krausz dimension* of G, could be accepted as a measure of complexity of the graph G. Note that the recognition problems '$G \in L_r$?' for $r \geq 4$ [29], '$G \in L_r^\ell$?' for $r \geq 3$ and the problem of recognizing line graphs of 3-uniform hypergraphs without multiple edges [15] are NP-complete.

Naik et al. [27] gave a finite induced subgraph characterization of graphs in L_3^ℓ whose vertex degrees are at least 69. We will see that this degree bound can be reduced to 19, and the corresponding list of subgraphs differs from the list in [27]. The proof is based on the theorems of Krausz [18] and Beineke [1], the Krausz-type characterization of the class L_r^ℓ [27] and the properties of graph cliques formulated in the previous section. We will discuss an analogous characterization of graphs in L_2^ℓ with vertex degrees at least 5. Also, we will show that a similar characterization is impossible for graphs in L_r^ℓ with vertex degrees at least c if $r>3$ and c is an arbitrary constant. In fact, this was conjectured in [27]. Nevertheless, such a finite characterization was obtained in [27] with a restriction on edge degrees. Moreover, we will describe a polynomial procedure for recognizing if $G \in L_3^\ell$ in the class of graphs with minimum degree $\delta(G) \geq 19$. Notice that efficient algorithms for recognizing the classes L_2 and L_2^ℓ are known [9, 19, 28, 31].

Let us remind ourselves of some basic definitions. An arbitrary complete subgraph of a graph is called a *clique*. The set of vertices of a clique may be called a *clique* as well. A maximal clique C is called *r-large* if

$$|C| \geq r^2 - r + 2.$$

Hence, 3- and 2-large cliques have orders at least 8 and 4, respectively.

A finite family $C = (C_i : i = 1, 2, \ldots, k)$ of cliques of a graph G is called a *covering* if G is the union of those cliques; each clique C_i is a *cluster* of the covering C. A covering is called *linear* if any two of its clusters have at most one common vertex. A covering is called an *r-covering* if each vertex of a graph belongs to at most r clusters. The neighbourhood of a vertex v in a graph G is denoted by $N_G(v)$, and $N_G[v] = N_G(v) \cup \{v\}$ is the closed neighbourhood of v. Also, the subgraph induced by the vertex set X is denoted by $G(X)$.

We will need the following theorems. Note that Theorem A is a particular case of Theorem 5.4 from the previous section.

Theorem A [34] *Any r-large clique of a graph is a cluster of each linear r-covering.*

Theorem B [27] *A graph belongs to the class L_r^ℓ if and only if the graph has a linear r-covering.*

5.2.2 Characterization of Graphs in L_3^ℓ with $\delta(G) \geq 19$

Let $V(G)$ and $E(G)$ denote the vertex and edge sets of a graph G, respectively. Define the sets $A_1, A_2, \ldots, A_8$ as follows.

- A_1 is the set of graphs of order 20 having a dominating vertex (i.e. a vertex adjacent to all other vertices) and not containing K_8.
- A_2 is the set of graphs of the form $K_9 - E(K_{1,m}), 1 \leq m \leq 4$ (m edges constituting a star are deleted from K_9).
- A_3 is the set of graphs that can be represented in the form

$$G = G_1 \cup G_2, \quad G_1 \simeq G_2 \simeq K_8, \quad 2 \leq |V(G_1) \cap V(G_2)| \leq 3.$$

- A_4 is the set of graphs that can be represented in the form

$$G = G_1 \cup G_2 \cup F, \quad G_1 \simeq G_2 \simeq K_8, \quad |V(G_1) \cap V(G_2)| = 2,$$

where $F = mK_2$ is the graph of order $2m$ with m pairwise disjoint edges $(1 \leq m \leq 6)$ and

$$V(F) \subseteq V(G_1 \cup G_2), \quad E(F) \cap E(G_1 \cup G_2) = \emptyset.$$

- $A_5 = \{K_{1,4}\}$.
- A_6, A_7 and A_8 are the sets of graphs that can be represented in the form of the union of graphs

$$G = B \cup G_1 \cup G_2 \cup \ldots \cup G_p \cup F$$

satisfying conditions I, II and III below, respectively, and such that any two of the graphs $B, G_1, G_2, \ldots, G_p, F$ have no common edge.

I. (a) G_i is a maximal clique of G, $i = 1, 2, \ldots, p$.
(b) $V(F) \subseteq V(G_1 \cup G_2 \cup \ldots \cup G_p)$.
(c) F does not contain an edge with both ends in B.
(d) $V(B) \cap V(G_i) \neq \emptyset$, $i = 1, 2, \ldots, p$.
(e) B is a Beineke graph (Figure 5.2).
(f) $8 \leq |G_i| \leq 11$, $i = 1, 2, \ldots, p$.
(g) Each vertex of B belongs to either one or two cliques G_i.

II. (a)–(b) The same as (a) and (b) from I.
(c) $B = P_3$ is the path with three vertices.
(d) $|G_i| = 8$, $i = 1, 2, 3$; $p = 3$.
(e) The central vertex of B belongs to G_1 and G_2 but not to G_3, and both endpoints belong to G_3 but not to G_1 and G_2.

III. (a)–(d) The same as (a)–(d) from I.
(e) B is one of the graphs shown in Figure 5.3 (if $B = R_j$ then we say that G has *type* j).
(f) $8 \leq |G_i| \leq 9$, $i = 1, 2, \ldots, p$.
(g) Each vertex of B belongs to one or two cliques G_i, and each vertex of degree 2 belongs to two cliques.

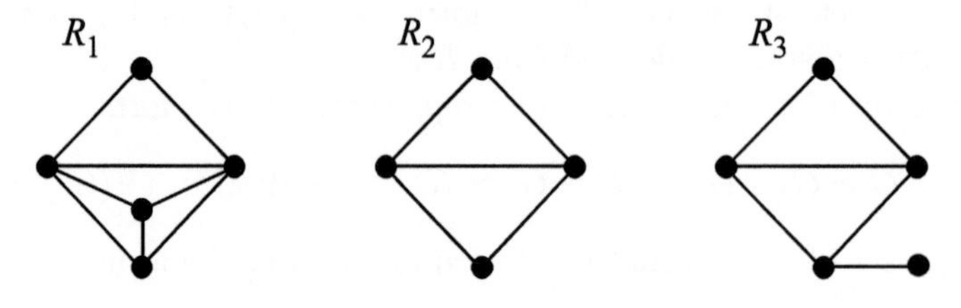

Figure 5.3 Graphs of item III (e).

For each $i = 1, 2, \ldots, 8$, the orders of graphs in A_i are restricted, hence A_i is finite. For example, $|G| = 14$ if $G \in A_4$. If $G \in A_6$, the conditions (d), (e) and (g) imply $p \leq 12$.

Taking the union of A_i, $1 \leq i \leq 8$, we obtain a finite list $\mathcal{A}$.

Theorem 5.6 [26] *For a graph G with $\delta(G) \geq 19$, the following two statements are equivalent:*

(i) $G \in L_3^\ell$;
(ii) None of the graphs from the list $\mathcal{A}$ is an induced subgraph of G.

Proof: First let us prove that (i) implies (ii). Obviously, L_3^ℓ is a hereditary class of graphs. Hence, it is sufficient to prove that

$$\mathcal{A} \cap L_3^\ell = \emptyset. \tag{5.18}$$

By Theorem B, the neighbourhood of every vertex in a graph in L_3^ℓ is divided into at most three cliques. Therefore,

$$(A_1 \cup A_5) \cap L_3^\ell = \emptyset.$$

Suppose that

$$G \in A_2 \cap L_3^\ell. \tag{5.19}$$

By Theorem B, there exists a linear 3-covering C of the graph G. Further, G contains a 3-large clique K_8 which must be a cluster of C by Theorem A. It is obvious that such a covering C cannot exist and, consequently, (5.19) is impossible.

In the same way, the cliques G_1 and G_2 must be clusters of every linear 3-covering for $G \in A_3 \cup A_4$. This is impossible because

$$|V(G_1) \cap V(G_2)| > 1.$$

Therefore,

$$(A_3 \cup A_4) \cap L_3^\ell = \emptyset.$$

Now, assume that

$$G = B \cup G_1 \cup G_2 \cup \ldots \cup G_p \cup F \in A_6 \cap L_3^\ell. \tag{5.20}$$

Fix a linear 3-covering C for G. By Theorem A, every clique G_i is a cluster of this covering and every vertex of G belongs to some cluster G_i. Let us delete the edges of each G_i and the resulting isolated vertices from G, and denote the constructed graph by H. The covering C becomes a linear 2-covering for H. By the Krausz theorem, $H \in L_2^\ell$, contrary to the Beineke theorem because B is an induced subgraph of H. Thus, (5.20) is impossible.

Finally, for $G \in A_7 \cup A_8$, every clique G_i must belong to each linear 3-covering. Obviously, such a covering cannot exist, and hence

$$(A_7 \cup A_8) \cap L_3^\ell = \emptyset.$$

Equality (5.18) is therefore proved.

Let us prove that (ii) implies (i). Suppose that $C = \{C_i : i = 1, 2, \ldots, k\}$ is the set of 3-large cliques of G, and denote $V_i = V(C_i)$. We proceed with the series of propositions.

Proposition 5.1 *Every vertex of G belongs to some clique C_i.*

Indeed, suppose that

$$v \in V(G), \quad U \subseteq N_G(v), \quad |U| = 19, \quad H = G(U \cup \{v\}).$$

Then $|H| = 20$ and v is a dominating vertex in H. Further, H contains the 3-large clique K_8 because $H \notin A_1$.

Proposition 5.2 *If $v \in V(G) - V_i$, then the vertex v has at most three neighbours in V_i.*

This is obvious, since otherwise G would contain an induced subgraph from A_2. In particular, Proposition 5.2 implies $|V_i \cap V_j| \leq 3$ for $i \neq j$.

Proposition 5.3 *We have $|V_i \cap V_j| \leq 1$ for $i \neq j$.*

This follows immediately from Proposition 5.2 because $G(V_i \cup V_j)$ contains an induced subgraph from A_3 or A_4 if

$$2 \leq |V_i \cap V_j| \leq 3.$$

Proposition 5.4 *If $v \in V_i \cap V_j$ for $i \neq j$, then*

$$N_G[v] = V_i \cup V_j \quad \text{or} \quad N_G[v] = V_i \cup V_j \cup F,$$

where F is a clique such that $F \cap (V_i \cup V_j) = \{v\}$.

Suppose to the contrary that av and bv are edges of G that do not belong to C_i and C_j, and a and b are not adjacent. By Proposition 5.2, there are vertices $c \in V_i$ and $d \in V_j$ which are adjacent to neither a, nor b nor each other. We have

$$G(\{v, a, b, c, d\}) = K_{1,4} \in A_5.$$

Proposition 5.4 implies, in particular, that each vertex of the graph G belongs to at most three cliques V_i. If there are exactly three such cliques, then the union of these cliques coincides with $N_G[v]$.

Now, delete from G the edges of all cliques C_i belonging to the set C, and delete also all isolated vertices if any appear. Denote the resulting graph by H.

Proposition 5.5 *Each vertex of H belongs to one or two cliques V_i.*

This follows from Propositions 5.1 and 5.4.

Proposition 5.6 *It is true that $H \in L_2^\ell$.*

Proof: It is sufficient to prove that H does not contain an induced subgraph forbidden by the Beineke theorem (see Figure 5.2). Suppose to the contrary that B is one of the Beineke graphs and

$$C_B = \{C_i : i = 1, 2, \ldots, q\}$$

is the set of cliques from C such that C_i contains vertices from B. Consider the set U_i of vertices $u \in V_i$ satisfying one of the following conditions:

- $u \in V(B)$;
- u belongs to two cliques from C_B.

Propositions 5.3 and 5.5 imply that $q \leq 12$ because $|B| \leq 6$. Hence $|U_i| \leq 11$ for $i = 1, 2, \ldots, q$. Consequently, for every C_i from C_B there is a clique G_i' with $8 \leq |G_i'| \leq 11$ containing the set U_i. Consider the induced subgraph

$$G' = G(V(G_1') \cup V(G_2') \cup \ldots \cup V(G_q')).$$

Obviously, it can be represented in the form

$$G' = B \cup G_1' \cup G_2' \cup \ldots \cup G_q' \cup F,$$

where

$$V(F) \subseteq V(G_1' \cup G_2' \cup \ldots \cup G_q'), \quad E(F) \cap E(B \cup G_1' \cup G_2' \cup \ldots \cup G_q') = \emptyset.$$

By Proposition 5.3, every clique G_i' is maximal in G_i' and $G' \in A_6$. Proposition 5.6 is proved. □

Proposition 5.7 *There exists a linear 2-covering D of the graph H such that $C \cup D$ is a linear 3-covering of G.*

Proof: Indeed, by the Krausz theorem and Proposition 5.6, there exists a linear 2-covering of the graph $H : D = (D_i : i = 1, 2, \ldots, s)$. Without loss of generality, suppose that D contains no one-vertex clusters. Taking Proposition 5.5 into account, we obtain that $Z = C \cup D$ is a linear 4-covering. Moreover, Z is a linear 3-covering if we choose D in a certain way. Indeed, suppose to the contrary that Z is not a linear 3-covering. This means that there is a vertex a belonging to four cliques, two from C, say C_1 and C_2, and two from D, say D_1 and D_2. Further, let $ab_1 \in E(D_1)$ and $ab_2 \in E(D_2)$. By Proposition 5.4, the vertices b_1 and b_2 are adjacent in G.

If $b_1 b_2 \notin E(H)$, then this edge belongs to some third clique $C_3 \in C$. Any of the three cliques C_i, $i = 1, 2, 3$, has at most one vertex in common with each of the remaining two cliques and contains at most two vertices from the set $\{a, b_1, b_2\}$. Therefore, each of the cliques C_i contains an 8-vertex clique G'_i such that

$$G(V(G'_1) \cup V(G'_2) \cup V(G'_3)) \in A_7,$$

a contradiction. Thus, $b_1 b_2 \in E(H)$. So, the cliques D_1 and D_2 together belong to some maximal clique K of H. If $|K| \geq 4$, then K is a cluster of the covering D by Theorem A. Therefore, $|K| = 3$, $V(D_i) = \{a, b_i\}$, $i = 1, 2$, and the edge $b_1 b_2$ belongs to some third cluster D_3 of the covering D.

Note that $|D_3| \leq 3$. If not, suppose that b_3 and b_4 are new vertices from D_3, $B = H(\{a, b_1, b_2, b_3, b_4\})$ and $C_B = \{C_1, C_2, \ldots, C_t\}$ is the set of cliques from C such that C_i contains vertices of the graph B. Using the same arguments as in the proof of Proposition 5.6, we obtain that each clique C_i from C_B contains a clique G'_i with $8 \leq |G'_i| \leq 9$ such that the induced subgraph

$$G(V(G'_1) \cup V(G'_2) \cup \ldots \cup V(G'_t))$$

is a graph of type 1 from the set A_8.

Now, let $V(D_3) = \{b_1, b_2, b_3\}$. Then, the vertex b_3 belongs to only one clique from the set C, since otherwise we can find in G an induced subgraph of type 2 from A_8 as above. In the same way, if H contains a new vertex b_4 adjacent to b_3, then, by the definition of the covering D, this vertex is adjacent to neither a, nor b_1, nor b_2. So, an induced subgraph of type 3 from A_8 can be found. Thus, b_3 belongs to exactly two clusters of the covering Z, namely, to D_3 and one cluster of C. Remove from D the clusters D_i, $i = 1, 2, 3$, and add the new ones K, $H(\{b_1, b_3\})$, $H(\{b_2, b_3\})$. We obtain a linear 2-covering D' of the graph H with a unique cluster containing the vertex a. Now, a belongs to exactly three clusters of the covering $C \cup D'$. Hence, the above operation 'corrected' the situation at the vertex a and did not 'worsen' the situation for any other vertex.

Analogously, for $V(D_3) = \{b_1, b_2\}$, eliminate from D the clusters D_i, $i = 1, 2, 3$, and add the new cluster K. The proof of Proposition 5.7 is complete. □

The truth of the implication (ii)⇒(i) in Theorem 5.6 follows from Proposition 5.7 and Theorem B. Theorem 5.6 is proved. □

5.2.3 Classes L_2^ℓ and $L_{r>3}^\ell$

Let us consider the class L_2^ℓ with vertex degrees at least 5. We will prove that this class can be characterized by a subset of the Beineke list.

Theorem 5.7 [26] *For a graph G with $\delta(G) \geq 5$, the following two statements are equivalent:*

(i) $G \in L_2^\ell$;

(ii) *G does not contain induced subgraphs isomorphic to the Beineke graphs B_1–B_6 in Figure 5.2.*

Proof: The fact that (i) implies (ii) is obvious, so we will prove that (ii) implies (i). Let $C = \{C_i : i = 1, 2, \ldots, k\}$ be the set of 2-large cliques of the graph G and $V_i = V(C_i)$. It is not difficult to see that the wheel W_5 (B_1 in Figure 5.2) is the only graph of order 6 that has a dominating vertex and does not contain the induced star $K_{1,3}$ (B_2 in Figure 5.2) and K_4. Hence, the following statement is true.

Proposition 5.8 *Every vertex of G belongs to some clique C_i.*

Since G does not contain the induced subgraphs B_3 and B_4, the following two propositions follow immediately.

Proposition 5.9 *If $v \in V(G) - V_i$, then the vertex v has at most two neighbours in V_i.*

Proposition 5.10 *We have $|V_i \cap V_j| \leq 1$ for $i \neq j$.*

Proposition 5.11 *If $v \in V_i$, then $N_G[v] = V_i$ or $N_G[v] = V_i \cup F$, where F is a clique such that $F \cap V_i = \{v\}$.*

Suppose to the contrary that it does not hold. Let $x, y \in N_G[v] - V_i$ and $xy \notin E(G)$. By Proposition 5.9, $|N(x) \cap V_i| \leq 2$ and $|N(y) \cap V_i| \leq 2$. Since $v \in N(x) \cap N(y)$, we obtain $|(N(x) \cup N(y)) \cap V_i| \leq 3$. It follows from the condition $|V_i| \geq 4$ that there exists a vertex $z \in V_i$ that is not adjacent to both x and y in G. Therefore, $G(v, x, y, z) \simeq K_{1,3}$.

If the union of all cliques C_i coincides with G, then C is a linear 2-covering of G. Otherwise, remove from G the edges of all cliques C_i and any resulting isolated vertices. Denote the constructed graph by H.

Proposition 5.12 *The graph H does not contain the path P_3 as an induced subgraph.*

Suppose that $V(P_3) = \{a, b, c\}$ and a is the central vertex of P_3 in H. By Proposition 5.10, $G(\{a, b, c\}) = K_3$. If $bc \notin E(H)$, then bc is an edge of some clique C_1 from C. The vertex a belongs to some other clique C_2 from C and, by Proposition 5.9, a is not adjacent to any vertex from C_1 differing from b and c. Again, by Proposition 5.9, since $|C_2| \geq 4$, C_2 contains the vertex f adjacent to neither b nor c. Let us fix two new vertices d and e in the clique C_1. At least one of these vertices must be adjacent to f, since otherwise $G(\{a, b, c, d, e, f\}) = B_5$. For instance, let f and d be adjacent. Then $G(\{a, b, c, d, f\}) = B_6$. Thus, $bc \in E(H)$.

By Proposition 5.12, each vertex of H belongs to exactly one maximal clique. Adding all such cliques to the set C, we obtain a linear 2-covering of the graph G. □

The following theorem confirms a conjecture from [27].

Theorem 5.8 [26] *For $r > 3$ and an arbitrary constant c, the class of all graphs G in L_r^ℓ with $\delta(G) \geq c$ cannot be characterized by a finite list of forbidden induced subgraphs.*

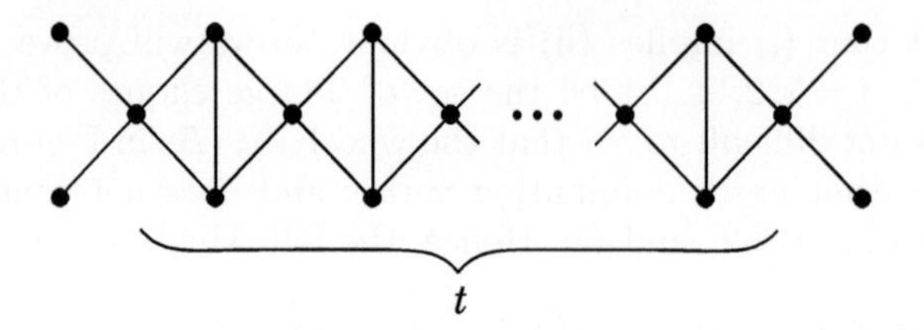

Figure 5.4 Graph G_t.

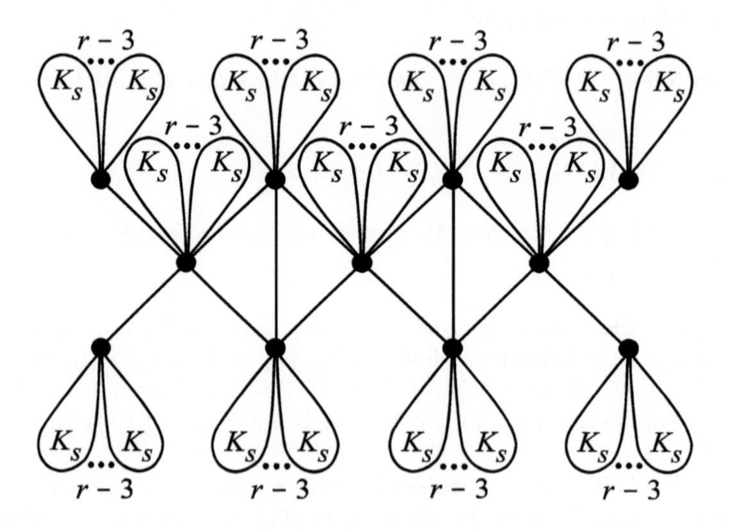

Figure 5.5 Graph H_2.

Proof: Suppose to the contrary that S is such a list. In [27], the infinite family of graphs shown in Figure 5.4 is given. In this figure, the graph $K_4 - e$ is repeated t times, where t is arbitrary. The graph G_t has the following property: $G_t \notin L_3^\ell$, but $G_t - v \in L_3^\ell$ for any vertex v.

Let us construct a new graph, which is based on G_t. We 'paste' $r-3$ pairwise disjoint copies of K_s, $s = \max\{r^2 - r + 2, c+1\}$, to each vertex v of G_t, so that v becomes the common vertex for all copies of K_s attached to it. Moreover, we require here that the copies of K_s 'pasted' to different vertices are pairwise disjoint (see Figure 5.5). Denote the resulting graph by H_t. Since $G_t \notin L_3^\ell$, it follows from Theorems A and B that $H_t \notin L_r^\ell$. But $\delta(H_t) \geq c$ and hence H_t contains the induced subgraph F_t from the list S. Further, if $v \in V(G_t)$, then $H_t - v \in L_r^\ell$ as $G_t - v \in L_3^\ell$. Thus, $V(G_t) \subseteq V(F_t)$ for each t. Since t is unbounded and the order of G_t grows with t, the list S cannot be finite. □

5.2.4 Recognition Algorithm

Let us consider the recognition problem

$$\text{`}G \in L_3^\ell\,?\text{'}, \quad \delta(G) \geq 19.$$

Algorithm 5.1: Recognition procedure for line graphs of linear 3-uniform hypergraphs with minimum degree at least 19.

Input: A graph G with $\delta(G) \geq 19$.

Output: $G \in L_3^\ell$ or $G \notin L_3^\ell$.

(1) First, construct a set $C \subseteq C^*$ of 3-large cliques of G covering $V(G)$; if $G \in L_3^\ell$ then such a set can be found. Delete the edges of the cliques belonging to C. Denote the resulting graph by G_1. Notice that every vertex belonging to exactly three cliques of C must be isolated in G_1.

(2) No four of the constructed cliques can have a common vertex, since otherwise we have a contradiction to Theorem A. If G_1 is the empty graph, then the set C is the required covering C^*, and we have $G \in L_3^\ell$.

(3) Consider the subset $V_1 \subseteq V(G)$ of vertices belonging to exactly two cliques from C. Let u be one of the vertices. The induced subgraph

$$G_u = G_1(N_{G_1}[u])$$

must be a clique. If $N_{G_1}(u) \neq \emptyset$, then add G_u to C as a new cluster and delete from G_1 the corresponding edges. Do the same for each u from V_1. Again, denote by C the constructed part of the sought-for covering C^* and by G_1 the remaining graph. If the new set C differs from the previous one, repeat steps (2)–(3).

(4) Each non-isolated vertex of the graph G_1 belongs to exactly one clique from C and, by Theorem A, this clique must be a cluster of any linear 3-covering. Therefore, $G \in L_3^\ell$ if and only if $G_1 \in L_2^\ell$. Delete all isolated vertices in G_1. Then, the sought-for covering C^* can be represented by the union $C^* = C \cup D$, where D is an arbitrary linear 2-covering of the graph G_1. The solution of the recognition problem '$G \in L_2^\ell$?' is known (see, for example, [31]).

(5) If one of the above conditions does not hold, then $G \notin L_3^\ell$.

By Theorem B, the answer is affirmative if we can construct a linear 3-covering C^* of the graph G. The construction procedure described above iteratively adds to the resulting covering just one cluster, simultaneously removing its edges from the graph. Eventually, it either constructs a Krausz 3-covering or produces an evidence that such covering does not exist. It is easy to see that this procedure is polynomial with respect to the order of a graph. Although this algorithm is generalized in the next section using a more sophisticated approach, it nicely illustrates the general idea.

5.3 Further Analysis of Krausz Coverings in L_3^ℓ

In this section, we will discuss an advanced machinery for analysing the geometry of Krausz coverings in the class L_3^ℓ of line graphs of linear 3-uniform hypergraphs.

This will enable us to show that the recognition problem for the class L_3^ℓ is polynomially solvable in the class of graphs with minimum degree $\delta(G) \geq 10$, thus extending the results described in the previous section. It is also proved that the class L_3^ℓ with $\delta(G) \geq 16$ can be characterized by a finite list of forbidden induced subgraphs. However, this list is implicit in contrast to the similar result from the previous section.

5.3.1 Reducing Families and Local Fragments

Let us remind ourselves of the basic notation: $u \sim v$ means that a vertex u is adjacent to a vertex v; and $u \sim X$ ($u \not\sim X$) means that a vertex u is adjacent to all (none) vertices of the set X. Also, $\mathrm{ecc}(v)$ is the eccentricity of a vertex v and $r(G)$ is the radius of the graph G. Denote by $B_k[a]$ and $S_k[a]$ the ball and the sphere in a graph H of radius k with centre in the vertex a. We will write B_k and S_k if it is clear which vertex is meant.

Let F be a family of cliques of a graph G. The cliques from F are called *clusters of F*. Denote by $V(F)$ and $E(F)$ the sets of vertices and edges covered by F, respectively. Also, denote by $\mathrm{cl}_F(v)$ the number of cliques from F covering the vertex v. For the sake of brevity, a *Krausz covering* further refers to a linear 3-covering.

A family F is called

(1) *a fragment of a Krausz covering* (or simply a *fragment*) if there exists a Krausz covering Q of G such that $F \subseteq Q$;
(2) *fundamental* if it satisfies the following condition: F is a fragment if and only if $G \in L_3^\ell$;
(3) *a reducing family* if it is fundamental and $\mathrm{cl}_F(v) = 1$ or $\mathrm{cl}_F(v) = 3$ for any vertex $v \in V(G)$.

Let F be a reducing family and $H = G - E(F)$. If there exists a vertex v such that $\mathrm{cl}_F(v) = 3$ and $N_H(v) \neq \emptyset$, then $G \notin L_3^\ell$. Otherwise, $G \in L_3^\ell$ if and only if $H \in L_2^\ell$, and if R is a Krausz 2-covering of H, then $F \cup H$ is a Krausz 3-covering of G. So, the reducing family indeed allows to reduce the problem '$G \in L_3^\ell$?' to the problem of recognizing line graphs, which is polynomially solvable.

The idea of the algorithm for solving the problem '$G \in L_3^\ell$?' for graphs with $\delta(G) \geq 10$ is the following. We will build a reducing family starting from some initial family of cliques by adding a new cluster at each iteration of the algorithm. The main purpose of further considerations is to show which cliques could be added. Recall that a maximal clique of order at least $k^2 - k + 2$ of a graph G is called a k-large clique; a 3-large clique will simply be called a *large clique*.

Let G be an arbitrary graph from L_3^ℓ, Q be its Krausz covering, $F \subset Q$ be some fragment ($F = \emptyset$ is possible) and $H = G - E(F)$. Denote by $E_k(a)$ the set of edges in H with at least one end in $B_{k-1}[a]$. If $\deg(v) \geq 19$ for some vertex $v \in V(H)$, then v should be contained in some large clique. According to Theorem A from Section 5.2, we may assume that all large cliques of G belong to F. This implies

$\omega(H) \leq 7$ and $\Delta(H) \leq 18$, where $\omega(H)$ is the density and $\Delta(H)$ is the maximum vertex degree of H.

A clique covering $F_k(c)$ of $E_k(c)$ is called a *local fragment of radius k with centre in c with respect to F*, or a *(c,k)-local fragment* or simply a *local fragment* if

(1) any two cliques from $F_k(c)$ have no common edge;
(2) every vertex $v \in B_k[c]$ is covered by at most 3 cliques from $F \cup F_k(c)$.

Sometimes, we will write F_k if the centre is clear from the context or is not important. The following properties of local fragments are evident:

(i) $\bigcup_{C_i \in F_k(c)} C_i = B_k[c]$.
(ii) For every vertex c and any $k \in \mathbb{N}$, the Krausz covering Q induces the local fragment $F_k(c)$.
(iii) If $k = \text{ecc}(c) + 1$, then $F \cup F_k(c)$ is a fragment.
(iv) Any local fragment $F_k(c)$ with radius k uniquely determines the decreasing sequence

$$F_k(c) \supset F_{k-1}(c) \supset \ldots \supset F_1(c) \tag{5.21}$$

of local fragments with radii $k, k-1, \ldots, 1$. We call $F_1(c)$ the *groundwork* of the local fragment $F_k(c)$.

Clusters of local fragments have the following property:

Lemma 5.6 *Let C be a cluster of a local fragment F_i from the sequence (5.21), $i < k$, and let v be a vertex such that $|C \cap N_H(v)| \geq 4$. Then $v \in C$. In particular, if $|C| \geq 4$, then the cluster C is a maximal clique.*

We call a clique $C \subseteq B_k[c]$ *special* if it is a cluster of any local fragment $F_{k+1}(c)$. The maximal cliques of order 6 and 7 are called *prelarge*. The clique $C \subset V(H)$ *touches* the fragment F if C contains a vertex covered by F. The above property (ii) implies the next lemma.

Lemma 5.7 *If C is a special clique, then $F \cup \{C\}$ is a fragment.*

Let us describe some special cliques.

Lemma 5.8 *The following cliques are special:*

(1) Any prelarge clique of H that touches the fragment F is special.

(2) Let C be a maximal clique of H and $v \in V(H) - C$ such that

$$|C \cap N_H(v)| \geq 1 \quad \textit{and} \quad |C - N_H(v)| \geq 5.$$

Then C is special.

Proof: We only prove (2), the proof of (1) is similar. Assume that there exists a local fragment F_k such that $C \notin F_k$. Then, a vertex $b \in C \cap N_H(v)$ should be covered by three clusters

$$C_1, C_2, C_3 \in F_k, \quad v \in C_3, \quad C_i \cap C \neq \emptyset, \quad i = 1, 2.$$

Without loss of generality, $|C_1| \geq 4$ and for any vertex x from $C_2 - \{b\}$, we have $|C_1 \cap N_H(x)| \geq 4$. This contradicts Lemma 5.6. □

The clique and the vertex in item (2) of Lemma 5.8 are called a *good clique* and a *good vertex*, respectively.

Suppose that $k \geq 3$ and there exist two sequences of local fragments:

$$F_{k+1} \supset F_k \supset \ldots \supset F_1, \quad F'_{k+1} \supset F'_k \supset \ldots \supset F'_1. \tag{5.22}$$

Let $W \subset H$. Denote by $D_W(v)$ ($D'_W(v)$) the number of clusters of F_{k+1} (F'_{k+1}) covering all edges between v and W. In particular, for $v \in S_i$,

$$D_-(v) = D_{S_{i-1}}(v), \quad D_+(v) = D_{S_{i+1}}(v).$$

The values $D'_-(v)$ and $D'_+(v)$ are defined analogously for F'_{k+1}.

Now, let S_{k-2} be the disjoint union of two sets:

$$S_{k-2} = P_{k-2} \cup Q_{k-2}, \quad P_{k-2} \cap Q_{k-2} = \emptyset.$$

Consider the following sets:

$$P_{k-1} = N_{S_{k-1}}(P_{k-2}), \quad Q_{k-1} = S_{k-1} - P_{k-1},$$

$$P_k = N_{S_k}(A_{k-1}) \quad \text{and} \quad Q_k = S_k - P_k.$$

We put

$$\widetilde{F}_k = F_{k-1} \cup \{C \in F_k : C \cap P_{k-1} \neq \emptyset\},$$

$$\widetilde{F}'_k = F'_{k-1} \cup \{C' \in F'_k : C' \cap P_{k-1} \neq \emptyset\}.$$

Lemma 5.9 *Let the following conditions hold:*

(1) $D_{P_k \cup Q_{k-1}}(v) \leq 1$ and $D'_{P_k \cup Q_{k-1}}(v) \leq 1$ for any vertex $v \in P_{k-1}$;

(2) $D_{Q_{k-1}}(u) \leq 1$ and $D'_{Q_{k-1}}(u) \leq 1$ for any vertex $u \in Q_{k-2}$;

(3) $\deg(\alpha) \geq 9$ for any vertex $\alpha \in P_k \cup Q_{k-1}$;

(4) there is neither large nor good clique in $S_{k-1} \cup S_k \cup S_{k+1}$.

Then, $F \cup \widetilde{F}_k$ *is a fragment if and only if* $F \cup \widetilde{F}'_k$ *is a fragment.*

Proof: Let $F \cup \widetilde{F}_k$ be a fragment and X be a list of cliques complementing it to some Krausz covering Q. Let us show that $F \cup \widetilde{F}'_k \cup X$ is also a Krausz covering of G. Evidently,

$$V(\widetilde{F}_k) = V(\widetilde{F}'_k) = B_{k-1} \cup P_k.$$

Let $xy \in E(H)$. The following three statements are true:

(a) If $x \in P_{k-1}$, then xy is covered by $\widetilde{F}_k$.

(b) If $x \in P_k$ and $y \in P_k \cup Q_{k-1}$, then xy is covered by $\widetilde{F}_k$ if and only if there exists a vertex $\alpha \in P_{k-1}$ such that a triple x, y, α induces a triangle in H. (This follows from the condition (1) of the lemma).

(c) If $x, y \in Q_{k-1}$, then xy is covered by $\widetilde{F}_k$ if and only if there exists a vertex $\beta \in P_{k-1} \cup Q_{k-2}$ such that x, y, β induces a triangle in H. (This follows from the conditions (1) and (2)).

These statements are also true for $\widetilde{F}'_k$. Thus, $\widetilde{F}_k$ and $\widetilde{F}'_k$ cover the same sets of vertices and edges.

Now, it is evident that the statement of the lemma could be false only if there exists a vertex $\alpha \in P_k \cup Q_{k-1}$ that belongs to exactly one cluster $C \in \widetilde{F}_k$ and exactly two clusters $C'_1, C'_2 \in \widetilde{F}'_k$. Clearly,

$$C = \{\alpha\} \cup \{x \in V(H) : \alpha x \in E(\widetilde{F}_k)\},$$

$$C'_1 \cup C'_2 = \{\alpha\} \cup \{x \in V(H) : \alpha x \in E(\widetilde{F}'_k)\}.$$

Then, the equality $E(\widetilde{F}_k) = E(\widetilde{F}'_k)$ implies $C = C'_1 \cup C'_2$. According to Lemma 5.6, the last equality implies $|C'_i| \leq 3$, $i = 1, 2$. Therefore, by virtue of the condition (3), there exists a third cluster C'_3 covering α, and $|C'_3| \geq 6$.

Let $u \in Q_{k-2} \cup P_{k-1}$ and $u \sim \alpha$. Without loss of generality, $u \in C'_1$. Then, the conditions (1) and (2) imply

$$u \nsim C'_3 - \{\alpha\}. \tag{5.23}$$

Because of the condition (4), the clique C'_3 cannot be large. Therefore, C'_3 is prelarge. But then (5.23) implies that C'_3 is a good clique. This contradicts (4). The lemma is proved. □

Assuming $Q_{k-2} = \emptyset$, we obtain:

Corollary 5.13 *Let the following conditions hold:*

(1) $D_+(v) \leq 1$ *and* $D'_+(v) \leq 1$ *for any vertex* $v \in S_{k-1}$;

(2) $\deg(\alpha) \geq 9$ *for any vertex* $\alpha \in S_k$;

(3) there are neither large nor good cliques in $S_{k-1} \cup S_k \cup S_{k+1}$.

Then, $F \cup F_k$ *is a fragment if and only if* $F \cup F'_k$ *is a fragment.*

5.3.2 Prelarge Cliques

Denote by $C(x_1, x_2, \ldots, x_r)$ the cluster of a local fragment $F_k(a)$ containing vertices $x_1, x_2, \ldots, x_r$ and, perhaps, some other vertices; for $F'_k(a)$, the cluster $C'(x_1, x_2, \ldots, x_r)$ is defined analogously.

Lemma 5.10 *Let C be some prelarge clique of H and $a \in C$. Assume that* $\deg(a) \geq 7$ *and there is no special clique in H. Then, for any $k \geq 2$, there exists an (a, k)-local fragment $F_k(a)$ containing C.*

Proof: Assume that $C \supseteq \{a, b, c, d, e, f\}$ and there is no local fragment containing C. Since there is no special clique in H, there exist two local fragments F_k and F'_k with different groundworks F_1 and F'_1. In these groundworks, C is partitioned differently into three clusters. Let us consider possible variants of such partitions.

(1)

$$F_1 = \{C(a,b,c), C(a,d,e), C(a,f)\},$$

$$F'_1 = \{C'(a,b,c), C'(a,d,e), C'(a,f)\}.$$

Since there are no special cliques, we may assume that $C(a,b,c) \neq C'(a,b,c)$. Suppose that there exists $x \in C'(a,b,c) - C(a,b,c)$. Then $x \in C(a,d,e) \cup C(a,f)$, say $x \in C(a,f)$. We have $f \sim \{a,b,c,x\}$, so f is adjacent to four vertices from $C'(a,b,c)$. But in this case $f \in C'(a,b,c)$, which contradicts the definition of the local fragment.

(2)

$$F_1 = \{C(a,b,c), C(a,d,e), C(a,f)\},$$

$$F'_1 = \{C'(a,d,c), C'(a,b,e), C'(a,f)\}.$$

Without loss of generality, assume that $C(a,f) \neq C'(a,f)$. Further arguments are the same as in (1).

(3)

$$F_1 = \{C(a,b,c), C(a,d,e), C(a,f)\},$$

$$F'_1 = \{C'(a,f,c), C'(a,d,e), C'(a,b)\}.$$

Without loss of generality, $C'(a,d,e) \neq C(a,d,e)$. Further consideration is similar to the case (1).

(4)

$$F_1 = \{C(a,b,c), C(a,d,e), C(a,f)\},$$

$$F'_1 = \{C'(a,d,c), C'(a,f,e), C'(a,b)\}.$$

Assume that there is $x \in C(a,b,c)$. It is evident that $x \notin C'(a,c,d)$. Indeed, if it is not true, then $d \sim \{a,b,c,x\}$. In other words,

$$|N(d) \cap C(a,b,c)| \geq 4,$$

which is impossible by Lemma 5.6. The vertex c is also covered by the clusters $C'(c,e)$ and $C'(c,f)$ from $F_2' - F_1'$. Therefore, x is contained in one of these clusters, say $x \in C'(c,e)$. But then $e \sim \{a,b,c,x\}$. This contradicts Lemma 5.6. So, $C(a,b,c) = \{a,b,c\}$. Analogously,

$$C(a,d,e) = \{a,d,e\}, \quad C'(a,d,c) = \{a,d,c\}, \quad C(a,f,e) = \{a,f,e\}.$$

Since $\deg(a) \geq 7$, there exist vertices $x, y \in N(a) - \{b,c,d,e,f\}$. We have

$$x, y \in C(a,f) \cap C'(a,b).$$

But in this case, b is adjacent to four vertices of $C(a,f)$. This contradicts Lemma 5.6.

Thus, the cases (1)–(4) are impossible. This proves Lemma 5.10. □

A vertex $v \in S_i$ is called a *deadlock* if $N(v) \cap S_{i+1} = \emptyset$. The following lemma is obvious.

Lemma 5.11 *Let (5.21) be the sequence of local fragments determined by the local fragment* F_k, *and* $i < k$. *If* $D_-(v) = 3$ *for the vertex* $v \in S_i$, *then* v *is a deadlock.*

Our next result is about cliques of size 7.

Lemma 5.12 *If* C *is a prelarge clique of* H *and* $|C| = 7$, *then* $F \cup \{C\}$ *is a fragment.*

Proof: The lemma is true if the clique C is either good, special or a connected component of H. Thus, taking into account Lemma 5.10, it is sufficient to consider the situation when C is not good and there are two different local fragments of radius 2 such that C is a cluster in exactly one of those fragments.

Let (5.22) be two sequences of local fragments with centre in $a \in C$ of radius 2, $C \in F_2$ and $C \notin F_2'$. It is easy to prove that

$$\mathrm{ecc}(a) = 2. \tag{5.24}$$

Indeed, we have $C \in F_1$ and $C \notin F_1'$. Then, by Lemma 5.6,

$$F_1' = \{C_1', C_2', C_3'\}; \;\; a \in C_i', \;\; |C_i' \cap C| = 3, \;\; i = 1,2,3; \;\; \bigcup_{i=1}^{3} C_i' = B_1. \tag{5.25}$$

Every vertex $s \in S_1$ is covered by one of the clusters C_i'. If, for example, $s \in C_1' = C'(a,b,c)$ and $a,b,c \in C$, then the edges sa, sb, sc belong to three different clusters

$$C(s,a), C(s,b), C(s,c) \in F_2.$$

Now, if $v \in S_2$ and $v \sim s$, then v is contained in one of the above clusters. Hence, for every vertex $v \in S_2$, we have $N(v) \cap C \neq \emptyset$. But the clique C is not good and $|C| = 7$. Therefore, $|N(v) \cap C| \geq 3$. Thus, $D_-(v) = 3$ and, by Lemma 5.11, the vertex v is a deadlock. The equality (5.24) is proved.

Thus, by property (iii) of local fragments, $F \cup F_2$ is a fragment. Therefore, $F \cup \{C\}$ is a fragment too. The lemma is proved. □

We will need the following technical lemma.

Lemma 5.13 *Let H contain no clique of size at least 7, and F_k, F'_k be local fragments. Further, let $C_1, C_2 \in F_k$ be a pair of cliques covering a vertex v with* $\deg(v) \geq 10$, *and $C' \in F'_k$. Then*

$$C_1 \cup C_2 \not\subseteq C'.$$

Proof: Let $C_1 \cup C_2 \subseteq C'$. Then $|C_i| \leq 3$, $i = 1, 2$. Therefore, the vertex v is covered by the cluster C_3 with $|C_3| \geq 7$, a contradiction. The lemma is proved. □

The following lemma is about cliques of size 6.

Lemma 5.14 *Let the following conditions hold:*

(1) for a vertex $a \in V(H)$, there exists a local fragment $F_4(a)$ that determines the sequence of local fragments

$$F_4(a) \supset F_3(a) \supset F_2(a) \supset F_1(a);$$

(2) there exists a cluster $C \in F_1(a)$ such that $|C| = 6$;

(3) in $B_4[a]$, there are neither good cliques nor cliques of size at least 7;

(4) the minimum degree $\delta(G) \geq 10$.

Then, $F \cup \{C\}$ is a fragment.

Proof: The required statement is evident if C is special. If C is not special, then there exists a local fragment F'_4 such that $C \notin F'_4$. Let $C = \{a, b, c, d, e, f\}$. By Lemma 5.6, without loss of generality, one may assume that the vertices of C belong to the following clusters:

$$C'(a, b, c), C'(a, d, e), C'(a, f) \in F'_3,$$

and there are also clusters

$$C'(f, b, d), C'(f, c, e), C'(b, e), C'(c, d) \in F'_3. \tag{5.26}$$

Some clusters are represented by bold lines in Figure 5.6.

Let us consider the following sets:

$$P_1 = C \cup N_{S_1}(C - \{f\}), \quad Q_1 = S_1 - P_1, \quad P_2 = N_{S_2}(P_1),$$
$$Q_2 = S_2 - P_2, \quad P_3 = N_{S_3}(P_2), \quad Q_3 = S_3 - P_3.$$

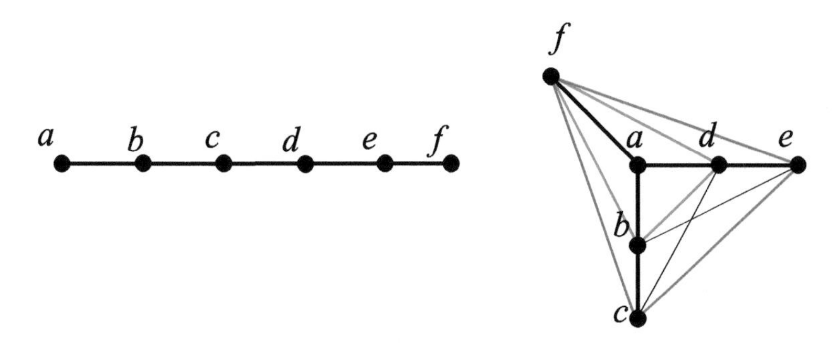

Figure 5.6 Clusters in the proof of Lemma 5.14.

It is sufficient to prove that these sets satisfy the conditions of Lemma 5.9.

It is easy to see that $P_2 = N_{S_2}(C)$. Indeed, let $v \in P_2$, $v \sim g \in P_1 - C$. Without loss of generality, $g \in C'(a, b, c)$. Then, g belongs to the clusters $C(g, a)$, $C(g, b)$, $C(g, c) \in F_2$ and, therefore, v belongs to one of these clusters.

Notice also that if $|N(v) \cap C| \geq 3$ for the vertex $v \in B_2 - C$, then $N_{P_3 \cup Q_2}(v) = \emptyset$. For example, if $v \sim \{b, d, f\}$, then v is covered by the clusters $C(v, b)$, $C(v, d)$ and $C(v, f)$. Hence, any vertex of $N_{P_3 \cup Q_2}(v)$ belongs to one of these clusters. This contradicts the definition of P_3 and Q_2.

Note the following facts:

Fact 1 *For any vertex* $v \in P_2$, $D_{P_3 \cup Q_2}(v) \leq 1$ *and* $D'_{P_3 \cup Q_2}(v) \leq 1$.

Proof of Fact 1: If $v \sim f$, then, without loss of generality, one may assume that $v \in C'(f, b, d)$. Hence, $|N(v) \cap C| \geq 3$ and, consequently, $N(v) \cap (P_3 \cup Q_2) = \emptyset$. If $v \nsim f$ and $v \in C'(b, e)$, then $v \in C(v, b) \cap C(v, e)$ and hence $D_{P_3 \cup Q_2}(v) \leq 1$.

Now, suppose that $D'_{P_3 \cup Q_2}(v) = 2$, in other words,

$$v \in C'(v, x) \cap C'(v, y), \quad x, y \in P_3 \cup Q_2.$$

Then, there exists a cluster $C(v, x, y) \in F_3$. By Lemma 5.6,

$$C'(v, x) \cup C'(v, y) \nsubseteq C(v, x, y).$$

Therefore, one may assume that there exists $z \in C'(v, x) \cap C(v, b)$. We obtain $z \in C'(a, b, c) \cup C'(b, d, f)$. So, z is adjacent to three vertices of C and hence $x \notin P_3 \cup Q_2$. This contradiction proves Fact 1. □

Fact 2 *For any vertex* $v \in Q_1$, $D_{Q_2}(v) \leq 1$ *and* $D'_{Q_2}(v) \leq 1$.

Proof of Fact 2: Evidently, $v \in C'(a, f)$ and $v \in C(v, a) \cap C(v, f)$. Therefore, $D_{Q_2}(v) \leq 1$. Now, assume that $D'_{P_3 \cup Q_2}(v) = 2$. Similar to the proof of Fact 1,

$$v \in C'(v, x) \cap C'(v, y), \quad v \in C(v, x, y), \quad x, y \in Q_2$$

and there exists a vertex $z \in C'(v,x) - C(v,x,y)$. Therefore, $z \in C(v,a) \cup C(v,f)$. This implies

$$z \in C'(f,b,d) \cup C'(f,c,e) \cup C'(a,b,c) \cup C'(a,d,e).$$

Therefore, $|N(z) \cap C| \geq 3$, and so $x \notin Q_2$, a contradiction. Fact 2 and at the same time Lemma 5.14 are proved. □

Thus, by the above lemmas, we can extend the family F if there exists a prelarge clique in the graph H. Hence, the problem is reduced to the case when $\omega(H) \leq 5$.

In what follows, the vertex a is covered by the fragment F.

Lemma 5.15 *Assume that $D_-(v) \geq 2$ for $v \in S_k$ and for any (a,k)-local fragment. If there is neither special nor prelarge clique in S_{k+2}, then $D_+(s) \leq 1$ for any vertex $s \in S_{k+1} \cap N(v)$ and for any $(a,k+1)$-local fragment.*

Proof: Let (5.22) be two sequences of local fragments. By assumption, v is covered by the clusters $C_1, C_2 \in F_{k-1}$ and $C_1', C_2' \in F_{k-1}'$. Suppose also that v is covered by the clusters C_3 and C_3' containing s. Since there is no special cliques, we can assume that $C_3 \neq C_3'$, that is, there exists $r \in C_3 \cap C_1'$. Hence, $s \in C_3' \cap C'(s,r)$ and $D_-'(s) \geq 2$. Consequently, $D_+'(s) \leq 1$.

It remains to show that the statement of the lemma holds for F_{k+1}. If there exists $u \in C_3' - C_3$, then $D_-(s) \geq 2$. Further, suppose that

$$C_3' \subseteq C_3. \tag{5.27}$$

Then $C_3' \supseteq N(r)$ and, by Lemma 5.6, $|C_3'| \leq 3$. Since $\deg(v) \geq 10$, we have

$$|C_1'| + |C_2'| + |C_3'| \geq 13.$$

Therefore, $|C_1'| + |C_2'| \geq 10$. By assumption, there are no prelarge cliques in H. Hence, $|C_1'| = |C_2'| = 5$. Now, suppose that the vertex s is also contained in clusters $C_4, C_5 \in F_{k+1}$ and $C_4' \in F_{k+1}'$. The same consideration as above implies $|C'(s,r)| = |C_4'| = 5$. If $C_4' = \{s,x,y,z,t\}$ and, for example, $x \in C_3$, then $x \sim v$. So, $x \in C_1' \cup C_2'$. Consequently, $C_4' \in F_k'$, that is, s is a deadlock. In this case, the statement of the lemma evidently is true.

Taking into account Lemma 5.6, we may assume that $x, y \in C_4$ and $z, t \in C_5$. Suppose also that $C_1' = \{v,r,b,c,d\}$ and $C_2' = \{e,f,g,h\}$ (see Figure 5.7). By Lemma 5.13, there exists $p \in (C_4 \cup C_5) - C_4'$ and, by (5.27), $p \notin C_3'$. Let us assume $p \in C'(s,r) \cap C_4$. If s is adjacent to some vertex of the set $\{b,c,d\}$, then s is a deadlock. The same is true if s is adjacent to any two vertices of the set $\{e,f,g,h\}$. Further, suppose that $N(s) \cap C_1' = \{r,v\}$ and $N(s) \cap C_2' = \{v\}$. Taking into account Lemma 5.6 and the considerations described, assume that $b,c,e \in C_1$. Then, because $r \sim \{b,c,p\}$, we can suppose that $p \in C(r,b)$. Therefore, $p \sim b$. For p, we know that

$$p \in C'(s,r,p) \cap C'(p,x) \cap C'(p,y).$$

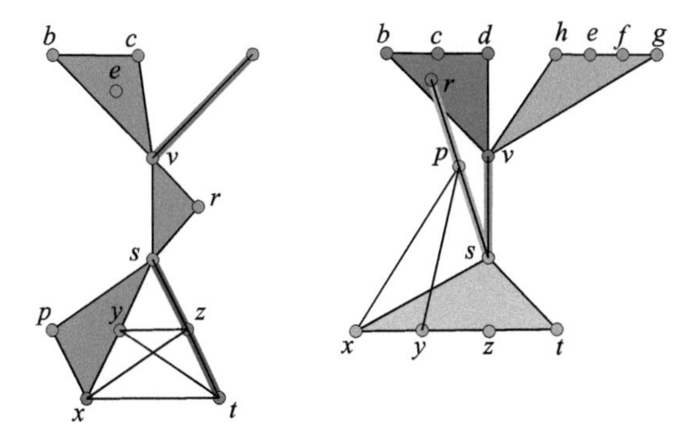

Figure 5.7 Clusters are represented as maximal line segments or filled polygons; not every vertex of each cluster is shown.

If $b \in C'(p, x)$, then $b \sim x$ and, consequently, one may assume that $b \in C(x, z)$, that is, $b \sim z$. Hence, we have

$$b \in C'(b, v) \cap C'(b, x) \cap C'(b, z).$$

Moreover, $b \sim e$. If, for example, $e \in C'(b, x)$, then $e \sim x$ and so $e \in C(x, t)$. Indeed, if $e \in C(x, z, b)$ then $C(x, z, b) \cap C_1 \supseteq \{b, e\}$. We obtain $x \sim b$, $z \sim b$, $x \sim e$ and $t \sim e$. Thus, $x, z, t \in B_{k+1}$. Therefore, the only vertex in $N(s) \cap S_{k+1}$ is y and the statement of the lemma is true. □

Lemma 5.16 [24] *Let $b \in S_1$ be not a good vertex and*

$$F_1 \subset F_2 \subset F_3 \tag{5.28}$$

be a sequence of local fragments. If there exists a cluster $C \in F_2 - F_1$ such that $b \in C$, $|C \cap S_2| \geq 4$ and C is not a special clique, then

$$D_-(s) \geq 2 \tag{5.29}$$

for any $s \in S_3 \cap N(v)$ where $v \in S_2 \cap N(b)$.

We say that the vertex $b \in S_1$ *satisfies the 7-condition* if $|N(b) \cap S_2| \geq 7$. The vertex $v \in S_2$ is called *appropriate* if it is adjacent to some vertex satisfying the 7-condition. Lemma 5.16 implies the following corollary.

Corollary 5.14 [24] *If there is no good vertex in B_1 and there is no special clique in B_2, then (5.29) is true for any vertex $s \in S_3$ adjacent to some appropriate vertex and for any sequence (5.28) of local fragments.*

Put $p = |B_1|$ and note that $p \leq 9$ if there are no good cliques in B_1.

Theorem 5.9 [32] *Let $G \in L_k^\ell$ and $\delta(G) \geq 10$. Assume that there are neither special nor prelarge cliques in $B_k[a]$, where*

$$k = \begin{cases} 3 & \text{if } p = 8 \text{ or } 9; \\ 4 & \text{otherwise.} \end{cases}$$

Then, for any $(a, k+1)$-local fragment $F_{k+1}(a)$ with the groundwork $F_1(a)$, the set $F_1(a) \cup F$ is a fragment.

Proof: Since no local fragment $F_1(a)$ contains cliques of size at least 6, we obtain for $p \geq 6$:

$$F_1(a) = \{C_1, C_2\}, \ |C_i| \leq 5, \ i = 1, 2, \ p \leq 9. \tag{5.30}$$

Because there are no special cliques in $B_2[a]$, there exist two local fragments of radius 3 with different groundworks. Let

$$F_1(a) \subset F_2(a) \subset F_3(a), \quad F_1'(a) \subset F_2'(a) \subset F_3'(a) \tag{5.31}$$

be two sequences of local fragments with $F_1(a) \neq F_1'(a)$, and

$$F_1'(a) = \{C_1', C_2'\}, \ |C_i'| = p_i', \ |C_i| = p_i, \ i = 1, 2. \tag{5.32}$$

We have $p_i \leq 5$ and $p_i' \leq 5$. Now, let v be an arbitrary vertex from $S_2(a)$. Let us consider the cases (1)–(6) corresponding to different values of p. We will show that $D_-(v) \geq 2$ for any vertex $v \in S_2(a)$ if $p = 8$ or 9. Then, the statement of the theorem will follow from Corollary 5.13 for $k = 3$.

(1) It is evident that if $p = 9$, then $(p_1, p_2) = (p_1', p_2') = (5, 5)$. Let

$$C_1 = \{a, b, c, d, e\}, \quad C_2 = \{a, f, g, h, i\}. \tag{5.33}$$

Taking into account Lemma 5.6, we may assume without loss of generality that

$$C_1' = \{a, b, c, f, g\}, \quad C_2' = \{a, d, e, h, i\}. \tag{5.34}$$

Further, let us turn to 2-local fragments. It follows from the definition of local fragments and (5.33) and (5.34) that $F_2'(a)$ covers all edges bd, be, cd, ce, fh, fi, gh, gi, whereas $F_2(a)$ covers the edges bf, bg, cf, cg, dh, di, eh, ei. Therefore, there are clusters

$$C(b,f),\ C(b,g),\ C(c,f),\ C(c,g),\ C(d,h),\ C(d,i),\ C(e,h),\ C(e,i);$$

$$C'(b,d),\ C'(b,e),\ C'(c,d),\ C'(c,e),\ C'(f,h),\ C'(f,i),\ C'(g,h),\ C'(g,i).$$

There are no other clusters in $F_2(a)$ and $F_2'(a)$ because every vertex of $S_1[a]$ is contained in exactly three clusters. Hence,

$$\begin{aligned} F_2(a) - F_1(a) &= \{C(b,f),\ C(b,g),\ C(c,f),\ C(c,g),\ C(d,h),\ C(d,i), \\ &\quad\ C(e,h), C(e,i)\}, \\ F_2'(a) - F_1'(a) &= \{C'(b,d),\ C'(b,e),\ C'(c,d),\ C'(c,e),\ C'(f,h), C'(f,i), \\ &\quad\ C'(g,h),\ C'(g,i)\}. \end{aligned}$$

The local fragments $F_2(a)$ and $F_2'(a)$ are shown in Figure 5.8. It is easy to see that

$$D_-(v) \geq 2 \quad \text{and} \quad D'_-(v) \geq 2. \tag{5.35}$$

Suppose that $v \in C(b,f)$. Because $\{b,f\} \subseteq C_1'$, we have $D'_-(v) \geq 2$. Further, v is contained in one of the clusters $C'(b,d)$, $C'(b,e)$, but $\{b,d\}$, $\{b,e\} \subseteq C_1$. Hence, $D_-(v) \geq 2$. Analogous reasonings can be used when v belongs to other clusters.

(2) In the case $p=8$, the arguments are similar (see Figure 5.9). If $v \sim h$, the vertex v belongs to one of the clusters $C(h,d)$, $C(h,e)$, and also to $C'(h,f)$ or $C'(h,g)$. If $v \in C(h,d) \cap C'(h,f)$, then the vertex v is covered by one of the clusters $C'(d,b)$, $C'(d,c)$, and by one of the clusters $C(f,b)$, $C(f,c)$. Hence, the inequality (5.35) holds. If $v \not\sim h$, one of the following statements is true: $v \sim d$, $v \sim e$, $v \sim f$, $v \sim g$. If, for example, $v \sim d$, then the vertex v is covered by one of the clusters $C'(d,b)$, $C'(d,c)$. If, for example, $v \in C'(d,b)$, then $v \sim b$ and hence $v \in C(b,f)$ or $v \in C(b,g)$. However, in $F_2(a) - F_1(a)$ there must exist the cluster $C(d,v)$ because $v \notin C(d,h)$. Therefore, $D_-(v) \geq 2$. Now, if $v \in C(b,f)$, then $v \sim f$. Consequently, $v \in C'(f,v)$ and $D'_-(v) \geq 2$.

Further, for $p \leq 7$, we will show that $D_+(s) \leq 1$ for any vertex $s \in S_3$ and for any local fragment. Then, the statement of the theorem will follow from Corollary 5.13.

(3) Assume that $p=7$. By Lemma 5.6, the pairs (p_1,p_2) and (p_1',p_2') are equal to (4,4) or (5,3).

(3.1) Let $(p_1,p_2) = (p_1',p_2') = (5,3)$. By Lemma 5.6, without loss of generality, we may assume that

$$C_1 = \{a,b,c,d,e\}, \ C_2 = \{a,f,g\}, \ C_1' = \{a,b,c,f,g\}, \ C_2' = \{a,d,e\}.$$

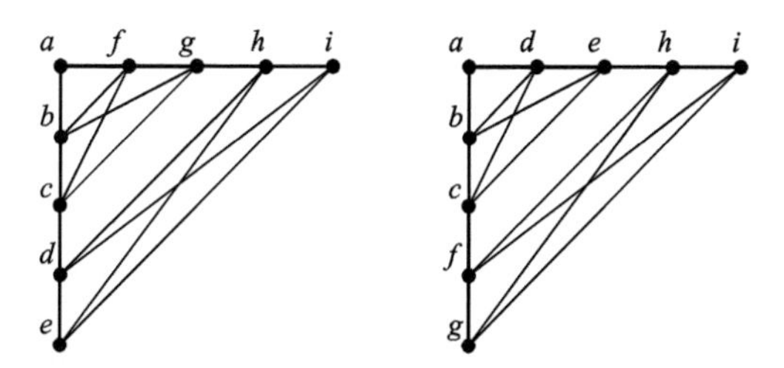

Figure 5.8 Local fragments in the case (1).

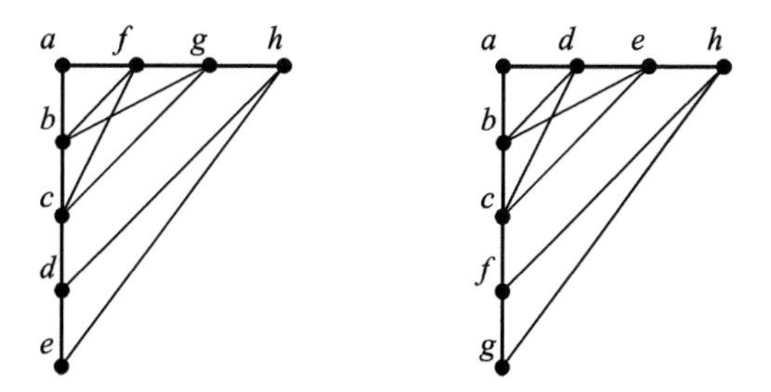

Figure 5.9 Local fragments in the case (2).

Then

$$F_2(a) - F_1(a) \supset \{C(b,f), C(b,g), C(c,f), C(c,g)\},$$
$$F_2'(a) - F_1'(a) \supset \{C'(b,d), C'(b,e), C'(c,d), C'(c,e)\},$$

see Figure 5.10. It is sufficient to prove that $D_-(v) \geq 2$ and $D'_-(v) \geq 2$ for any vertex $v \in S_2(a)$. Then, the required statement will follow from Lemma 5.15.

It is easy to see that v is adjacent to b or c. If, for instance, $v \sim b$ then, without loss of generality, $v \in C(b,f) \cap C'(b,d)$ and, evidently, $D_-(v) \geq 2$ and $D'_-(v) \geq 2$. In the case $v \sim c$, the reasonings are the same.

(3.2) Let $(p_1, p_2) = (5,3)$ and $(p_1', p_2') = (4,4)$. Without loss of generality,

$$C_1 = \{a,b,c,d,e\}, \;\; C_2 = \{a,f,g\}, \;\; C_1' = \{a,b,c,f\}, \;\; C_2' = \{a,d,e,g\}.$$

Now, we have the situation shown in Figure 5.10.

Let us show that s is adjacent to some vertex $u \in S_2(a)$, which is adjacent to f or g. If, for example $u \sim f$ then, without loss of generality, $u \in C(b,f) \cap C'(b,d)$. Therefore, $D_-(u) \geq 2$, $D'_-(u) \geq 2$ and the required statement follows from Lemma 5.15.

Let s be adjacent to $v \in S_2(a)$ and let $v \nsim \{f,g\}$. Then, without loss of generality, $v \in C'(b,d)$ and

$$v \in C(v,b) \cap C(v,d) \cap C(v,s).$$

It is evident that $|C(b,f)| = |C(d,g)| = 5$. Indeed, if $|C(b,f)| \leq 4$, then $|C(c,f)| \geq 6$ because $\deg(f) \geq 10$. Hence, $C(c,f)$ is a prelarge clique. Also, it is easy to see that $\deg(b) \leq 11$ and $\deg(d) \leq 11$. Indeed, if $\deg(b) \geq 12$, then

$$|C'(b,d)| + |C'(b,e)| + |C'(a,b,c,f)| = |C'(b,d)| + |C'(b,e)| + 4 \geq 15,$$

and so $C'(b,d)$ or $C'(b,e)$ is a prelarge clique. Hence,

$$|C(b,v)| \leq 4 \tag{5.36}$$

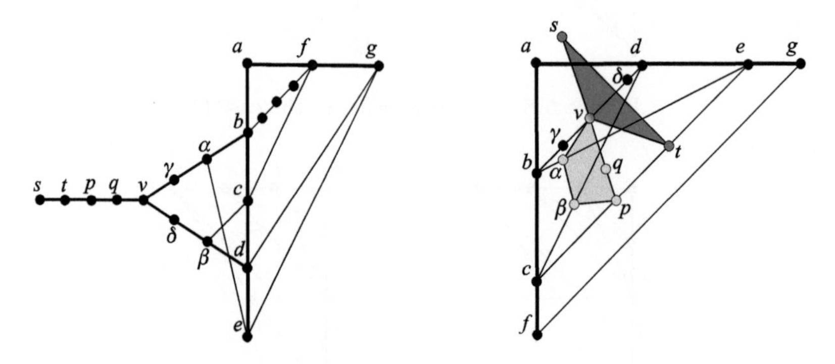

Figure 5.10 Local fragments in the case (3).

and

$$|C(d,v)| \le 4. \tag{5.37}$$

On the other hand, $|C(v,s)| \le 5$. Therefore,

$$|C(v,b)| + |C(v,d)| \ge 8. \tag{5.38}$$

Thus, from the inequalities above, we obtain

$$|C(b,v)| = 4, \quad |C(d,v)| = 4, \quad |C(v,s)| = 5.$$

Also, let v belong to the clusters $C_1'(v)$ and $C_2'(v)$ from $F_3'(a)$. Let $s \in C_2'(v)$ and

$$C(v,b) = \{v,b,\alpha,\beta\}, \quad C(v,s) = \{v,p,q,t,s\}.$$

If $p \in C(b,f)$, then the vertex p will play the same role as u; for example, $p \in C'(b,d)$. So, let $p,q \in C_1'(v)$ and $s,t \in C_2'(v)$. Let $C(b,f) = \{b,f,x,y,z\}$. If, for example, $y,z \in C'(b,d)$, then $v \sim \{y,z\}$. Hence, without loss of generality, $s \sim y$.

Let $y,z \in C'(b,e) = C'(b,y,z,e)$. If $\alpha,\beta \in C'(b,d)$, then d is adjacent to four vertices of $C(v,b)$. So, let $\alpha \in C'(b,y,z,e)$, that is, $\alpha \sim \{y,z,e\}$. Now, assume that α is covered by the clusters $C(\alpha,y,e)$ and $C(\alpha,z)$. Also, we have $\alpha \sim v$. If $\alpha \in C_2'(v)$, then $\alpha \sim s$ and so $s \in C(\alpha,z)$, that is, $\alpha \sim z$.

If $\alpha \in C_1'(v)$ then $\alpha \sim \{p,q\}$. Assume without loss of generality that $p \in C(\alpha,y,e)$. Then, p is covered by the clusters $C_1'(v), C'(p,e), C'(p,y)$. But $p \sim s$, therefore, $s \in C'(p,y)$. Thus, in any case, s is adjacent to one of the vertices of S_2, which is adjacent to f.

(3.3) Assume that $(p_1,p_2) = (p_1',p_2') = (4,4)$ and

$$C_1 = \{a,b,c,d\}, \quad C_2 = \{a,e,f,g\}, \quad C_1' = \{a,b,c,e\}, \quad C_2' = \{a,d,f,g\}.$$

Then, we have the situation shown in Figure 5.11.

Suppose that there is one of the edges bf, bg, cf, cg in the graph H; for example, $bf \in E(H)$. Then $C(b,f) \in F_2 - F_1$ and $C'(b,f) \in F_2' - F_1'$. Since $C(b,f)$ is not a special clique, we may assume that $C(b,f) \ne C'(b,f)$. If $A' = C'(b,f) - C(b,f)$, then $A' \subseteq C(b,e)$ and $A' \subset C(d,f)$. So, we have $|A'| \le 1$. Analogously, $|A| \le 1$ for $A = C(b,f) - C'(b,f)$.

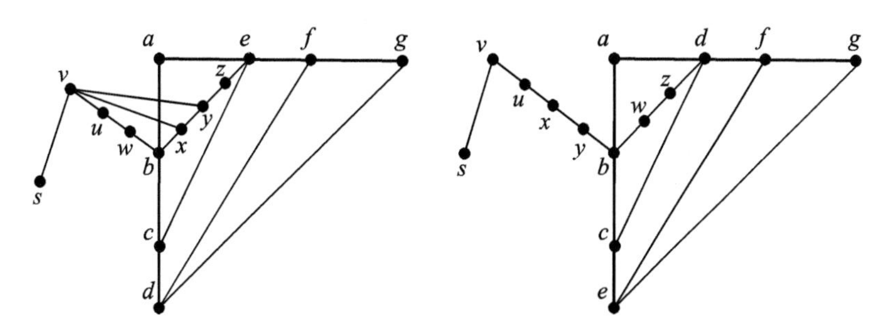

Figure 5.11 Local fragments in the case (3.3).

Let $A = \{\alpha\}$. We have $\alpha \in C(b,f)$ and $\alpha \in C'(f,e)$. By Lemma 5.6, $|C'(b,f) \cap C(b,f)| \leq 3$. Hence,

$$|C(b,f)| = |C'(b,f) \cap C(b,f)| + |A| \leq 4.$$

In a similar way, $|C'(b,f)| \leq 4$. Further,

$$|C'(f,e)| = (\deg(f) + 3) - 4 - |C'(b,f)| \geq 5.$$

Analogously, $|C(f,d)| \geq 5$. So, we obtain

$$|C(f,d) \cap C'(f,e)| = |C(f,d)| - |A'| \geq 4.$$

Therefore, $\alpha \in C(b,f)$ and α is adjacent to at least four vertices of $C(f,d)$. This is a contradiction. Hence, there are no edges bf, bg, cf, cg in H.

For the same reason as in (3.2), it is sufficient to show that s is adjacent to some vertex of S_2, which is adjacent to e or d. Let $s \sim v$, $v \sim b$ and $v \nsim \{d,e\}$. Let us suppose that $\deg(b) = 10$; otherwise, b satisfies the 7-condition and the required statement follows from Corollary 5.14. It is evident that

$$|C(b,e)| + |C(b,v)| = 9.$$

There are three subcases to consider.

(3.3.1) $|C(v,b)| = |C'(v,b)| = 5$. In this case, the inequalities $D_-(s) \geq 2$ and $D'_-(s) \geq 2$ follow immediately from Lemma 5.16.

(3.3.2) $|C(v,b)| = |C'(b,d)| = 4$ and $|C(b,e)| = |C'(v,b)| = 5$. Let

$$C(v,b) = \{v,b,u,w\}, \quad C(b,e) = \{b,e,x,y,z\},$$

$$C'(v,b) = \{v,u,x,y,b\}, \quad C'(b,d) = \{b,d,w,z\},$$

see Figure 5.11. We have $v \in C(v,x) \cap C(v,y)$, so $s \sim x$ or $s \sim y$.

(3.3.3) $|C(v,b)| = |C'(v,b)| = 4$ and $|C(b,e)| = |C'(b,d)| = 5$. If v is adjacent to some three vertices of $C(b,e)$, then the considerations are the same as in (3.3.2). Taking into account Lemma 5.6, we may assume without loss of generality that

$$C(v,b) = \{v,b,u,w\}, \quad C(b,e) = \{b,e,x,y,z\},$$

$$C'(v,b) = \{v,b,u,x\}, \quad C'(b,d) = \{b,d,y,z,w\},$$

see Figure 5.12. Since there are no special cliques in H, we may assume that there exists a vertex $p \in C(v,s) - C'(v,s)$. So, $p \in C'(v,w) = C'(v,w,p)$. Moreover, $|C(v,s)| \geq 4$. This follows from the evident inequality

$$|C(v,s)| + |C(v,x)| \geq 9$$

and from the fact that there are no prelarge cliques in H. Therefore, we have $C(v,s) \supset \{p,q\}$. If $q \in C'(v,w,p)$ then $w \sim \{p,q\}$. Consequently, w is covered by the clusters $C(w,p)$ and $C(w,q)$. Now, because $w \sim d$, the vertex d is contained in some of these clusters. Hence, $p \sim d$ or $q \sim d$.

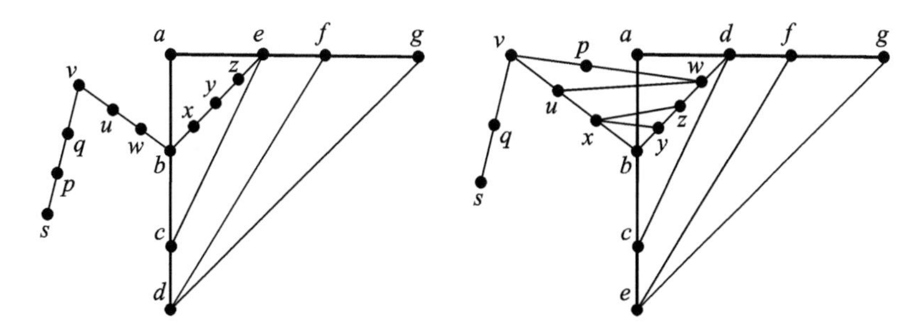

Figure 5.12 Local fragments in the case (3.3.3).

Further, let $q \in C'(v,s)$. Since $p \sim w$, we may assume that $p \in C(w,y)$, that is, $p \sim y$. Moreover,

$$p \in C'(v,p,w) \cap C'(p,q) \cap C'(p,s).$$

If $y \in C'(p,q)$ then $y \in C(y,p) \cap C(y,q)$, and hence d belongs to one of these clusters. If $y \notin C'(p,q)$ then $y \in C'(p,s)$. Thus, s is adjacent to some vertex of $S_2 \cap N(d)$.

(4) Suppose that $p = 6$. Note that there are no local fragments with groundwork that has a cluster of order 2. Indeed, if there exists a cluster $C = \{a,x\} \in F_1(a)$, then the vertex x belongs to a prelarge clique because $\deg(x) \geq 10$. Therefore, $(p_1, p_2) = (p_1', p_2') = (4,3)$.

There are two subcases to consider.

(4.1) $C_1 = \{a,b,c,d\}$, $C_2 = \{a,e,f\}$, $C_1' = \{a,b,e,f\}$, $C_2' = \{a,c,d\}$. Let us show that $ce, cf, de, df \notin E(H)$. Assume, for example, that $df \in E(H)$. We have

$$|C_2| + |C(f,b)| + |C(f,d)| \geq 13.$$

Therefore, $|C(f,b)| + |C(f,d)| \geq 10$ and, hence, $|C(f,d)| = 5$. Analogously, $|C'(f,d)| = 5$. Let $C(f,d) = \{f,d,x,y,z\}$. Since there are no special cliques in H, we may assume that $x \in C(f,d) - C'(f,d)$. Then $x \in C'(d,b)$. If $y,z \in C'(d,f)$, then x is adjacent to $C'(d,f)$. If $y \in C'(b,d)$, then b is adjacent to four vertices of $C(d,f)$. In both cases, we have a contradiction. Hence, there are no edges ce, cf, de, df in H and the vertices c,d,e,f satisfy the 7-condition. Every vertex of S_2 is adjacent to one of these vertices and, therefore, is appropriate. Thus, the required statement follows from Corollary 5.14.

(4.2) $C_1 = \{a,b,c,d\}$, $C_2 = \{a,e,f\}$, $C_1' = \{a,b,c,e\}$, $C_2' = \{a,f,d\}$. Let $s \sim v$, $v \in S_2$. Assume that $v \sim b$ and $v \not\sim \{e,d\}$. The same consideration as in (3.3.3) shows that s is adjacent to some vertex, which is adjacent to d or e. Now, let $v \sim e$ and $v \nsim f$. Consider the vertex f. Since $\deg(f) \geq 10$, f is covered by the clusters

$$C(f,d),\ C(f) \in F_2 - F_1 \quad \text{and} \quad C'(f,e),\ C'(f) \in F_2' - F_1'$$

in addition to C_2 and C_2'. Moreover, by virtue of the inequalities

$$|C(f,d)| + |C(f)| \geq 10 \quad \text{and} \quad |C'(f,e)| + |C'(f)| \geq 10,$$

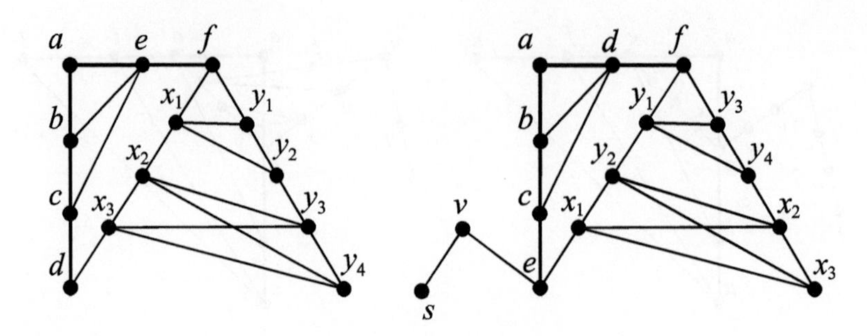

Figure 5.13 Local fragments in the case (4.2).

we have

$$|C(f,d)| = |C(f)| = |C'(f,e)| = |C'(f)| = 5.$$

Let $C(f,d) = \{f, d, x_1, x_2, x_3\}$ and $C(f) = \{f, y_1, y_2, y_3, y_4\}$. Also, taking into account Lemma 5.6, assume without loss of generality that $C'(f,e) = \{f, e, x_1, y_1, y_2\}$ and $C'(f) = \{f, x_2, x_3, y_3, y_4\}$, see Figure 5.13. We have $e \sim \{x_1, y_1, y_2\}$. Then $e \in C(e, y_1) \cap C(e, y_2)$. If $v \in C(e, y_1)$ then $v \sim y_1$, therefore, without loss of generality, $v \in C'(y_1, y_3)$. We may assume now that $v \in C(y_3, x_3)$. So, we have $v \in C'(v, y_3, y_1) \cap C'(v, x_3)$. It is easy to see that e does not belong to any of these clusters. Indeed, if $e \in C'(v, x_3)$ then $v \sim x_3$. Hence, without loss of generality, $e \in C(x_3, y_4)$. So, we obtain $|N(y_4) \cap C'(f,e)| \geq 4$.

Thus, $v \in C'(v, y_3, y_1) \cap C'(v, x_3) \cap C'(v, e)$. Therefore, $s \in C'(v, y_3, y_1) \cap C'(v, x_3)$. So, our considerations are reduced to the case when s is adjacent to some vertex of S_2, which is adjacent to f. Without loss of generality, assume that $s \sim y_1$ and $s \in C'(y_1, y_3) \cap C(y_3, x_3)$. We have $D_{-}(s) \geq 2$ and $D'_{-}(s) \geq 2$.

(5) Suppose that $p = 5$. Similar to (4), the case $(p_1, p_2) = (2, 4)$ is impossible. Thus, the only possible case is $C_1 = \{a, b, c\}$, $C_2 = \{a, d, e\}$, $C'_1 = \{a, b, d\}$, $C'_2 = \{a, c, e\}$. For the same reasons as in (4.1), we obtain $be, dc \notin E(H)$. Therefore, b, c, d and e satisfy the 7-condition and, hence, every vertex of S_2 is appropriate. Again, the required statement follows from Corollary 5.14.

(6) If $p \leq 4$, then every vertex of S_1 satisfies the 7-condition. Thus, the theorem is true for the same reason as in (5). □

5.3.3 Recognition Algorithm

The algorithm for solving the problem '$G \in L_3^\ell$?' for graphs with $\delta(G) \geq 10$ consists of three stages:

(A) *Construct the initial family of cliques F_0.*

Choose an arbitrary vertex $z \in V(G)$. If $\deg(z) \geq 19$, then consider the set $S \subseteq B_1[z]$ with $|S| \geq 20$. The set S must contain a clique of size at least 8. Extend it to the large clique C'. If $\deg(z) \leq 18$, then for C' we take an arbitrary maximal clique from $B_1[z]$ satisfying the following conditions:

$$z \in C', \quad |C'| \geq 5. \tag{5.39}$$

We set $F_0 = \{C'\}$. If Algorithm 5.2 cannot extend F_0 to the reducing family F, or the graph $H = G - E(F)$ is not a line graph, then there is no fragment containing C' as a cluster. Hence, if C' is a large clique, then $G \notin L_3^\ell$. In the case when C' is not a large clique, choose another clique satisfying (5.39). In the worst case, we should test all cliques from $B_1[z]$ satisfying (5.39), where $|B_1[z]| \leq 19$.

(B) *Extend* F_0 *to the reducing family* F *(Algorithm 5.2).*

(C) *Check whether* $H = G - E(F) \in L_2^\ell$.

Algorithm 5.2: Extending the initial family of cliques F_0 of the graph G to the reducing family F.

Input: A connected graph G with $\delta(G) \geq 10$, and some family of cliques F_0 of the graph G.

Output: One of the variants (a) or (b):

(a) A reducing family F such that $F_0 \subseteq F$.

(b) The answer that there is no fragment containing F_0. If F_0 contains a large clique of G, then the answer is $G \notin L_3^\ell$.

(1) Initialize $F = F_0$ and $H = G - E(F)$. For an arbitrary vertex v, the label $\mathrm{cl}(v)$ is the number of clusters of F covering v. If one of the conditions below fails, then there is no fragment containing F and the algorithm stops.

(2) While there are vertices in H with label 0 or 2, do:

(2.1) If v is a vertex with label $\mathrm{cl}(v) = 2$, then $B_1[v]$ should be a clique. In this case, put $C = B_1[v]$ and go to Step (2.5).

(2.2) Let a be a vertex of H such that $\mathrm{cl}(a) > 0$, and in the neighbourhood of a there exists a vertex, say w, with $\mathrm{cl}(w) = 0$. Then, the inequality $|S_i(a)| \leq 18 \cdot |S_{i-1}(a)|$ holds.

(2.3) For i from $i = 1$ to $i = \min\{9, \mathrm{ecc}(a)\}$, do:

(2.3.1) Search for a large clique C in $B_i[a]$ and if found, go to (2.5).

(2.3.2) Search for a special clique C in $B_i[a]$ and if found, go to (2.5).

(2.4) At this stage, there are neither special nor good cliques in $B_k[a]$, where $k = \min\{9, \mathrm{ecc}(a)\}$, and $|B_k[a]| \leq \mathrm{const}$.

(2.4.1) Search for a 7-clique C in $B_4[a]$ and if it exists, go to (2.5).
(2.4.2) Search for a 6-clique C in $B_4[a]$ and if it exists, go to (2.5).
(2.4.3) At this stage, there are no prelarge cliques in $B_4[a]$. Take for C an arbitrary cluster of any groundwork F_1 of an arbitrary local fragment F_k and go to (2.5).

(2.5) Extend the family F, update the graph H and recount the labels of the vertices: $F = F \cup \{C\}; H = H - E(C); \text{cl}(v) = \text{cl}(v) + 1$ for all $v \in C$.

(3) At this stage, there are no vertices with labels 0 or 2 in the graph H. If there exists a vertex v in H such that (a) $\text{cl}(v) > 3$ or (b) $\text{cl}(v) = 3$ and $S_1(v) \neq \emptyset$, then there is no fragment containing F_0. If there are no such vertices, then F is a reducing family.

Theorem 5.10 [32] *There exists an algorithm with complexity $O(nm)$ for solving the recognition problem '$G \in L_3^\ell$?' in the class of graphs with minimum vertex degree $\delta(G) \geq 10$.*

Proof: The existence of a polynomial algorithm follows from Algorithm 5.2 and the previous discussion. The complexity of Algorithm 5.2 is determined by the complexity of finding a large clique containing a fixed vertex x. This can be done in time $O(m)$, where m is the number of edges in G. Hence, the complexity of Algorithm 5.2 is equal to $O(nm)$, where n is the order of G. Thus, the complexity of the algorithm for solving the problem '$G \in L_3^\ell$?' is the same because Algorithm 5.2 is repeated at most $\binom{18}{4}$ times. □

5.3.4 Forbidden Induced Subgraphs

It is known that the class of line graphs of simple graphs, L_2^ℓ, is characterized by a finite list of forbidden induced subgraphs, see Section 5.2.1. Let $\mathcal{A}$ be the set of all graphs of order $\leq 10^{11}$ not belonging to L_3^ℓ. Further, let G be an arbitrary graph, F_0 be the set of all large cliques of G and $H_0 = G - E(F_0)$.

Lemma 5.17 *If G contains no graph from $\mathcal{A}$ as an induced subgraph, then the cliques in F_0 have no common edges.*

Proof: Assume that there exist two cliques C_1 and C_2 in F_0 such that $|C_1 \cap C_2| > 1$. Now, suppose that $|C_1 \cap C_2| \geq 4$ and

$$v_1, v_2, \ldots, v_8 \in C_1, \quad v_1, v_2, v_3, v_4 \in C_1 \cap C_2, \quad u \in C_2 - C_1.$$

Consider the induced subgraph $G_1 = G(v_1, v_2, \ldots, v_8, u)$. By Theorem A (Section 5.2.1) and Lemma 5.6, $G \in \mathcal{A}$, which is impossible. Therefore, $2 \leq |C_1 \cap C_2| \leq 3$. Let

$$v_1, v_2 \in C_1 \cap C_2, \quad u_1, u_2, \ldots, u_6 \in C_1 - C_2, \quad w_1, w_2, \ldots, w_6 \in C_2 - C_1.$$

According to Theorem A and Lemma 5.6, the induced subgraph

$$G_2 = G(v_1, v_2, u_1, u_2, \ldots, u_6, w_1, w_2, \ldots, w_6)$$

belongs to $\mathcal{A}$. This contradiction proves the lemma. □

Lemma 5.18 *If G contains no graph from $\mathcal{A}$ as an induced subgraph, then every vertex $v \in V(G)$ belongs to at most three cliques from F_0.*

Proof: Suppose that there exists a vertex $v \in V(G)$ such that $v \in C_1, C_2, C_3, C_4$ and $C_i \in F_0$, $i = 1, 2, 3, 4$. Further, let $v_1^i, v_2^i, \ldots, v_7^i \in C_i$, $i = 1, 2, 3, 4$. By Lemma 5.17,

$$v_j^i \in C_i - \bigcup_{\substack{1 \le k \le 4 \\ k \ne i}} C_k, \quad 1 \le i \le 4, \quad 1 \le j \le 7.$$

Consider the induced subgraph $G(v, v_1^1, v_2^1, \ldots, v_7^1, v_1^2, v_2^2, \ldots, v_7^4)$. By Theorem A, it belongs to $\mathcal{A}$. □

Lemma 5.19 *If G contains no graph from $\mathcal{A}$ as an induced subgraph, then for every $a \in V(G)$ there exists an $(a, 4)$-local fragment with respect to F_0.*

Proof: Note that if $\deg(v) \ge 19$, then v belongs to some clique from F_0. Indeed, let $v_1, v_2, \ldots, v_{19} \in N(v)$. Since $G(v, v_1, v_2, \ldots, v_{19}) \notin \mathcal{A}$, there exists a clique $C \subseteq \{v, v_1, v_2, \ldots, v_{19}\}$ of size at least 8. Hence, $\Delta(H_0) \le 18$. Choose a vertex $a \in V(G)$ and consider the ball $B = B_4[a]$ in H_0. Evidently, $|B| \le 18 + 18^2 + 18^3 + 18^4 \le 105{,}000$. Let $C_1, C_2, \ldots, C_k$ be the large cliques from F_0 having a non-empty intersection with the set B; it is possible that there are no such cliques. By Lemma 5.18, $k \le 3|B| \le 315{,}000$. Let us choose the sets $C_i' \subseteq C_i$, $i = 1, 2, \ldots, k$, such that

(1) $C_i \cap B = C_i' \cap B$;
(2) $C_i' - (B \cup (\bigcup_{j \ne i} C_j')) \le 8$;
(3) $|C_i'| \ge 8$, $i = 1, 2, \ldots, k$.

Let $Y = B \cup C_1' \cup C_2' \cup \ldots \cup C_k'$. By Lemmas 5.17 and 5.18, we have

$$|Y| \le |B| + \sum_{i \ne j} |C_i' \cap C_j'| + \sum_i \left| C_i' - \left(B \cup \left(\bigcup_{j \ne i} C_j' \right) \right) \right| \le 105{,}000 + k^2 + 8k \le 10^{11}.$$

Let us consider the induced subgraph $G_1 = G(Y)$. We have $G_1 \notin \mathcal{A}$ and, by Theorem A, there exists a Krausz 3-covering Q of G_1 such that $C_i' \in Q$, $i = 1, 2, \ldots, k$. The clique covering $Q' = Q - \{C_1', C_2', \ldots, C_k'\}$ of $B_4[a]$ evidently induces the required local fragment. □

Theorem 5.11 [32] *Let G be a graph such that*

(1) $\delta(G) \geq 16$*;*
(2) any two cliques from F_0 have no common edges;
(3) every vertex of G is covered by at most three cliques from F_0;
(4) for any $a \in V(G)$, there exists a local fragment with centre in a of radius 4 *with respect to F_0.*

Then $G \in L_3^\ell$.

Proof: Let us build the Krausz 3-covering of G by adding new clusters to the initial family of cliques F. At the beginning, put $F = F_0$ (it is correct by the conditions (2) and (3)) and $H = G - E(F)$. Next, assume that there exists a connected component H' of the graph H with radius $r(H') \leq 2$; that is, there exists a vertex $a_{H'} \in V(H')$ such that $\operatorname{ecc}(a_{H'}) \leq 2$. Then, a local fragment $F_4(a_{H'})$ covers all vertices of H'. Hence, we can choose a vertex $a_{H'}$ in every such component H' and put $F = F \bigcup_{r(H') \leq 2} F_4(a_{H'})$.

Further, assume that $\operatorname{ecc}(v) \geq 3$ for all $v \in V(H)$ and let $a \in V(H)$ be a vertex not covered by F.

Proposition 5.13 *There exists a unique groundwork $F_1(a)$ of every local fragment $F_k(a)$ with centre in a.*

Proof: By the conditions of the theorem, there exists a local fragment $F_4(a)$ and the corresponding sequence $F_4(a) \supset F_3(a) \supset F_2(a) \supset F_1(a)$ of local fragments. Since $\deg_G(a) \geq 16$ and a is not covered by F, there exists a 7-clique $C \in F_1(a)$. The clique C is special. Indeed, in Lemma 5.12 it is proved that if there exists some other sequence $F_2'(a) \supset F_1'(a)$, then $\operatorname{ecc}(a) \leq 2$. Let $F_1(a) = \{C, C_1, C_2\}$ and $|C_1| \geq 5$. If there exists a local fragment $F_1'(a) \not\ni C_1$, then the vertices from C_1 are contained in exactly two clusters of F_1'. This is impossible by Lemma 5.6. Thus, the clique C_1 is special and, therefore, C_2 is also special. The proposition is proved. □

Now, consider the set $X = \bigcup F_1(a)$, where $F_1(a) \subset F_4(a)$ and the union is taken over all a not covered by F.

Proposition 5.14 *We have*

(i) The clusters of X have no common edges.
(ii) Every vertex of G is covered by at most three cliques of $F \cup X$.
(iii) If some vertex v is covered by exactly three cliques $C_1(v), C_2(v), C_3(v) \in F \cup X$, then $N_G(v) - (C_1(v) \cup C_2(v) \cup C_3(v)) = \emptyset$.
(iv) If some vertex $v \in V(G)$ is covered by exactly two cliques $C_1(v), C_2(v) \in F \cup X$, then $N_G(v) - (C_1(v) \cup C_2(v))$ is a clique.

Proof:

(i) Let there exist cliques $C_1, C_2 \in X$ such that $|C_1 \cap C_2| \geq 2$. Then $C_1 \in F_1(a), C_2 \in F_1(b), a \neq b$ and a, b are not covered by F. Also,

$$F_1(a) \subset F_2(a) \subset F_3(a) \subset F_4(a) \quad \text{and} \quad F_1(b) \subset F_2(b) \subset F_3(b) \subset F_4(b).$$

For the distance between the vertices a and b, we have $d(a,b) \leq 2$. Hence, the local fragment $F_4(a)$ induces the sequence of local fragments $F_2'(b) \supset F_1'(b)$ and $F_2'(b) \subset F_4(a)$. By Proposition 5.13, $F_1(b) = F_1'(b)$, so $C_2 \in F_4(a)$. But then two clusters C_1, C_2 of the local fragment $F_4(a)$ have a common edge. This contradicts the definition of the local fragment.

(ii) Let there exist a vertex $v \in V(G)$ that belongs to the cliques $C_1, C_2, C_3, C_4 \in F \cup X$. By the condition (3) of the theorem, assume that $C_1 \in F_1(a) \subset F_4(a)$ and a is not covered by F_0. Using the same arguments as in (i), we can prove that if $C_i \in X$ for some $i \in \{2,3,4\}$, then $C_i \in F_4(a)$. Hence, $v \in B_1[a]$, and v is covered by four cliques of $F_4(a) \cup F$. This contradicts the definition of the local fragment.

(iii) Let there exist a vertex $v \in V(G)$ that belongs to the cliques $C_1, C_2, C_3 \in F \cup X$, and $C = N_G(v) - (C_1 \cup C_2 \cup C_3) \neq \emptyset$. There exists a local fragment with centre in v with respect to F. Hence, without loss of generality, C_1 is not a large clique and $C_1 \in F_1(a) \subset F_4(a)$. Similar to the arguments above, if some $C_i \in X$ for $i \in \{2,3\}$, then $C_i \in F_4(a)$. Therefore, $v \in B_1[a]$, and v is covered by more than three clusters of $F_4(a) \cup F$; that is, v is covered by C_1, C_2, C_3 and some extra clusters covering vertices from the set C, which is impossible.

(iv) Let a vertex $v \in V(G)$ be covered by cliques $C_1, C_2 \in F \cup X$. If $C_1, C_2 \in F$, then the required statement is true because there exists a local fragment with respect to F and with centre in v. Let $C_1 \in X$ and $C_1 \in F_1(a) \subset F_4(a)$. If $C_2 \in X$ then $C_2 \in F_4(a)$, so $v \in B_1[a]$, and v is covered by at most three clusters of $F_4(a) \cup F$. That is, v is covered by C_1, C_2 and $C_3 = N_G(v) - (C_1 \cup C_2)$. The proposition is proved. □

Thus, we can put $F = F \cup X$ and, by Proposition 5.14, a reducing family of cliques will be obtained. Now, in the graph $H = G - E(F)$, every vertex is covered by some large clique.

Proposition 5.15 *We have* $H \in L_2^\ell$.

Proof: Let us prove that H contains no Beineke graph as a forbidden induced subgraph (see Section 5.2.1). Suppose to the contrary that there exists an induced subgraph B *of* H isomorphic to some Beineke graph. It is easy to see that $r(B) \leq 2$. Choose a vertex $a \in V(B)$ with $\mathrm{ecc}(a) \leq 2$. By statement (iv) of Proposition 5.14, a is covered by exactly one clique $C \in F$, $|C| \geq 8$. By the conditions of the theorem, there exists a local fragment $F_4(a)$ and $V(B) \subseteq B_2[a]$. Hence, $F_4(a)$ induces a Krausz 2-covering of B, which is impossible. The proposition is proved. □

Proposition 5.16 *There exists a Krausz 2-covering* Y *of* H *such that every vertex* $a \in V(H)$*, which is covered by exactly two clusters of* F*, is covered by exactly one cluster of* Y*.*

Proof: Let Z be an arbitrary Krausz 2-covering of H, and a vertex $a \in V(H)$ be covered by exactly two clusters of F. Then, by statement (iv) of Proposition 5.14, $C = N_H(v) \cup \{v\}$ is a clique. Let $C \notin Z, C = C_1 \cup C_2$ and $C_1, C_2 \in Z$. Then $|C| = 3$, otherwise either $|C| = 2$ or C is a 2-large clique, and so $C \in Z$. Hence, $C_1 = \{a, b\}$ and $C_2 = \{a, c\}$. We will prove that $Z' = (Z - \{C_1, C_2\}) \cup \{C\}$ is a Krausz 2-covering. The only thing that could impede this is the following: there exist vertices $d, e \in V(H)$ such that $bd, cd, ed \in E(H)$. Let us prove that this is impossible.

By the conditions of the theorem, there exists a local fragment $F_4(a)$ with respect to the set of all large cliques F_0. It is easy to see that $C \in F_4(a)$, the reasoning is the same as in the proof of Proposition 5.14. Hence, the vertex d is covered by the clusters $C(b, d), C(c, d) \in F_4(a)$. Since the vertex d is covered by some large clique and $d \sim e$, we may assume without loss of generality that $e \in C(b, d)$, which implies $be \in E(H)$. Now, consider the covering Z. If $ec \notin E(H)$, then either $C(b, c, d), C(e, b) \in Z$ or $C(b, d, e), C(b, c) \in Z$. In both cases, b is covered by three clusters of Z, which is impossible. Thus, $ec \in E(H)$. Consequently, $C(b, c, d, e) \in F_4(a)$ and $|C(b, c, d, e) \cap C| \geq 2$, a contradiction. The proposition is proved. □

Let us consider $F = F \cup Y$. By the above propositions, F is a Krausz 3-covering. □

Thus, Lemmas 5.17–5.19 and Theorem 5.11 imply the following:

Theorem 5.12 [32] *If a graph G with $\delta(G) \geq 16$ contains no graph from the set $\mathcal{A}$ as an induced subgraph, then $G \in L_3^\ell$.*

5.4 Line Graphs of Helly Hypergraphs

In this section, we consider the class of r-minoes, which is a natural generalization of the notion of domino. We will show that this class coincides with the class of line graphs of Helly hypergraphs with rank at most r. For an arbitrary r, we will prove the existence of a finite list of forbidden induced subgraphs that characterizes r-minoes, and an explicit finite characterization will be given for 3-minoes. Also, it will be shown that the GRAPH 3-COLOURABILITY PROBLEM remains NP-complete when restricted to linear dominoes with vertex degrees at most 4.

5.4.1 Preliminaries

A graph is called a *domino* if each of its vertices belongs to at most two maximal cliques. The class of dominoes was introduced by Kloks et al. [17]. In a similar way, a graph is called an *r-mino* if each of its vertices belongs to at most r maximal cliques. The class of r-minoes is denoted by M_r; thus M_2 is the class of dominoes. Obviously, each graph is an r-mino for an appropriate r, and we have the following strictly increasing sequence:

$$M_1 \subset M_2 \subset \ldots \subset M_r \subset \ldots$$

Two characterizations of the class M_r will be given. The first one states that M_r is the class of line graphs of Helly hypergraphs with rank at most r. In particular, M_r is a *hereditary class*; this means that any induced subgraph of a graph in M_r is also in $\mathcal{M}_r$. It is well known that every hereditary class of graphs P can theoretically be characterized by means of a list of forbidden induced subgraphs. If $\mathcal{F}$ is such a list, then we write $P = \text{Forb}(\mathcal{F})$. If, in addition, $\mathcal{F}$ is finite, then we have a *finite characterization* of P. We will recursively define a finite list of forbidden induced subgraphs $\mathcal{F}_r$ for the class M_r for every $r \geq 1$. The existence of a finite characterization implies that, for every fixed r, there is a polynomial time algorithm for determining if a given graph belongs to M_r.

For a fixed constant r, let us consider the class of line graphs of hypergraphs with rank at most r. It is easy to see that this class is exactly the class L_r of line graphs of r-uniform hypergraphs. A non-trivial characterization of L_r is known only for $r \leq 2$ (see Bermond and Meyer [7]). Poljak et al. [29] proved that the problem of determining if a graph belongs to L_r is NP-complete for an arbitrary $r \geq 4$.

Let us consider the following two graph-theoretic invariants:

- *Rank dimension* $\dim_R(G) = \min\{r : G \in L_r\}$;
- *Helly rank dimension* $\dim_{HR}(G) = \min\{r : G \in M_r\}$.

The characterization of the class M_r below implies the strict inclusion

$$M_r \subset L_r,$$

hence

$$\dim_R(G) \leq \dim_{HR}(G).$$

However, the difference $\dim_{HR}(G) - \dim_R(G)$ can be arbitrarily large. The complexity of determining $\dim_{HR}(G)$ is still unknown. The dimensionality of graphs will be further explored in Chapter 6.

We only consider finite hypergraphs in which every vertex is contained in some edge. For a hypergraph H with the incidence matrix M, the *dual hypergraph* H^* is the hypergraph with the transposed incidence matrix M^{T}. For an arbitrary clique covering $Q = \{C_i : i \in I\}$ of G, we define the hypergraph F_Q as follows: the vertices of F_Q are the vertices of G, and the edges are the clusters of Q. The edges C_i and C_j are different for $i \neq j$ even if the sets C_i and C_j coincide. The dual hypergraph $(F_Q)^*$ is called the *canonical hypergraph* and is denoted by $C(Q)$.

A hypergraph H is called a *root* of a graph G if the line graph $L(H)$ is isomorphic to G. Berge [2] described all roots for an arbitrary graph in terms of clique coverings, see Theorem 5.13. Note that this theorem was generalized in Theorem 5.2 from Section 5.1.4.

Theorem 5.13 [2] *The roots of a graph G are exactly the canonical hypergraphs $C(Q)$, where Q runs over all clique coverings of G.*

Recall some definitions and notation from Section 5.1.5. Let P be a *hypergraph-theoretic property*; that is, a class of hypergraphs distinguished up to isomorphism. We say that a clique covering Q of a graph G *has the property* P if $F_Q \in P$. Put

$$P^* = \{H^* : H \in P\}, \quad L(P) = \{L(H) : H \in P\}.$$

Theorem 5.13 (or Theorem 5.3) immediately implies

Corollary 5.15 *Let P be an arbitrary hypergraph-theoretic property and G be a graph. Then $G \in L(P)$ if and only if G has a clique covering with the property P^*.*

A hypergraph H is a *Helly hypergraph* if the family $\mathcal{E}$ of its edges satisfies the following *Helly condition*: for each subfamily $\mathcal{E}' \subseteq \mathcal{E}$ of pairwise intersecting edges, there is a vertex which belongs to all edges in $\mathcal{E}'$. The *2-section graph* $[H]_2$ of a hypergraph H is the graph whose vertices are the vertices of H, and two different vertices are adjacent if and only if they belong to the same edge of H. A hypergraph H is called *conformal* if the vertex set of each clique in $[H]_2$ is contained in some edge of H.

Lemma 5.20 [2] *A hypergraph H is a Helly hypergraph if and only if H^* is conformal.*

A clique covering Q of a graph G is called *conformal* if the hypergraph F_Q is conformal.

Lemma 5.21 *For any graph G, the set of maximal cliques is the unique minimal (with respect to inclusion) conformal covering of G.*

Proof: Let Q be a clique covering of G and $H = F_Q$. Obviously, $[H]_2 = G$. Therefore, H is conformal if and only if each maximal clique of G is an edge of H, that is, a cluster of the covering Q. □

Corollary 5.16 [23] *Each graph is the line graph of a Helly hypergraph.*

5.4.2 r-Minoes

Let us remind ourselves that a graph G is an *r-mino* if each vertex of G belongs to at most r maximal cliques. If, in addition, every edge of G belongs to exactly one maximal clique, then we have a *linear r-mino*. The sets of r-minoes and of linear r-minoes are denoted by M_r and M_r^ℓ, respectively.

A hypergraph is called *linear* if every pair of edges has at most one common vertex. A clique covering Q of a graph is called *linear* if F_Q is linear. A clique covering is called an *r-covering* if each vertex of the graph belongs to at most r clusters.

Theorem 5.14 [25] *The following statements are true:*

(i) M_r coincides with the set of line graphs of Helly hypergraphs with rank at most r.

(ii) M_r^ℓ *coincides with the set of line graphs of linear Helly hypergraphs with rank at most* r.

Proof: If $G \in M_r$, then the set Q of maximal cliques is an r-covering of G. By Lemma 5.21, Q is conformal. Then, the canonical hypergraph $C(Q)$ is a Helly hypergraph by Lemma 5.20, and rank $C(Q) \leq r$. By Theorem 5.13, $L(C(Q)) \simeq G$.

Conversely, let H be a Helly hypergraph, rank $H \leq r$ and $L(H) \simeq G$. Then, by Corollary 5.15 and Lemma 5.20, there is a conformal r-covering of G. By Lemma 5.21, this covering contains all maximal cliques of G. Therefore, the set of maximal cliques is also an r-covering of G. Thus, $G \in M_r$ and (i) is proved.

Obviously, the linearity condition is self-dual, so (ii) holds. □

Corollary 5.17 [17] *The following statements hold:*

(i) M_2 *coincides with the set of line graphs of multigraphs without triangles.*

(ii) M_2^ℓ *coincides with the set of line graphs of simple graphs without triangles.*

Corollary 5.18 M_r *and* M_r^ℓ *are hereditary classes of graphs.*

The proof follows from the fact that the property 'being a Helly hypergraph with rank at most r' is hereditary with respect to deleting edges.

Corollary 5.19 *The following statements are true:*

(i) *For a constant* $r \geq 2$, *we have*

$$M_r \subset L_r. \tag{5.40}$$

(ii) *For a graph* G,

$$\dim_R(G) \leq \dim_{HR}(G). \tag{5.41}$$

Equality in (5.41) is possible, but the difference $\dim_{HR}(G) - \dim_R(G)$ *can be arbitrarily large.*

Proof: The inclusion (5.40) follows immediately from Theorem 5.14. Let $W_r, r > 3$, denote the $(r+1)$-vertex wheel W_r; W_4 is shown in Figure 5.14. For the wheel W_r, we have

$$\dim_{HR}(W_r) = r, \quad \dim_R(W_r) = \lceil r/2 \rceil .$$

Hence, the inclusion (5.40) is strict and the difference $\dim_{HR}(G) - \dim_R(G)$ can be arbitrarily large. The inclusion (5.40) immediately implies (5.41). Finally, for the star $K_{1,r}$,

$$\dim_R(K_{1,r}) = r = \dim_{HR}(K_{1,r}).$$

□

Forbidden Induced Subgraph Characterizations

Let us turn to the problem of characterizing the class of r-minoes.

Theorem 5.15 [25] *For any constant r, there exists a finite characterization of the class M_r.*

Proof: The proof is by induction on r. It is obvious that $M_1 = \text{Forb}(\{P_3\})$; that is, $\mathcal{F}_1 = \{P_3\}$.

Now let $r \geq 2$. Denote by $s(G)$ the number of maximal cliques in G. By hypothesis, $M_{r-1} = \text{Forb}(\mathcal{F}_{r-1})$ and $\mathcal{F}_{r-1}$ is finite. Without loss of generality, suppose that the list $\mathcal{F}_{r-1}$ is minimal. Evidently, if $G \in \mathcal{F}_{r-1}$, then G has a unique dominating vertex $v(G)$ and it is the union of $s(G)$ maximal complete subgraphs, $s(G) \geq r$.

We shall construct the list $\mathcal{F}_r$, which provides a finite characterization of the class M_r. Any graph G satisfying the conditions

$$G \in \mathcal{F}_{r-1}, \quad s(G) \geq r+1,$$

is included in the list $\mathcal{F}_r$. Next, let

$$G \in \mathcal{F}_{r-1}, \quad s(G) = r, \quad v(G) = v, \tag{5.42}$$

and let

$$C_1, C_2, \ldots, C_r \tag{5.43}$$

be the list of maximal cliques of G.

Define two supergraphs F and H for the graph G as follows:

(1) $G = F - A$, where A is a clique in F, $1 \leq |A| \leq r$, $A \sim v, s(F) \geq r+1$.
(2) $G = H - a - b$, $a \not\sim b$, $a \sim C_i$, $b \sim C_i$ for some index i.

The remaining adjacencies in the graphs F and H can be arbitrary. Thus, the conditions (1) and (2) determine the sets of graphs of two types, F and H. For every graph G satisfying the conditions in (5.42), we now include all graphs of type F and H into $\mathcal{F}_r$. The list $\mathcal{F}_r$ is constructed. If m_i is the maximal order of the graphs in $\mathcal{F}_i$, then $m_r \leq m_{r-1} + r$. Hence, the set $\mathcal{F}_r$ is finite.

We shall prove that

$$M_r = \text{Forb}(\mathcal{F}_r). \tag{5.44}$$

Obviously, $\mathcal{F}_r \cap M_r = \emptyset$ because the vertex $v(G)$ of any graph G in $\mathcal{F}_r$ belongs to at least $r+1$ maximal cliques. The induced subgraphs of the graphs in M_r do not belong to the list $\mathcal{F}_r$, since they belong to M_r. Hence,

$$M_r \subseteq \text{Forb}(\mathcal{F}_r).$$

Now, let $B \in \text{Forb}(\mathcal{F}_r)$. Consider the subgraph $U = B(N[u])$ for a vertex $u \in V(B)$, where $N[u] = N(u) \cup \{u\}$. Each clique of B containing u is a clique of U. If $U \in \text{Forb}(\mathcal{F}_{r-1})$, then u belongs to at most $r-1$ maximal cliques of U. If some graph G in $\mathcal{F}_{r-1}$ is an induced subgraph of U, then G satisfies the conditions in (5.42). Without loss of generality, suppose that $u = v(G) = v$. Let (5.43) be the list of maximal cliques in G. Every clique C_i is contained in one maximal clique D_i of $U, i = 1, 2, \ldots, r$, because B has no induced subgraphs of type H. It remains to prove that

$$D_1, D_2, \ldots, D_r \tag{5.45}$$

is the complete list of maximal cliques of U. Let D be an arbitrary clique of $U, v \in D$ and $|D - v| = k$. Then

$$D \subseteq D_i \tag{5.46}$$

for some index i. In fact, (5.46) holds for $k \leq r$ because U contains no induced subgraphs of type F. Let $k \geq r+1$, and let (5.46) hold if $|D - v| < k$. Consider the $(k-1)$-subsets of the clique $D - v$. The number of these subsets is equal to k, and each one is contained in the corresponding clique D_i. Because $k > r$, there exist two such subsets, for instance X_1 and X_2, both contained in the same clique D_i. Also, $D - v = X_1 \cup X_2$. Therefore, $D \subseteq D_i$. Hence, (5.45) is the complete list of maximal cliques of B containing the vertex u. Thus, $B \in M_r$, and the equality (5.44) is proved. □

Applying the recursive procedure from the proof of Theorem 5.15 to the list $\mathcal{F}_1 = \{P_3\}$, one can obtain the following result.

Theorem 5.16 [33, 17] $M_2 = \text{Forb}(K_{1,3}, W_4, W_4')$ *(see Figure 5.14).*

In Section 5.4.4, we will give a finite characterization of the class M_3. In this case, applying the above recursive procedure to the list of forbidden graphs in Theorem 5.16 is rather tedious. Hence, this characterization will be produced in a different way.

Adding all odd simple cycles of length at least 5 to the list in Figure 5.14, we obtain the following characterization of the class L_{bm} of line graphs of bipartite multigraphs.

Corollary 5.20 [33] $L_{\text{bm}} = \text{Forb}(K_{1,3}, W_4, W_4', C_{2n+1} : n \geq 2)$.

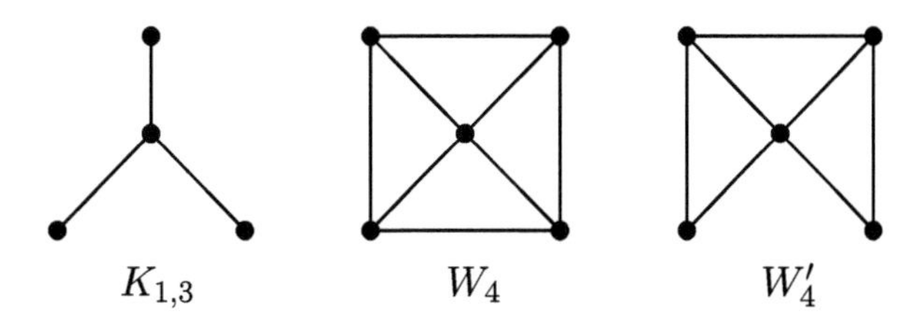

Figure 5.14 List of forbidden subgraphs for the class M_2.

Notice that $L_{\rm bm}$ is exactly the class of graphs with the equivalence covering number at most 2 [33]. Now, adding all simple cycles of length at least 4 to the list in Figure 5.14, we obtain the characterization of the class $\mathrm{Ch_d}$ of chordal dominoes.

Corollary 5.21 [17] *The following statements are true:*

(i) $\mathrm{Ch_d}$ *coincides with the class of line graphs of acyclic multigraphs.*
(ii) $\mathrm{Ch_d} = \mathrm{Forb}(K_{1,3}, W_4, W_4', C_n : n \geq 4)$.

5.4.3 Linear r-Minoes

Let us explore some properties of linear r-minoes.

Proposition 5.17 [25] *The list of maximal cliques and Helly rank dimension of a graph can be found in polynomial time for the class of linear r-minoes.*

Proof: Let G be a linear r-mino, $v \in V(G)$, and let A be a connected component of the subgraph $G(N(v))$. Then, $A \cup \{v\}$ is a maximal clique of G, and each maximal clique can be obtained analogously. Furthermore, $\dim_{HR}(G)$ is the maximal number of connected components of the graphs $G(N(v))$ for all $v \in V(G)$. □

Proposition 5.18 [25] $M_r^\ell = \mathrm{Forb}(K_{1,r+1}, K_4 - e)$.

Proof: Obviously, every edge of a graph belongs to exactly one maximal clique if and only if this graph is $(K_4 - e)$-free. Now, let G be a $(K_4 - e)$-free graph, and let its vertex v belong to exactly p maximal cliques

$$C_1, C_2, \ldots, C_p.$$

Taking an arbitrary vertex $v_i \neq v$ in $C_i, i = 1, 2, \ldots, p$, we obtain the induced star

$$G(v, v_1, v_2, \ldots, v_p) = K_{1,p}.$$

□

Corollary 5.22 *For a linear r-mino G,*

$$\dim_{HR}(G) = \dim_R(G) = \max\{p : G \textit{ contains an induced } K_{1,p}\}.$$

Now, let us consider linear dominoes. Denote $\chi(G)$ and $\chi'(G)$ the chromatic number and the chromatic index (the edge chromatic number) of a graph G, respectively. Theorem 5.17 answers the question posed in [17].

Theorem 5.17 [25] *The decision problem '$\chi(G) \leq 3$?' is NP-complete for linear dominoes with* $\Delta(G) \leq 4$.

Proof: First, we consider the following two decision problems:

$$\chi'(G) \leq 3 \text{ for a graph } G \text{ with } \Delta(G) \leq 3; \tag{5.47}$$

$$\chi'(G) \leq 3 \text{ for a triangle-free graph } G \text{ with } \Delta(G) \leq 3. \tag{5.48}$$

Holyer [16] proved that the problem (5.47) is NP-complete. We shall show that the problem (5.47) can be reduced to the problem (5.48) in polynomial time. This would mean that the problem (5.48) is NP-complete.

Let F be a graph with $\Delta(F) \leq 3$. Consider a triangle with the vertex set $\{a, b, c\}$ in F. Replace this triangle in F by the 7-vertex graph shown in Figure 5.15. Let $\widetilde{F}$ denote the resulting graph. The graph $\widetilde{F}$ has fewer triangles than F. Obviously, the following implication is true (see Figure 5.16):

$$\chi'(F) \leq 3 \Rightarrow \chi'(\widetilde{F}) \leq 3. \tag{5.49}$$

Conversely, assume that $\chi'(\widetilde{F}) \leq 3$. Let φ be a proper 3-colouring of the edges of $\widetilde{F}$, and let us associate with the vertex a the 2-element set $\{a_1, a_2\}$ of colours of the edges aa', ac'. Analogously, define the sets $\{b_1, b_2\}$ and $\{c_1, c_2\}$ for the vertices b and c. The correspondence

$$x \rightarrow \{x_1, x_2\}, \quad x = a, b, c,$$

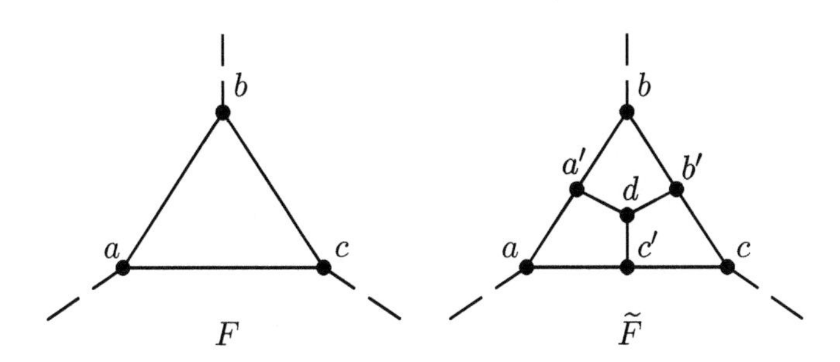

Figure 5.15 Graphs F and $\widetilde{F}$.

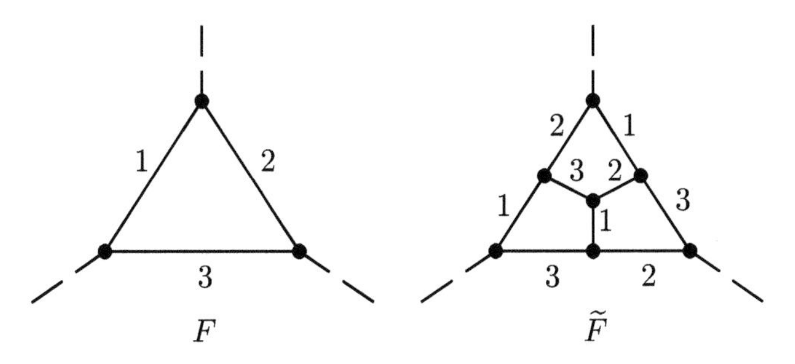

Figure 5.16 Illustration of the implication (5.49).

is injective. Indeed, suppose $\{a_1, a_2\} = \{b_1, b_2\}$ and, without loss of generality,

$$\varphi(aa') = \varphi(bb') = 1, \quad \varphi(ac') = \varphi(a'b) = 2.$$

Then $\varphi(a'd) = 3$, $\varphi(b'd) = 2$, $\varphi(c'd) = 1$. We have $\varphi(cc') = \varphi(b'c) = 3$, a contradiction.

Without loss of generality, let

$$\{a_1, a_2\} = \{1, 2\}, \quad \{b_1, b_2\} = \{1, 3\}, \quad \{c_1, c_2\} = \{2, 3\}.$$

Put

$$\varphi(ab) = 1, \quad \varphi(bc) = 3, \quad \varphi(ac) = 2.$$

The colours of all the other edges in F are the same as in $\widetilde{F}$. Thus, the following implication is true:

$$\chi'(\widetilde{F}) \leq 3 \Rightarrow \chi'(F) \leq 3.$$

Applying the above transformation, we eliminate all triangles in F one after another. Let H denote the resulting graph. If the graph F has no triangles, then put $H = F$. Obviously, the correspondence $F \to H$ is a polynomial reduction of the problem (5.47) to the problem (5.48).

Now, we shall give a polynomial reduction of the NP-complete problem (5.48) to the problem in the statement of Theorem 5.17. Let H be a triangle-free graph with $\Delta(H) \leq 3$ and $G = L(H)$. We have

$$\chi'(H) = \chi(G), \quad \Delta(G) \leq 4.$$

Moreover, G is a linear domino by Corollary 5.17. Obviously, the correspondence $H \to G$ is the required polynomial-time reduction. □

5.4.4 Finite Characterization of the Class of 3-Minoes

The following theorem provides a finite induced subgraph characterization of the class of 3-minoes.

Theorem 5.18 [25] $M_3 = \mathrm{Forb}(\{G_1, G_2, \ldots, G_8\})$, *where the graphs* G_i *are shown in Figure 5.17.*

Proof: One can easily see that

$$M_3 \subseteq \mathrm{Forb}(\{G_1, G_2, \ldots, G_8\}).$$

Now, let $G \in \mathrm{Forb}(\{G_1, G_2, \ldots, G_8\})$. Take an arbitrary vertex $u \in V(G)$ such that $H = G(N(u))$ is not a complete graph. Without loss of generality, suppose that

$$H \text{ has no dominating vertices.} \tag{5.50}$$

Put $G_i' = G_i - v(G_i)$, where $v(G_i)$ is the unique dominating vertex of G_i.

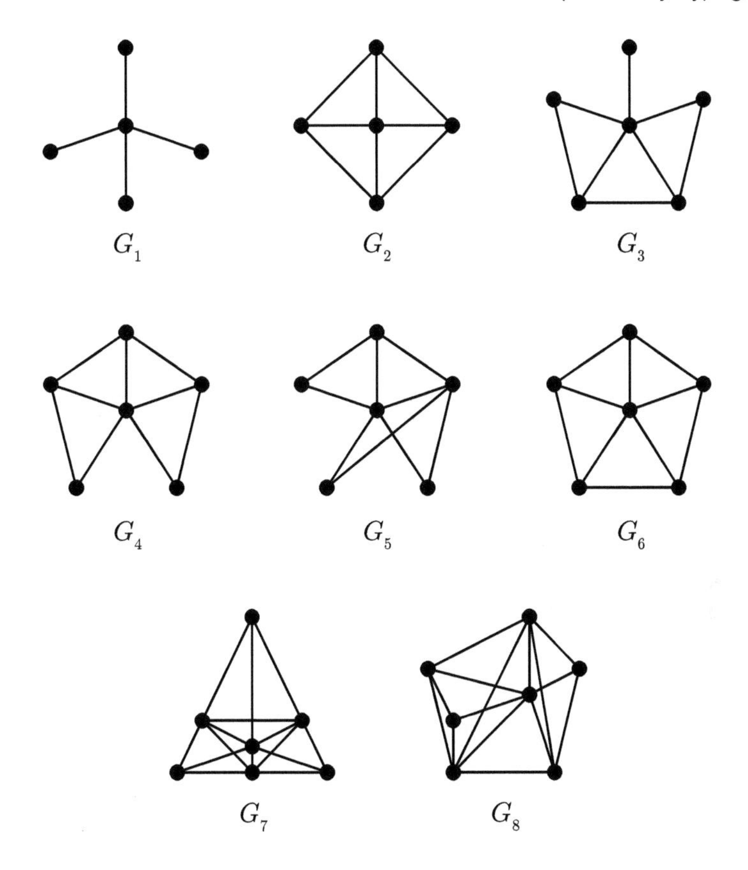

Figure 5.17 List of forbidden subgraphs for the class M_3.

Case 1: Suppose H contains three pairwise non-adjacent vertices x, y, z. Then

$$V(H) = N[x] \cup N[y] \cup N[z] \tag{5.51}$$

and

$$X = N[x],\ Y = N[y],\ Z = N[z] \text{ are cliques in } H. \tag{5.52}$$

The equality (5.51) holds because H does not contain an induced G_1'. Suppose that there exist $x_1, x_2 \in N(x)$ such that $x_1 \not\sim x_2$. The graphs $H(x, x_1, x_2, y)$ and $H(x, x_1, x_2, z)$ are not isomorphic to G_2', and the graph $H(x_1, x_2, y, z)$ is not empty. Therefore, $H(x, x_1, x_2, y, z) \simeq G_3', G_4'$ or G_5', a contradiction.

Thus, (5.51) and (5.52) hold for any pairwise non-adjacent $x, y, z \in V(H)$. Let K be a maximal clique in $H, K \neq X, Y, Z$. For any $a, b \in K$, we have

$$a, b \in X, \quad a, b \in Y \quad \text{or} \quad a, b \in Z. \tag{5.53}$$

In fact, if $a \in X - Y, b \in Y - X$ and $a \notin Z$, then a, y, z are pairwise non-adjacent in H. By (5.52), $N[a]$ is a clique. But $x, b \in N[a]$ and $x \not\sim b$, a contradiction.

By (5.50) and (5.52), each vertex of K belongs to exactly two of the cliques X, Y, Z. Let $a \in K$ and $a \in (X \cap Y) - Z$. Since K is maximal, there exist $b, c \in K$ such that $b \in (X \cap Z) - Y$ and $c \in (Y \cap Z) - X$. We have $H(a, b, c, x, y, z) \simeq G_7'$, a contradiction.

Thus, H has no maximal cliques different from X, Y, Z; that is, the number of maximal cliques $s(H)$ is 3.

Case 2: H does not contain three pairwise non-adjacent vertices. Let x and y be two non-adjacent vertices in H. Clearly, $N[x] \cup N[y] = V(H)$. We set

$$A = N(x) - N(y), \quad B = N(y) - N(x), \quad C = N(x) \cap N(y).$$

Observe that A, B, C are cliques. In fact, $A \cup \{y\}$ and $B \cup \{x\}$ do not contain three pairwise non-adjacent vertices, and $H(C \cup \{x, y\})$ has no induced G_2'. Without loss of generality, assume that no two of the sets A, B, C are empty. Otherwise, $s(H) \leq 2$.

Subcase 2a: $C = \emptyset$.

Suppose that H contains maximal cliques C_1, C_2 different from $A \cup \{x\}$ and $B \cup \{y\}$. Then $C_1, C_2 \subseteq A \cup B$, and there are $a \in A$, $b \in B$, $a \not\sim b$ such that $a \in C_1 - C_2$ and $b \in C_2 - C_1$. Obviously,

$$C_i \cap A \neq \emptyset, \quad C_i \cap B \neq \emptyset, \quad i = 1, 2.$$

Hence, there exist $a_1 \in A$ and $b_1 \in B$ such that $a_1 \sim b$, $b_1 \sim a$. We have $a_1 \sim b_1$ because $H(a, a_1, b, b_1) \not\simeq G_2'$. Therefore, $H(x, y, a, b, a_1, b_1) \simeq G_8'$, a contradiction. Thus, $s(H) \leq 3$.

Subcase 2b: $C \neq \emptyset$.

Lemma 5.22 *If C has no dominating vertices of the graph*

$$F_A = H(A \cup C \cup \{x\}),$$

then $s(F_A) = 2$. If C has no dominating vertices of the graph

$$F_B = H(B \cup C \cup \{y\}),$$

then $s(F_B) = 2$.

Proof: Let C have no dominating vertices of F_A. Then $A \neq \emptyset$. If $A = \{a\}$, then $a \not\sim C$ and F_A contains exactly two maximal cliques: $A \cup \{x\}$ and $C \cup \{x\}$. If $C = \{c\}$, then the only such cliques are $N[c] - (B \cup \{y\})$ and $A \cup \{x\}$.

Let $|A| \geq 2$ and $|C| \geq 2$. Divide A into subsets $A_1 = \{a \in A : a \sim C\}$ and $A_2 = A - A_1$. By the assumptions, $A_2 \neq \emptyset$. Suppose that F_A contains a maximal clique K different from $A \cup \{x\}$ and $C \cup A_1 \cup \{x\}$. Then, there exist $a \in A_2$ and $c \in C$ such that $a, c \in K$. By the definition of A_2, there exists $c' \in C - \{c\}, c' \not\sim a$. By the assumptions, there exists $a' \in A_2, c \not\sim a'$. Since $H(a, c, a', c') \not\simeq G_2'$, we obtain $a' \not\sim c'$. Hence, $H(a, c, a', c', x, y) \simeq G_8'$, a contradiction. Thus, $s(F_A) = 2$. This finishes the proof of the lemma. □

Taking into account (5.50), it follows from Lemma 5.22 that $s(H) \leq 3$ if $A = \emptyset$ or $B = \emptyset$. Hence, we may assume that $A \neq \emptyset$ and $B \neq \emptyset$. Next, we prove that

$$A \sim C \text{ or } B \sim C. \tag{5.54}$$

Otherwise, if there exist $a \in A$, $b \in B$ and $c \in C$ such that $c \not\sim \{a, b\}$, then $H(a, b, c, x, y) \simeq G_4'$ or G_6', a contradiction. If there are $c_1, c_2 \in C$, $a \in A$ and $b \in B$ such that

$$c_1 \not\sim a, \quad c_2 \not\sim b, \quad c_1 \sim b, \quad c_2 \sim a,$$

then $a \not\sim b$ because $H(a, b, c_1, c_2) \not\simeq G_2'$. Hence, $H(a, b, c_1, c_2, x, y) \simeq G_8'$, a contradiction.

Thus, (5.54) is true. Taking (5.50) into account, we conclude that $C \sim A \cup B$ is impossible. Without loss of generality, let $A \sim C$. By Lemma 5.22, $s(F_B) = 2$. The inequality $s(H) \leq 3$ will be proved if we show that $A \not\sim B$. Let $b \in B, c \in C$ and $b \not\sim c$. If there exists $a \in A, a \sim b$, then $H(a, b, c, y) \simeq G_2'$. Otherwise, if $a \in A, b' \in B - \{b\}, a \sim b'$ and $a \not\sim b$, then $b' \sim c$ because $H(a, b', c, y) \not\simeq G_2'$. We have $H(a, b, b', c, x, y) \simeq G_8'$. Thus, $A \not\sim B$ and $s(H) \leq 3$. □

Acknowledgements

This chapter is based on the following publications: (A) Reprinted by permission from Elsevier: Discrete Mathematics, 161, R. I. Tyshkevich and V. E. Zverovich, Line hypergraphs, (1–3), 265–283, ©1996. Some results of Section 5.1 in a less general form were obtained in [20]. (B) Reprinted by permission from John Wiley & Sons: Journal of Graph Theory, 25, Yu. M. Metelsky, R. I. Tyshkevich, On line graphs of linear 3-uniform hypergraphs, (4), 243–251, ©1997. (C) Reprinted by permission from Elsevier: Discrete Mathematics, 309, P. V. Skums, S. V. Suzdal and R. I. Tyshkevich, Edge intersection graphs of linear 3-uniform hypergraphs, (11), 3500–3517, ©2009. (D) Reprinted by permission from Society for Industrial and Applied Mathematics: SIAM Journal on Discrete Mathematics, 16, Yu. M. Metelsky and R. I. Tyshkevich, Line graphs of Helly hypergraphs, (3), 438–448, ©2003.

5.5 References

[1] L. W. Beineke, Derived graphs and digraphs, in H. Sachs, H. Voss and H. Walter (eds), *Beiträge zur Graphentheorie*, Teubner, Leipzig, 1968, 17–33.

[2] C. Berge, *Graphs and Hypergraphs*, Amsterdam, The Netherlands: North-Holland Publishing Company, 1973.

[3] C. Berge, *Hypergraphs: Combinatorics of Finite Sets*, Amsterdam, The Netherlands: Elsevier Science Publishers B.V., 1989.

[4] C. Berge, Isomorphism problems for hypergraphs, *Lecture Notes in Mathematics*, **411** (1974), 1–12.

[5] C. Berge, Une condition pour qu'un hypergraphe soit fortement isomorphe a un hypergraphe complet ou multiparti, *Comptes Rendus de l'Académie des Sciences Paris, Ser. A–B*, **274** (1972), 1783–1786.

[6] C. Berge and R. Rado, Note on isomorphic hypergraphs and some extensions of Whitney's theorem to families of sets, *Journal of Combinatorial Theory, Ser. B*, **13** (1972), 226–241.

[7] J. C. Bermond and J. C. Meyer, Graphe représentaif des arêtes d'un multigraphe, *Journal de Mathématiques Pures et Appliquées*, **52** (9)(1973), 299–308.

[8] W. H. Cunningham, On theorems of Berge and Fournier, *Lecture Notes in Mathematics*, **411** (1974), 67–74.

[9] D. G. Degiorgi and K. Simon, A dynamic algorithm for line graph recognition, International Workshop on Graph-Theoretic Concepts in Computer Science, *Lecture Notes in Computer Science*, **1017** (1995), 37–48.

[10] V. Faber, Hypergraph reconstruction, *Lecture Notes in Mathematics*, **411** (1974), 85–94.

[11] J.-C. Fournier, Sur les isomorphismes d'hypergraphes, *Comptes Rendus de l'Académie des Sciences Paris, Ser. A–B*, **274** (1972), 1612–1614.

[12] J.-C. Fournier, Une condition pour qu'un hypergraphe, ou son complementaire, soit fortement isomorphe a un hypergraphe complet, *Lecture Notes in Mathematics*, **411** (1974), 95–98.

[13] M. L. Gardner, Forbidden configurations in intersection graphs of r-graphs, *Discrete Mathematics*, **31** (1980), 85–88.

[14] M. L. Gardner, Hypergraphs and Whitney theorem on edge isomorphisms of graphs, *Discrete Mathematics*, **51** (1984), 1–9.

[15] P. Hliněný and J. Kratochvíl, Computational complexity of the Krausz dimension of graphs, International Workshop on Graph-Theoretic Concepts in Computer Science, *Lecture Notes in Computer Science*, **1335** (1997), 214–228.

[16] I. Holyer, The NP-completeness of edge-coloring, *SIAM Journal on Computing*, **10** (1981), 718–720.

[17] T. Kloks, D. Kratsch and H. Müller, Dominoes, International Workshop on Graph-Theoretic Concepts in Computer Science, *Lecture Notes in Computer Science*, **903** (1995), 106–120.

[18] J. Krausz, Démonstration nouvelle d'une theoreme de Whitney sur les réseaux, *Matematikai és Fizikai Lapok*, **50** (1943), 75–85.

[19] P. G. H. Lehot, An optimal algorithm to detect a line graph and output its root graph, *Journal of the ACM*, **21** (4)(1974), 569–575.

[20] A. G. Levin and R. I. Tyshkevich, Line hypergraphs, *Discrete Mathematics and Applications*, **3** (4)(1993), 407–428.

[21] L. Lovász, *Combinatorial Problems and Exercises*, Budapest: Akadémiai Kiadó, 1979.

[22] L. Lovász, Problem 9, in *Beiträge zur Graphentheorie und deren Anwendungen.* Vorgetragen auf dem International Kolloquium in Oberhof (DDR), 1977, 313.

[23] T. A. McKee and F. R. McMorris, *Topics in Intersection Graph Theory*, Philadelphia, PA: SIAM, 1999.

[24] Yu. M. Metelsky, S. Suzdal and R. I. Tyshkevich, Recognizing edge intersection graphs of linear 3-uniform hypergraphs, *Proceeding of the Institute of Mathematics of the NAS of Belarus*, **8** (2001), 76–92.

[25] Yu. M. Metelsky and R. I. Tyshkevich, Line graphs of Helly hypergraphs, *SIAM Journal on Discrete Mathematics*, **16** (3)(2003), 438–448.

[26] Yu. M. Metelsky and R. I. Tyshkevich, On line graphs of linear 3-uniform hypergraphs, *Journal of Graph Theory*, **25** (4)(1997), 243–251.

[27] R. N. Naik, S. B. Rao, S. S. Shrikhande and N. M. Singhi, Intersection graphs of k-uniform linear hypergraphs, *European Journal of Combinatorics*, **3** (2)(1982), 159–172.

[28] J. Naor and M. B. Novick, An efficient reconstruction of a graph from its line graph in parallel, *Journal of Algorithms*, **11** (1)(1990), 132–143.

[29] S. Poljak, V. Rödl and D. Turzik, Complexity of representation of graphs by set systems, *Discrete Applied Mathematics*, **3** (1981), 301–312.

[30] R. Rado, Reconstruction theorems for infinite hypergraphs, *Lecture Notes in Mathematics*, **411** (1974), 140–146.

[31] N. D. Roussopoulos, A $\max\{m, n\}$ algorithm for determining the graph H from its line graph G, *Information Processing Letters*, **2** (4)(1973), 108–112.

[32] P. V. Skums, S. V. Suzdal and R. I. Tyshkevich, Edge intersection graphs of linear 3-uniform hypergraphs, *Discrete Mathematics*, **309** (11)(2009), 3500–3517.

[33] R. I. Tyshkevich and O. P. Urbanovich, Graphs with matroidal number 2, *Izvestiya Akademii Nauk BSSR*, Ser. fiz.-mat. nauk, (3)(1989), 13–17.

[34] R. I. Tyshkevich and V. E. Zverovich, Line hypergraphs, *Discrete Mathematics*, **161** (1–3)(1996), 265–283.

[35] H. Whitney, Congruent graphs and the connectivity of graphs, *American Journal of Mathematics*, **54** (1)(1932), 150–168.

6
Dimensionality of Graphs

P. Skums

In this chapter, we will explore the notions of graph dimensions that arise from the theory of intersection graphs described in the previous chapters. We will exploit the duality between decompositions of graphs into cliques and representation of graphs as set systems established by the Berge theorem—a powerful tool that allows to demonstrate interesting connections between combinatorics, general topology and information theory. First, we will discuss two graph dimensions associated with representation of graphs as intersection graphs of hypergraphs: rank dimension and product dimension. Next, we will demonstrate how and why these dimensions can be considered as discrete analogues of the classical topological dimensions: Lebesgue dimension and Hausdorff dimension. We will link these dimensions to graph homomorphisms and show how they measure graph self-similarity. Out of these considerations, an interesting class of graph fractals naturally emerges—a graph-theoretic analogue of topological fractals. We will look at graph fractals in more detail and study their basic properties. Finally, we will establish a link between graph dimensions, graph fractals and graph Kolmogorov complexity, which is one of the basic concepts of information theory and often studied in association with fractal and chaotic systems.

6.1 Preliminaries

6.1.1 Graph Dimensions and Intersection Graphs

Let $G = (V(G), E(G))$ be a simple graph. A family of cliques $\mathcal{C} = (C_1, C_2, \ldots, C_m)$ of G is a *clique covering* if every edge $uv \in E(G)$ is contained in at least one clique from $\mathcal{C}$. The subgraphs that form the covering are referred to as its *clusters*. A covering $\mathcal{C}$ is *k-covering* if every vertex $v \in V(G)$ belongs to at most k clusters. A cluster $C \in \mathcal{C}$ *separates* vertices $u, v \in V(G)$ if $|C \cap \{u, v\}| = 1$. Also, a covering is *separating* if every two distinct vertices are separated by some cluster.

P. Skums, *Dimensionality of Graphs.* In: *Methods of Graph Decompositions.* Edited by: Vadim Zverovich and Pavel Skums, Oxford University Press. © Vadim Zverovich & Pavel Skums (2024). DOI: 10.1093/oso/9780198882091.003.0006

Now, let us consider a hypergraph $\mathcal{H} = (\mathcal{V}(\mathcal{H}), \mathcal{E}(\mathcal{H}))$. The *rank* of $\mathcal{H}$, denoted $\text{rank}(\mathcal{H})$, is the maximal size of an edge of $\mathcal{H}$. The following theorem straightforwardly follows from the Berge theorem and establishes a connection between intersection graphs and clique k-coverings:

Theorem 6.1 [5] *A graph G is the intersection graph of a hypergraph of rank at most k if and only if it has a clique k-covering.*

Rank dimension [29] of a graph G, denoted $\dim_R(G)$, is the minimal k such that G satisfies Theorem 6.1. In particular, graphs with $\dim_R(G) = 1$ are disjoint unions of cliques; such graphs are called *equivalence graphs* [2] or *M-graphs* [38].

A hypergraph $\mathcal{H}$ is *strongly k-colourable* if one can assign colours from the set $\{1, 2, \ldots, k\}$ to its vertices in such a way that vertices in every edge receive different colours. The vertices of the same colour form a *colour class.* For example, strongly 2-colourable simple graphs are *bipartite graphs.* The *edge k-colouring* and *edge colour classes* of a hypergraph are defined analogously, with the condition that the edges that share a vertex receive different colours. The *chromatic number* $\chi(H)$ and the *edge chromatic number* $\chi'(H)$ are the minimal numbers of colours required to colour vertices and edges of a hypergraph, respectively.

Analogously to the rank dimension, we can define a new graph 'colour dimension' (not to be confused with the graph chromatic number) associated with hypergraph colourings as the minimal k such that a graph is the intersection graph of a strongly k-colourable hypergraph. However, we can approach this dimension in a more subtle way. Namely, we will introduce another, seemingly unrelated but a more natural definition, and then demonstrate that in reality it leads to the same parameter.

Categorical product of graphs G_1 and G_2 is the graph $G_1 \times G_2$ with the vertex set

$$V(G_1 \times G_2) = V(G_1) \times V(G_2),$$

where two vertices (u_1, u_2) and (v_1, v_2) are adjacent whenever $u_1v_1 \in E(G_1)$ and $u_2v_2 \in E(G_2)$. Readers familiar with abstract algebra can easily check that this operation indeed is a categorical product in the category theory sense.

Product dimension (also called *Prague dimension* or *Nešetřil-Rödl dimension*), $\dim_P(G)$, is the minimal integer d such that G is an induced subgraph of the categorical product

$$K^1_{n_1} \times K^2_{n_2} \times \ldots \times K^d_{n_d}$$

of d complete graphs [21].

Now, let us return to clique coverings. The *equivalent k-covering* of a graph G is a covering of its edges by equivalence graphs. It can be equally defined as a clique covering $\mathcal{C}$ such that the hypergraph $\mathcal{H}(\mathcal{C}) = (V(G), \mathcal{C})$ is edge k-colourable. Relations between product dimension, clique coverings and intersection graphs are described in the following theorem:

Theorem 6.2 [4, 21] *The following statements are equivalent:*

(1) $\dim_P(\overline{G}) \leq k$.

(2) There exists a separating equivalent k-covering of G.

(3) G is the intersection graph of a strongly k-colourable hypergraph without multiple edges.

(4) There exists an injective mapping

$$\phi : V(G) \to \mathbb{Z}^k, \quad v \mapsto (\phi_1(v), \phi_2(v), \ldots, \phi_k(v))$$

such that $uv \in E(G)$ whenever $\phi_j(u) = \phi_j(v)$ for some $j \in \{1, 2, \ldots, k\}$.

Proof: The equivalence of (1) and (4) follows directly from the definition if we assume that the vertices of complete graphs are labelled by distinct integer numbers. The equivalence of (2) and (3) follows from Theorem 5.3. Thus, it remains to prove the equivalence of (1) and (2). Suppose that $\overline{G}$ is isomorphic to an induced subgraph of the product of k complete graphs:

$$H = K^1_{n_1} \times K^2_{n_2} \times \ldots \times K^k_{n_k}.$$

Then, by definition, two vertices u and v are adjacent in G if and only if their isomorphic images in H have the same i-th coordinate for some $i \in \{1, 2, \ldots, k\}$. Thus, all vertices that share the element x at the i-th coordinate form a clique $C_{i,x}$. It is now straightforward to check that

$$\mathcal{C} = \{C_{i,x} \, : \, i = 1, 2, \ldots, k \text{ and } x \in V(K^i_{n_i})\}$$

is a separating equivalent k-covering of G. Conversely, let $\mathcal{C}$ be a separating equivalent k-covering whose cliques are separated into k colour classes:

$$\mathcal{C} = \mathcal{C}_1 \cup \mathcal{C}_2 \cup \ldots \cup \mathcal{C}_k,$$

and where the colour class $\mathcal{C}_i$ consists of n_i cliques:

$$\mathcal{C}_i = \{C_{i,1}, C_{i,2}, \ldots, C_{i,n_i}\}.$$

Then, we can embed G into the product of k complete graphs

$$K^1_{n_1} \times K^2_{n_2} \times \ldots \times K^k_{n_k}$$

with $V(K^i_{n_i}) = \{1, 2, \ldots, n_i\}$ as follows: set the i-th coordinate of the vertex v as j if $v \in C_{i,j}$. □

In light of Theorem 6.2, let us refer to $\dim_P(\overline{G})$ as *co-product dimension* of G. A simple corollary of Theorem 6.2 is that the product and co-product dimensions are well defined for all graphs.

Corollary 6.1 *Every graph G is isomorphic to an induced subgraph of the categorical product of complete graphs.*

In particular, graphs whose complements have product dimension of 1 or 2 are exactly the line graphs of bipartite graphs.

6.1.2 Some Definitions from General Topology

Since our goal is to establish connections between the graph-theoretic dimensions described in the previous section and several classical dimensions from general topology, we need to introduce some topological definitions. Let X be a compact metric space. A family $\mathcal{C} = \{C_\alpha : \alpha \in A\}$ of open subsets of X is a *covering* if $X = \bigcup_{\alpha \in A} C_\alpha$. A covering $\mathcal{C}$ is a *k-covering* if every $x \in X$ belongs to at most k sets from $\mathcal{C}$, and an *ϵ-covering* if for every set $C_i \in \mathcal{C}$ its diameter, $\mathrm{diam}(C_i)$, does not exceed ϵ. A covering is called an *(ϵ, k)-covering* if it is both an ϵ-covering and a k-covering. *Lebesgue dimension* of X, $\dim_L(X)$, is the minimal integer k such that for every $\epsilon > 0$ there exists an $(\epsilon, k+1)$-covering of X.

Let $\mathcal{F}$ be a semiring of subsets of a set X. A function $m : \mathcal{F} \to \mathbb{R}_0^+$ is a *measure* if $m(\emptyset) = 0$ and for any countable collection of pairwise disjoint sets

$$A_i \in \mathcal{F}, \quad i = 1, 2, \ldots, \infty,$$

one has

$$m\left(\bigcup_{i=1}^{\infty} A_i\right) = \sum_{i=1}^{\infty} m(A_i).$$

Let now X be a subspace of an Euclidean space $\mathbb{R}^d$. A *hyper-rectangle* R is the Cartesian product of semi-open intervals:

$$R = [a_1, b_1) \times [a_2, b_2) \times \ldots \times [a_d, b_d),$$

where $a_i, b_i \in \mathbb{R}$. The *volume* of the hyper-rectangle R is equal to

$$\mathrm{vol}(R) = \prod_{i=1}^{d} (b_i - a_i).$$

The *d-dimensional Jordan measure* of the set X is defined as

$$\mathcal{J}^d(X) = \inf\left\{\sum_{R \in C} \mathrm{vol}(R)\right\},$$

where the infimum is taken over all finite coverings $\mathcal{C}$ of X by disjoint hyper-rectangles. The *d-dimensional Lebesgue measure* of a measurable set $\mathcal{L}^d(X)$ is defined analogously, with the infimum taken over all countable coverings $\mathcal{C}$ of X by (not necessarily disjoint) hyper-rectangles. Finally, the *d-dimensional Hausdorff measure* of the set X is defined as

$$\mathcal{H}^d(X) = \lim_{\epsilon \to 0} \mathcal{H}^d_\epsilon(X),$$

where

$$\mathcal{H}^d_\epsilon(X) = \inf\left\{\sum_{C \in \mathcal{C}} \mathrm{diam}(C)^d\right\}$$

and the infimum is taken over all ϵ-coverings of X. These three measures are related: the Jordan and Lebesgue measures of the set X are equal if the former exists, whereas the Lebesgue and Hausdorff measures of Borel sets differ only by a multiplicative constant.

Hausdorff dimension of the set X, $\dim_H(X)$, is the value

$$\dim_H(X) = \inf\{s \geq 0 : \mathcal{H}^s(X) < \infty\}. \tag{6.1}$$

Notice that Lebesgue and Hausdorff dimensions of X are related as follows:

$$\dim_L(X) \leq \dim_H(X). \tag{6.2}$$

By Mandelbrot's definition, the set X is a *fractal* [14] if inequality (6.2) is strict.

6.2 Dimensions

6.2.1 Rank Dimension as Lebesgue Dimension of Graphs

Lebesgue dimension of a metric space is defined through k-coverings by sets of arbitrary small diameter. It is natural to transfer this definition to graphs using graph k-coverings by subgraphs of smallest possible diameter, i.e. by cliques. Thus, in light of Theorem 6.1, we define Lebesgue dimension of a graph through its rank dimension:

$$\dim_L(G) = \dim_R(G) - 1. \tag{6.3}$$

An analogy between Lebesgue dimension of a metric space and rank dimension of a graph is reinforced by Theorem 6.3. This theorem basically extends the analogy from graph theory back to general topology by stating that any compact metric spaces of bounded Lebesgue measure could be approximated by intersection graphs of (infinite) hypergraphs of bounded rank. To prove this, we will use the following fact from general topology (without proof):

Lemma 6.1 [14] *Let X be a compact metric space and $\mathcal{U}$ be its open covering. Then, there exists $\delta > 0$ (called a* Lebesgue number *of $\mathcal{U}$) such that for every subset $A \subseteq X$ with* $\operatorname{diam}(A) < \delta$ *there is a set $U \in \mathcal{U}$ such that $A \subseteq U$.*

The following theorem establishes that $\dim_L(X) \leq k$ whenever for any $\epsilon > 0$ there is a hypergraph $\mathcal{H}(\epsilon)$ of rank at most k with edges in bijective correspondence with points of X such that two points are close if and only if the corresponding edges intersect.

Theorem 6.3 [34] *Let X be a compact metric space with metric ρ. Then, $\dim_L(X) \leq k - 1$ if and only if for any $\epsilon > 0$ there exist a number δ, $0 < \delta < \epsilon$, and a hypergraph $\mathcal{H}(\epsilon)$ on a finite vertex set $V(\mathcal{H}(\epsilon))$ with the edge set $E(\mathcal{H}(\epsilon)) = \{e_x : x \in X\}$ satisfying the following conditions:*

(1) $\operatorname{rank}(\mathcal{H}(\epsilon)) \leq k$,

(2) $e_x \cap e_y \neq \emptyset$ *for every $x, y \in X$ such that $\rho(x, y) < \delta$,*

(3) $\rho(x, y) < \epsilon$ *for every* $x, y \in X$ *such that* $e_x \cap e_y \neq \emptyset$,
(4) *for every* $v \in V(\mathcal{H}(\epsilon))$, *the set* $X_v = \{x \in X : v \in e_x\}$ *is open.*

Proof: Suppose that $\dim_L(X) \leq k-1$ and $\epsilon > 0$, and let $\mathcal{C}$ be the corresponding (ϵ, k)-covering of X. Since X is compact, we can assume that the covering $\mathcal{C}$ is finite, i.e. $\mathcal{C} = \{C_1, C_2, \ldots, C_m\}$. Let δ be the Lebesgue number of $\mathcal{C}$.

For a point $x \in X$, let $e_x = \{i \in [m] : x \in C_i\}$, where $[m] = \{1, 2, \ldots, m\}$. Consider the hypergraph $\mathcal{H}$ with $V(\mathcal{H}) = [m]$ and $E(\mathcal{H}) = \{e_x : x \in X\}$. Then, $\mathcal{H}$ satisfies the above conditions (1)–(4). Indeed, $\text{rank}(\mathcal{H}) \leq k$ because $\mathcal{C}$ is a k-covering. If $\rho(x, y) < \delta$ then, by Lemma 6.1, there is $i \in [m]$ such that $\{x, y\} \in C_i$, i.e. $i \in e_x \cap e_y$. The condition $j \in e_x \cap e_y$ means that $x, y \in C_j$, and so $\rho(x, y) < \epsilon$, since $\text{diam}(C_j) < \epsilon$. Finally, for every $i \in V(\mathcal{H})$ we have $X_v = C_v$, and thus X_v is open.

Conversely, let $\mathcal{H}$ be a hypergraph with $V(\mathcal{H}) = [m]$ that satisfies the conditions (1)–(4). Then, it is straightforward to check that $\mathcal{C} = \{X_1, X_2, \ldots, X_m\}$ is an open (ϵ, k)-covering of X. □

It is easy to see that the clique covering consisting of all edges of G is a $\Delta(G)$-covering, where $\Delta(G)$ is the maximum vertex degree of G. Such a covering can be used to obtain the following upper bound for $\dim_L(G)$:

Proposition 6.1 [34] $\dim_L(G) \leq \Delta(G) - 1$, *and the equality holds if* G *is triangle-free.*

6.2.2 Co-product Dimension as Hausdorff Dimension of Graphs

Now, we will demonstrate that co-product dimension is a graph-theoretic analogue of Hausdorff dimension. First, we establish a formal connection by proving that co-product dimension is associated with a graph measure analogous to the Hausdorff measure of topological spaces. Second, we demonstrate that co-product dimension is also associated with graph self-similarity.

Graph Measure

In order to rigorously define a graph analogue of Hausdorff dimension, we need to define the corresponding measure first. Note that in any meaningful finite graph topology, every set is a Borel set. As mentioned above, for measurable Borel sets in $\mathbb{R}^n$, the Jordan, Lebesgue and Hausdorff measures are equivalent. Thus, further we will consider the graph analogue of the Jordan measure. We will discuss a parameter that is aimed to serve as the graph analogue of the Jordan measure and prove that it indeed satisfies the axioms of measure. Finally, based on this parameter, Hausdorff dimension of a graph will be defined.

By Corollary 6.1, every graph G is isomorphic to an induced subgraph G' of the categorical product

$$K^1_{n_1} \times K^2_{n_2} \times \ldots \times K^d_{n_d}$$

of complete graphs [21]. Without loss of generality, we may assume that

$$n_1 = n_2 = \ldots = n_d = n;$$

that is, G' is an induced subgraph of the graph $\mathbb{S}_d = (K_n)^d$. The graph $\mathbb{S}_d$ will be referred to as the *space of dimension* d and G' as an *embedding* of G into $\mathbb{S}_d$. After assuming that $V(K_n) = [n]$, we may say that every vertex $v \in \mathbb{S}_d$ is a vector

$$v = (v_1, v_2, \ldots, v_d) \in [n]^d,$$

and two vertices v and u are adjacent if and only if $v_r \neq u_r$ for every $r \in [d]$.

The *hyper-rectangle* $R = R(J_1, J_2, \ldots, J_d)$ is a subgraph of $\mathbb{S}_d$ that is defined as follows:

$$R = K_n(J_1) \times K_n(J_2) \times \ldots \times K_n(J_d),$$

where the set $J_i \subseteq [n]$ is non-empty for every $i \in [d]$. The *volume* of R is naturally defined as

$$\text{vol}(R) = |V(R)| = \prod_{i=1}^{d} |J_i|.$$

The family $\mathcal{R} = \{R^1, R^2, \ldots, R^m\}$ of hyper-rectangles is a *rectangle co-covering* of G' if the subgraphs R^i are pairwise vertex-disjoint, $V(G') \subseteq \bigcup_{i=1}^{m} V(R^i)$ and $\mathcal{R}$ covers all non-edges of G'; that is, for every pair of non-adjacent vertices $x, y \in V(G')$ there exists $j \in [m]$ such that $x, y \in V(R^j)$. The *d-volume* of a graph G is defined as follows:

$$\text{vol}^d(G) = \min_{G'} \min_{\mathcal{R}} \sum_{R \in \mathcal{R}} \text{vol}(R), \tag{6.4}$$

where the first minimum is taken over all embeddings G' of G into d-dimensional spaces $\mathbb{S}_d$ and the second minimum is taken over all rectangle co-coverings of G'. For example, Figure 6.1 (top) demonstrates that the 2-dimensional volume of the path P_4 is equal to 6.

Based on the definition of d-volume, we define the *d-measure* of a graph F as follows:

$$\mathcal{H}^d(F) = \begin{cases} \text{vol}^d(\overline{F}) \text{ if } \overline{F} \text{ has an embedding into } \mathbb{S}_d; \\ +\infty \text{ otherwise.} \end{cases} \tag{6.5}$$

The main theorem of this section confirms that $\mathcal{H}^d$ indeed satisfies the axioms of measure:

Theorem 6.4 [34] *Let F_1 and F_2 be two graphs, and $F^1 \cup F^2$ is their disjoint union. Then*

$$\mathcal{H}^d(F^1 \cup F^2) = \mathcal{H}^d(F^1) + \mathcal{H}^d(F^2). \tag{6.6}$$

Proof: It can be shown [3] that $\overline{F^1 \cup F^2}$ can be embedded into the categorical product of d complete graphs if and only if both $\overline{F_1}$ and $\overline{F_2}$ have such embeddings.

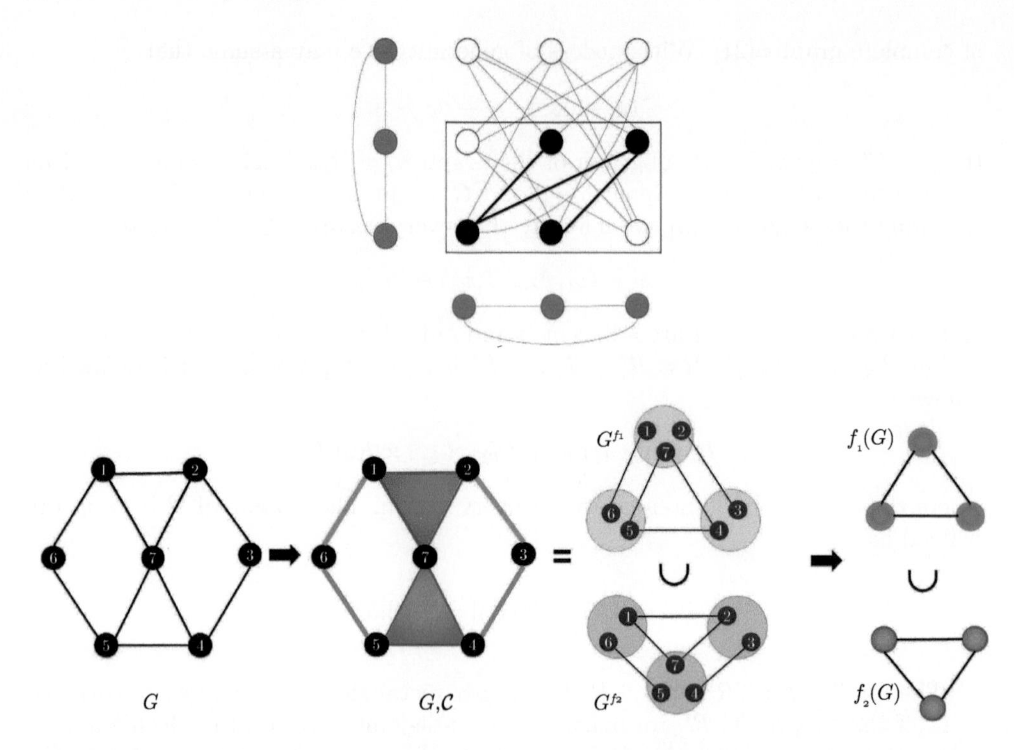

Figure 6.1 *Top*: Embedding of P_4 into the 2-dimensional space $\mathbb{S}_2 = (K_3)^2$ and its rectangle co-covering by a hyper-rectangle of volume 6. *Bottom*: Equivalent k-covering that determines self-similarity of a graph G. *From left to right*: graph G; equivalent 2-covering $\mathcal{C}$ of G with clusters of the same colour highlighted in brown and green; subgraphs G^{f_1} and G^{f_2} such that $G = G^{f_1} \cup G^{f_2}$ for the contracting family (f_1, f_2) defined by $\mathcal{C}$; contractions $f_1(G)$ and $f_2(G)$.

Therefore, relation (6.6) holds if some of the terms are equal to $+\infty$. If all $\overline{F_1}$, $\overline{F_2}$ and $\overline{F^1 \cup F^2}$ can be embedded into $\mathbb{S}_d$, then we proceed with a series of claims.

Let $W_1, W_2 \subseteq V(\mathbb{S}_d)$. We write $W_1 \sim W_2$ if every vertex from W_1 is adjacent to every vertex from W_2. Denote by $P_k(W_1)$ *the k-th projection of* W_1; that is, the set of all k-coordinates of vertices of W_1:

$$P_k(W_1) = \{v_k : v \in W_1\}.$$

In particular,

$$P_k(R(J_1, J_2, \dots, J_d)) = J_k.$$

The definition of $\mathbb{S}_d$ immediately implies the following claim:

Claim 6.1 $W_1 \sim W_2$ *if and only if* $P_k(W_1) \cap P_k(W_2) = \emptyset$ *for every* $k \in [d]$.

Assume that $\mathcal{R} = \{R^1, R^2, \ldots, R^m\}$ is the minimal rectangle co-covering of the minimal embedding G', i.e.

$$\mathrm{vol}^d(G) = \sum_{R \in \mathcal{R}} \mathrm{vol}(R).$$

We will demonstrate that $\mathcal{R}$ has a rather simple structure.

Claim 6.2 *Let $J_k^i = P_k(R^i)$. Then $J_k^i \cap J_k^j = \emptyset$ for every $i, j \in [m]$, $i \neq j$ and every $k \in [d]$.*

Proof: First note that for every hyper-rectangle $R^i = R(J_1^i, J_2^i, \ldots, J_d^i)$, every coordinate $k \in [d]$ and every $l \in J_k^i$, there exists $v \in V(G') \cap V(R^i)$ such that $v_k = l$. Indeed, suppose that it does not hold for some $l \in J_k^i$. If $|J_k^i| = 1$, this means that $V(G') \cap V(R^i) = \emptyset$. Hence, $\mathcal{R}' = \mathcal{R} - \{R^i\}$ is a rectangle co-covering, which contradicts the minimality of $\mathcal{R}$. If $|J_k^i| > 1$, consider the hyper-rectangle

$$(R^i)' = R(J_1^i, \ldots, J_{k-1}^i, J_k^i - \{l\}, J_{k+1}^i, \ldots, J_d^i).$$

The set $\mathcal{R}' = \mathcal{R} - \{R^i\} \cup \{(R^i)'\}$ is a rectangle co-covering, and the d-volume of $(R^i)'$ is smaller than the d-volume of R^i. Again, this contradicts the minimality of $\mathcal{R}$.

Now, assume that for some distinct $i, j \in [m]$ and $k \in [d]$ we have $J_k^i \cap J_k^j \supseteq \{l\}$. Then, there exist $u \in V(G') \cap V(R^i)$ and $v \in V(G') \cap V(R^j)$ such that $u_k = v_k = l$. So, $uv \notin E(G')$ and therefore, by definition, uv is covered by some $R^h \in \mathcal{R}$. The hyper-rectangle R^h intersects both R^i and R^j, which contradicts the definition of a rectangle co-covering. □

Claim 6.3 *Let $U^i = V(G') \cap V(R^i)$, $i = 1, 2, \ldots, m$. Then, the set $U = \{U^1, U^2, \ldots, U^m\}$ coincides with the set of co-connected components of G'.*

Proof: Claims 6.1 and 6.2 imply that the vertices of distinct hyper-rectangles from the co-covering $\mathcal{R}$ are pairwise adjacent. So, $U^i \sim U^j$ for every $i, j \in [m]$, $i \neq j$.

Let $\mathcal{C} = \{C^1, C^2, \ldots, C^r\}$ be the set of co-connected components of G', so that

$$V(G') = \bigsqcup_{l=1}^{r} C^l = \bigsqcup_{i=1}^{m} U^i,$$

where $\sqcup$ denotes the disjoint union. Consider the component $C^l \in \mathcal{C}$ and the sets $C_i^l = C^l \cap U^i$, $i = 1, 2, \ldots, m$. We have, $C^l = \bigsqcup_{i=1}^m C_i^l$ and $C_i^l \sim C_j^l$ for all $i \neq j$. Therefore, due to co-connectedness of C^l, exactly one of the sets C_i^l is non-empty.

Thus, we showed that every co-connected component C^l is contained in some of the sets U^i. Now, let some U_i consist of several components; that is, without loss of generality,

$$U_i = C^1 \sqcup C^2 \sqcup \ldots \sqcup C^q, \quad q \geq 2.$$

Denote $I_k^j = P_k(C^j)$, $j = 1, 2, \ldots, q$. By Claim 6.2, we have $I_k^{j_1} \cap I_k^{j_2} = \emptyset$ for all $j_1 \neq j_2$, $k = 1, 2, \ldots, d$. Let us consider the following hyper-rectangles:

$$R^{i,1} = R(I_1^1, I_2^1, \ldots, I_d^1), \; \ldots\ldots \; , R^{i,q} = R(I_1^q, I_2^q, \ldots, I_d^q).$$

These hyper-rectangles are pairwise vertex-disjoint, and $C^j \subseteq V(R^{i,j})$ for all $j \in [q]$. Since every pair of non-connected vertices in G' is contained in some of its co-connected components, we conclude that the set

$$\mathcal{R}' = \mathcal{R} - \{R^i\} \cup \{R^{i,1}, R^{i,2}, \ldots, R^{i,q}\}$$

is a rectangle co-covering. Moreover,

$$V(R^{i,1}) \sqcup V(R^{i,2}) \sqcup \ldots \sqcup V(R^{i,q}) \subsetneq V(R^i)$$

and, therefore,

$$\sum_{j=1}^{q} \mathrm{vol}(R^{i,j}) < \mathrm{vol}(R^i),$$

which contradicts the minimality of $\mathcal{R}$. □

Claim 6.4 *Let $\mathcal{C} = \{C^1, C^2, \ldots, C^m\}$ be the set of co-connected components of G'. Then*

$$R^i = R(P_1(C^i), P_2(C^i), \ldots, P_d(C^i)) \supseteq C^i.$$

Proof: By Claim 6.3, $|\mathcal{R}| = |\mathcal{C}|$, and every component $C^i \in \mathcal{C}$ is contained in the unique hyper-rectangle $R^i \in \mathcal{R}$, $i = 1, 2, \ldots, m$. Every pair of non-adjacent vertices of G' belongs to some of its co-connected components. This fact, together with the minimality of $\mathcal{R}$, implies that R^i is the minimal hyper-rectangle that contains C^i. Thus,

$$R^i = R(P_1(C^i), P_2(C^i), \ldots, P_d(C^i)).$$

□

Claim 6.5 *If $\overline{G}$ is connected, then $\mathrm{vol}^d(G)$ is the minimal volume of all d-dimensional spaces $\mathbb{S}_d$, where G can be embedded.*

Claim 6.6 *Let $\mathcal{D} = \{D^1, D^2, \ldots, D^m\}$ be the set of co-connected components of G. Then*

$$\mathrm{vol}^d(G) = \sum_{i=1}^{m} \mathrm{vol}^d(G(D^i)).$$

Proof: Suppose that $\{C^1, C^2, \ldots, C^m\}$ is the set of co-connected components of G', and $G(D^i) \simeq G'(C^i)$. Claims 6.3 and 6.4 imply that $\{R_i\}$ is a rectangle co-covering of an embedding of $G(D^i)$. Therefore, we have

$$\text{vol}(R^i) \geq \text{vol}^d(G(D^i)),$$

and thus

$$\text{vol}^d(G) = \sum_{i=1}^{m} \text{vol}(R^i) \geq \sum_{i=1}^{m} \text{vol}^d(G(D^i)).$$

So, it remains to prove the inverse inequality.

Let G'^i be a minimal embedding of $G(D^i)$ into $\mathbb{S}_d$. By Claim 6.3, every minimal hyper-rectangle co-covering of G'^i consists of a single hyper-rectangle $R^i = R(J_1^i, J_2^i, \ldots, J_d^i)$ or, in other words, $G(D^i)$ is embedded into R^i. Now, we can construct the embedding G' of G into $\mathbb{S}_d = (K_{nm})^d$ and its hyper-rectangle co-covering. This can be done as follows. Let us consider

$$I_k^i = \{(i-1)m + l : l \in J_k^i\}, \quad k = 1, 2, \ldots, d,$$

and let

$$Q^i = K_{mn}(I_1^i) \times K_{mn}(I_2^i) \times \ldots \times K_{mn}(I_d^i).$$

Obviously, $Q^i \simeq R^i$. Now, let us embed $G(D^i)$ into Q^i, and let G'^i be those embeddings. By Claim 6.1, $V(G'^i) \sim V(G'^j)$, so

$$G' = G'^1 \cup G'^2 \cup \ldots \cup G'^m$$

is indeed an embedding of G.

All hyper-rectangles Q^i are pairwise disjoint. Also, $\mathcal{Q} = \{Q^1, Q^2, \ldots, Q^m\}$ is a hyper-rectangle co-covering of G' because every pair of non-adjacent vertices belongs to some subgraph $G(D^i)$. Therefore,

$$\text{vol}^d(G) \leq \sum_{i=1}^{m} \text{vol}(Q^i) = \sum_{i=1}^{m} \text{vol}^d(G(D^i)).$$

□

Now, equality (6.6) follows from Claim 6.6. This concludes the proof of the theorem. □

Following the analogy with Hausdorff dimension of topological spaces (6.1), we define *Hausdorff dimension of a graph* G as follows:

$$\dim_H(G) = \min\{s \geq 0 : \mathcal{H}^s(G) < \infty\} - 1. \tag{6.7}$$

Thus, Hausdorff dimension of a graph can be identified with product dimension of its complement.

Self-similarity

The self-similarity of a compact metric space (X, d) is defined using the notion of *contraction* [14]. An open mapping $f : X \to X$ is a *similarity mapping* if

$$d(f(u), f(v)) \leq \alpha d(u, v) \quad \text{for all} \quad u, v \in X,$$

where α is called its *similarity ratio.* Notice that such a mapping is obviously continuous. If $\alpha < 1$, then it is a *contraction.* The space X is *self-similar* if there exists a family of contractions $f_1, f_2, \ldots, f_k$ such that

$$X = \bigcup_{i=1}^{k} f_i(X).$$

This definition cannot be directly applied to discrete metric spaces such as graphs because for them contractions in the strict sense do not exist. To define formally and rigorously the self-similarity of graphs, we proceed as follows. It is convenient to assume that every vertex is adjacent to itself. For two graphs G and H, a *homomorphism* [21] is a mapping $f : V(G) \to V(H)$ which maps adjacent vertices to adjacent vertices, i.e. $f(u)f(v) \in E(H)$ for every $uv \in E(G)$. A homomorphism f is a *similarity mapping* if inverse images of adjacent vertices are also adjacent, i.e. $uv \in E(G)$ whenever $f(u)f(v) \in E(H)$ (it is possible that $f(u) = f(v)$). In other words, for a similarity mapping, images and inverse images of cliques are cliques. With a similarity mapping f we can associate a subgraph G^f of G, which is formed by all edges uv such that $f(u) \neq f(v)$ (see Figure 6.1).

A family of graph similarity mappings

$$f_i : V(G) \to V(G_i), \quad i = 1, 2, \ldots, k,$$

is a *contracting family* if every edge of G is contracted by some mapping; that is, for every $uv \in E(G)$ there exists $i \in \{1, 2, \ldots, k\}$ such that $f_i(u) = f_i(v)$. The graphs G_i are *contractions* of G. Finally, the graph G is called *self-similar* if $G = \bigcup_{i=1}^{k} G^{f_i}$ (see Figure 6.1).

Proposition 6.2 [34] *A graph G is self-similar with a contracting family $f_1, f_2, \ldots, f_k$ if and only if there is an equivalent separating k-covering of G.*

Proof: For the given contracting family $f_1, f_2, \ldots, f_k$ and any $i \in \{1, 2, \ldots, k\}$, the sets

$$\mathcal{C}_i = \{f_i^{-1}(v) : v \in V(G_i)\}$$

consist of disjoint cliques. By definition, every edge of G is covered by one of these cliques. Therefore, $\mathcal{C} = \mathcal{C}_1 \cup \mathcal{C}_2 \cup \ldots \cup \mathcal{C}_k$ is an equivalent k-covering of G. Furthermore, due to the self-similarity of G, for every edge $xy \in E(G)$ there is a mapping f_i that does not contract it, i.e. $f_i(x) = u \neq v = f_i(y)$. Thus, x and y are separated by the cliques $f_i^{-1}(u)$ and $f_i^{-1}(v)$, and therefore $\mathcal{C}$ is a separating covering.

Conversely, let $\mathcal{C} = \mathcal{C}_1 \cup \mathcal{C}_2 \cup \ldots \cup \mathcal{C}_k$ be a separating equivalent k-covering, where

$$\mathcal{C}_i = (C_i^1, C_i^2, \ldots, C_i^{r_i})$$

is the set of connected components of the i-th equivalence graph (some of them may consist of a single vertex). Now, let us construct the graph G_i by contracting every clique C_i^j into a single vertex v_i^j and define the mapping f_i by setting $f_i(C_i^j) = v_i^j$. Then, the collection $(f_1, f_2, \ldots, f_k)$ is a contracting family. □

According to Proposition 6.2, all graphs could be considered as self-similar—for example, we can construct $|E(G)|$ trivial similarity mappings by contracting every edge individually. Thus, it is natural to concentrate our attention on non-trivial similarity mappings and measure the degree of graph self-similarity by the minimal number of similarity mappings in a contracting family, i.e. by its Hausdorff dimension. The smaller number of similarity mappings indicates the denser packing of a graph by its contraction subgraphs, i.e. the higher self-similarity degree. In particular, the *normalized Hausdorff dimension*

$$\overline{\dim}_H(G) = \frac{\dim_H(G)}{|V(G)|}$$

could serve as a measure of self-similarity.

6.2.3 Lebesgue and Hausdorff Dimensions of Random Graphs

Let us establish upper and lower bounds on graph dimensions for two widely used models of random graphs: scale-free graphs and Erdős–Rényi graphs.

Scale-free graphs are defined in the literature as graphs whose degree distribution (asymptotically) follows the power law. This means that the probability that a given vertex has degree d could be approximated by the function $cd^{-\alpha}$, where c is a constant and α is a *scaling exponent*. There is a number of models of scale-free networks of different degree of mathematical rigour. Here, we will consider the following formal probabilistic model [23,12]. For each vertex $i \in \{1, 2, \ldots, n\}$, we assign the weight $w_i = \left(\frac{n}{i}\right)^{1/\alpha}$. Then, we construct a graph $G(n, \alpha)$ by independently connecting any pair of vertices i, j by an edge with probability $p_{ij} = 1 - e^{-\lambda_{ij}}$, where $\lambda_{ij} = b\frac{w_i w_j}{n}$ and b is a constant.

From now on, we will use the following standard nomenclature [9]. An induced subgraph isomorphic to a cycle is a *hole*, a hole with the odd number of vertices is an *odd hole*. The star $K_{1,3}$ is the *claw*, the 4-vertex graph consisting of two triangles with a common edge is the *diamond* and the 5-vertex graph consisting of two triangles with a common vertex is the *butterfly*.

Theorem 6.5 [34] *For graphs $G = G(n, \alpha)$ with $\alpha > \frac{12}{5}$ and maximum degree $\Delta = \Delta(G)$, with high probability*

$$\dim_L(G) \in \{\Delta - 2, \Delta - 1\}$$

and

$$\dim_H(G) \in \{\Delta - 2, \Delta - 1, \Delta, \Delta + 1\}.$$

Proof: It was proved in [23] that the clique number of a graph $G(n, \alpha)$ with the scaling exponent $\alpha > 2$ is either 2 or 3 with high probability:

$$\mathbb{P}[\omega(G(n, \alpha)) \in \{2, 3\}] \to 1 \quad \text{as} \quad n \to \infty$$

for the n-vertex scale-free graph $G(n, \alpha)$ that has the power-law degree distribution with the exponent α. Lemma 6.2 complements this fact.

Lemma 6.2 [34] *The following statements are true:*

(1) For $\alpha > 2.4$, graphs $G(n, \alpha)$ with high probability do not contain diamonds as induced subgraphs.

(2) For $\alpha > 2$, graphs $G(n, \alpha)$ with high probability do not contain butterflies as induced subgraphs.

Proof: **(1)** Let vertices i, j, k, l form a diamond, where i and j are non-adjacent. For the probability of this event, we have

$$\begin{aligned} p' &= p_{ik} p_{jk} p_{il} p_{jl} p_{kl} (1 - p_{ij}) \\ &\leq p_{ik} p_{jk} p_{il} p_{jl} p_{kl} \\ &\leq \lambda_{ik} \lambda_{jk} \lambda_{il} \lambda_{jl} \lambda_{kl} \\ &= \frac{b^5}{n^5} w_i^2 w_j^2 w_k^3 w_l^3. \end{aligned}$$

Thus, the total probability that those vertices form a diamond (i.e. any two selected distinct vertices may be non-adjacent) can be estimated as

$$\begin{aligned} p &\leq \frac{b^5}{n^5} w_i^2 w_j^2 w_k^2 w_l^2 (w_i w_j + w_i w_k + w_i w_l + w_j w_k + w_j w_l + w_k w_l) \\ &\leq 6 \frac{b^5}{n^5} w_i^3 w_j^3 w_k^3 w_l^3. \end{aligned}$$

Let X_D be the number of diamonds in $G(n, \alpha)$. For the expected value of this random variable, we have

$$\begin{aligned} \mathbb{E}[X_D] &\leq 6 \frac{b^5}{n^5} \sum_{\{i,j,k,l\} \subseteq [n]} w_i^3 w_j^3 w_k^3 w_l^3 \\ &\leq 6 \frac{b^5}{n^5} \left(\sum_{i=1}^{n} w_i^3 \right)^4 \\ &= 6 \frac{b^5}{n^5} \left(\sum_{i=1}^{n} \left(\frac{n}{i} \right)^{\frac{3}{\alpha}} \right)^4. \end{aligned}$$

Using an integral upper bound, it is easy to see that

$$g_r(n) = \sum_{i=1}^{n} \left(\frac{n}{i}\right)^r = \begin{cases} O(n^r) \text{ if } r > 1; \\ O(n\log(n)) \text{ if } r = 1; \\ O(n) \text{ if } r < 1. \end{cases} \tag{6.8}$$

Furthermore, $g_r(n) \leq g_s(n)$ whenever $r \leq s$.

Let us select $\beta \in (0, \frac{1}{4})$ such that $\alpha \geq p = \frac{12}{5}(1+\beta)$. Then, we have

$$g_{\frac{3}{\alpha}}(n) \leq g_{\frac{3}{p}}(n) = O\left(n^{\frac{3}{p}}\right).$$

Thus,

$$\mathbb{E}[X_D] = O\left(n^{\frac{12}{p}-5}\right) = O\left(n^{-5\left(1-\frac{1}{1+\beta}\right)}\right) = o(1).$$

Finally, by Markov's inequality, we have

$$\mathbb{P}[X_D \geq 1] \;\leq\; \mathbb{E}[X_D] \to 0 \quad \text{as} \quad n \to +\infty.$$

(2) Similar to (1), for the probability p' that vertices i, j, k, l, r form a butterfly with the centre i, we have

$$p' \leq \frac{b^6}{n^6} w_i^4 w_j^2 w_k^2 w_l^2 w_r^2.$$

Thus, for the number of butterflies X_B, its expectation satisfies the following chain of inequalities:

$$\begin{aligned} \mathbb{E}[X_D] &\leq \frac{b^6}{n^6} \sum_{i=1}^{n} w_i^4 \sum_{\{j,k,l,r\}\subseteq[n]} w_j^2 w_k^2 w_l^2 w_r^2 \\ &\leq \frac{b^6}{n^6} \sum_{i=1}^{n} w_i^4 \left(\sum_{i=1}^{n} w_i^2\right)^4 \\ &= \frac{b^6}{n^6} \sum_{i=1}^{n} \left(\frac{n}{i}\right)^{\frac{4}{\alpha}} \left(\sum_{i=1}^{n} \left(\frac{n}{i}\right)^{\frac{2}{\alpha}}\right)^4 \\ &= \frac{b^6}{n^6} g_{\frac{4}{\alpha}}(n) \left(g_{\frac{2}{\alpha}}(n)\right)^4. \end{aligned}$$

Now, let us select a small number $\beta > 0$ such that $\alpha \geq p = 2(1+\beta)$. Then, by (6.8), we obtain

$$g_{\frac{4}{\alpha}}(n) \leq g_{\frac{4}{p}}(n) = O\left(n^{\frac{4}{p}}\right)$$

and

$$g_{\frac{2}{\alpha}}(n) = O(n).$$

Thus,

$$\mathbb{E}[X_D] = O\left(n^{4+\frac{4}{p}-6}\right) = O\left(n^{-2\left(1-\frac{1}{1+\beta}\right)}\right) = o(1).$$

After applying Markov's inequality, statement (2) of the lemma follows.

Thus, a typical scale-free graph with the scaling exponent $\alpha > 2.4$ has only cliques of size 2 and 3, and every vertex belongs to at most one triangle. For such graphs, the minimal clique covering that determines $\dim_L(G)$ consists of all triangles and the edges not belonging to triangles. Let $\operatorname{tr}(v) \in \{0, 1\}$ be the number of triangles which include a vertex v. Then, with high probability,

$$\dim_L(G) = \max_v(\deg(v) - \operatorname{tr}(v)) - 1,$$

and the statement about $\dim_L(G)$ follows. By the Vizing theorem, two-vertex clusters of the above covering could be coloured by at most $\Delta + 1$ colours, and one additional colour could be used to colour the triangles. Thus,

$$\dim_L(G) \leq \dim_H(G) \leq \Delta + 1,$$

which proves the statement for $\dim_H(G)$. □

The *Erdős–Rényi graph* $G(n, p)$ is a random graph, where each possible edge is selected independently with fixed probability p. It turned out that properties of dimensions similar to those of scale-free graphs hold for sparse Erdős–Rényi graphs with $p = \frac{1}{n^\alpha}$:

Theorem 6.6. [34] *For Erdős–Rényi graphs $G = G(n, p)$ with $\alpha > \frac{5}{6}$ and $\Delta = \Delta(G)$, with high probability*

$$\dim_L(G) \in \{\Delta - 2, \Delta - 1\}$$

and

$$\dim_H(G) \in \{\Delta - 2, \Delta - 1, \Delta, \Delta + 1\}.$$

Indeed, it is implied by the following simple statement and considerations analogous to the ones in Theorem 6.5.

Lemma 6.3 [34] *The following statements are true:*

(1) *For $\alpha > \frac{2}{3}$, graphs $G(n, p)$ with high probability do not contain K_4.*
(2) *For $\alpha > \frac{4}{5}$, graphs $G(n, p)$ with high probability do not contain diamonds as induced subgraphs.*
(3) *For $\alpha > \frac{5}{6}$, graphs $G(n, p)$ with high probability do not contain butterflies as induced subgraphs.*

Proof: Let X_D be the number of diamonds in $G(n,p)$. Then,

$$\mathbb{E}[X_D] = \binom{n}{4} p^5(1-p) = O(n^{4-5\alpha}) = o(1),$$

and (2) follows from Markov's inequality. Other statements can be proved analogously. □

6.3 Fractals

Topological fractals appear in many research domains, including dynamical systems, physics, biology and behavioural sciences [17]. By Mandelbrot's classical definition, the geometric fractal is a topological space (usually a subspace of the Euclidean space) whose Lebesgue dimension is strictly smaller than Hausdorff dimension. It is also usually assumed that fractals have some form of geometric or statistical self-similarity [17].

Self-similarity and fractal properties of complex networks have been subjects of a number of studies, largely inspired by various applications [31, 33, 37]. Although such studies are at least partially based on genuine ideas from graph theory and general topology, they are often not supported by a rigorous mathematical framework. As a result, such methods may not be directly applicable to many important classes of graphs and networks [26, 40].

6.3.1 Graph Fractals

Fractal graphs naturally emerge under the combinatorial framework described above. As will be shown next, fractality is closely related to edge colourings, and separation of graphs into fractals and non-fractals could be considered as a generalization of one of the most renowned dichotomies in graph theory—the separation of graphs into class 1 and class 2 [39]; that is, graphs whose edge chromatic number is equal to Δ or $\Delta + 1$, where Δ is the maximum vertex degree of a graph.

Formally, let G be a connected graph. Our starting point will be a simple fact that relation (6.2) between Lebesgue and Hausdorff dimensions of topological spaces remains true for graphs.

Proposition 6.3 [34] *For any graph G,*

$$\dim_R(G) - 1 = \dim_L(G) \leq \dim_H(G) = \dim_P(\overline{G}) - 1.$$

Proof: Let product dimension of the graph $\overline{G}$ be equal to k. Then, by Theorem 6.2, G is the intersection graph of a strongly k-colourable hypergraph. Since the rank of every such hypergraph obviously does not exceed k, Theorem 6.1 implies $\dim_R(G) \leq k$. □

Thus, Proposition 6.3 allows us to define graph fractals analogously to the definition of fractals for topological spaces: a graph G is a *fractal* if $\dim_L(G) < \dim_H(G)$, i.e. $\dim_R(G) < \dim_P(\overline{G})$. In particular, we say that a fractal graph G is a *k-fractal* if $\dim_L(G) = k$. For example, the graph G in Figure 6.1 (bottom) is self-similar, but not a fractal because

$$\dim_L(G) = \dim_H(G) = 1.$$

As the first example of a fractal graph, let us consider the so-called *Sierpiński gasket graphs* S_n [24]. They are associated with the Sierpiński gasket—the well-known topological fractal with Hausdorff dimension $\log(3)/\log(2) \approx 1.585$. The edges of S_n are line segments of the n-th approximation of the Sierpiński gasket, and the vertices are intersection points of these segments (Figure 6.2). Formally, the Sierpiński gasket graphs are recursively defined as follows. Consider the tetrads $T_n = (S_n, x_1, x_2, x_3)$, where x_1, x_2, x_3 are distinct vertices of S_n called *contact vertices.* The first Sierpiński gasket graph S_1 is a triangle K_3 with vertices x_1, x_2, x_3, and the first tetrad is defined as $T_1 = (S_1, x_1, x_2, x_3)$. The $(n+1)$-th Sierpiński gasket graph S_{n+1} is constructed from three disjoint copies (S_n, x_1, x_2, x_3), (S'_n, x'_1, x'_2, x'_3), $(S''_n, x''_1, x''_2, x''_3)$ of the n-th tetrad T_n by gluing together x_2 with x'_1, x'_3 with x''_2 and x_3 with x''_1; the corresponding $(n+1)$-th tetrad is $T_{n+1} = (S_{n+1}, x_1, x'_2, x''_3)$.

Figure 6.2 demonstrates that the Sierpiński gasket graph S_3 is a 1-fractal. In fact, as the following theorem indicates, all Sierpiński gasket graphs are fractals.

Theorem 6.7 [34] *For every $n \geq 2$, the Sierpiński gasket graph S_n is a fractal with*

$$\dim_L(S_n) = 1 \quad \textit{and} \quad \dim_H(S_n) = 2.$$

Proof: First, we will prove that for every $n \geq 2$,

$$\dim_R(S_n) \leq 2 \quad \text{and} \quad \dim_P(\overline{S_n}) \leq 3. \tag{6.9}$$

We will show this by induction on n. In fact, we will prove the slightly stronger fact that for any $n \geq 2$ there exists a clique covering $\mathcal{C} = \{C_1, C_2, \ldots, C_m\}$ such that:

(i) Every non-contact vertex is covered by two cliques from $\mathcal{C}$.
(ii) Every contact vertex is covered by one clique from $\mathcal{C}$.

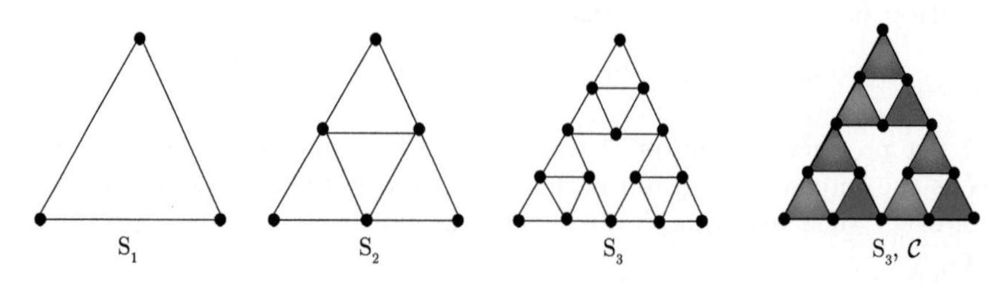

Figure 6.2 Sierpiński gasket graphs S_1–S_3 and the optimal equivalent separating 3-covering of S_3. Clusters of the same colour are highlighted in brown, green and blue. Graph S_3 is a fractal: every vertex is covered by two clusters, while the clusters can be coloured using three colours.

(iii) Cliques from $\mathcal{C}$ can be coloured using three colours in such a way that intersecting cliques receive different colours and cliques containing different contact vertices also receive different colours.
(iv) Every two distinct vertices are separated by some clique from $\mathcal{C}$.

For $n=2$, the clique covering $\mathcal{C}$, which consists of three cliques that contain contact vertices, obviously satisfies conditions (i)–(iv). Now, suppose that $\mathcal{C}$, $\mathcal{C}'$ and $\mathcal{C}''$ are clique coverings of S_n, S_n' and S_n'' with properties (i)–(iv). Assume that $x_i \in C_i$, $x_i' \in C_i'$, $x_i'' \in C_i''$ and C_i, C_i', C_i'' have colours i, $i = 1, 2, 3$. Then, it is straightforward to check that $\mathcal{C} \cup \mathcal{C}' \cup \mathcal{C}''$ with all cliques keeping their colours is a clique covering of S_{n+1} that satisfies (i)–(iv). So, (6.9) is proved.

Finally, note that for every $n \geq 2$, the graph S_n contains graphs $K_{1,2}$ and $K_4 - e$ as induced subgraphs. Since $\dim_R(K_{1,2}) = 2$ and $\dim_P(\overline{K_4 - e}) = 3$ (the latter is easy to see using the definition of clique coverings), we have equalities in (6.9). □

Now, let us turn to more substantial classes of graphs. Let $\chi'(G)$ denote the edge chromatic number of a graph G. The classical Vizing theorem [39] states that

$$\Delta(G) \leq \chi'(G) \leq \Delta(G) + 1.$$

This means that the set of all graphs can be partitioned into two classes: graphs for which $\chi'(G) = \Delta(G)$ (class 1), and graphs for which $\chi'(G) = \Delta(G) + 1$ (class 2).

By Proposition 6.1,

$$\dim_L(G) = \dim_R(G) - 1 = \Delta(G) - 1$$

if G contains no triangles. For such graphs, we have:

Proposition 6.4 [34] *Triangle-free fractals are exactly triangle-free graphs of class 2.*

Proof: The statement holds if $G = K_2$. Suppose that G has $n \geq 3$ vertices. For such graphs, every clique covering is a collection of its edges and vertices. However, because G is connected, for every pair of vertices there is an edge that separates them. Therefore, we may assume that the clique covering consists only of edges, and a feasible assignment of colours to the cliques is an edge colouring. Thus, it is true that $\dim_P(\overline{G}) = \chi'(G)$ (i.e. $\dim_H(G) = \chi'(G) - 1$), and the statement of the proposition follows. □

In particular, bipartite graphs are triangle-free graphs of class 1 [25]. Hence, bipartite graphs (and trees in particular) are not fractals, even though some of them may have high degree of self-similarity (e.g. binary trees). Furthermore, a typical Erdős–Rényi graph with high probability has a unique vertex of the maximum degree. For such graphs, it is also known that $\Delta(G) = \chi'(G)$. Therefore, with high probability, sparse Erdős–Rényi random graphs are not fractals either.

A graph is *cubic* if all its vertices have degree three. Among cubic graphs, the remarkable class of so-called *snarks* is distinguished in graph theory. The most famous snark, the Petersen graph, is shown in Figure 6.3. A *snark* is defined as a biconnected cubic graph of class 2 [11]. A snark is non-trivial if it is triangle-free

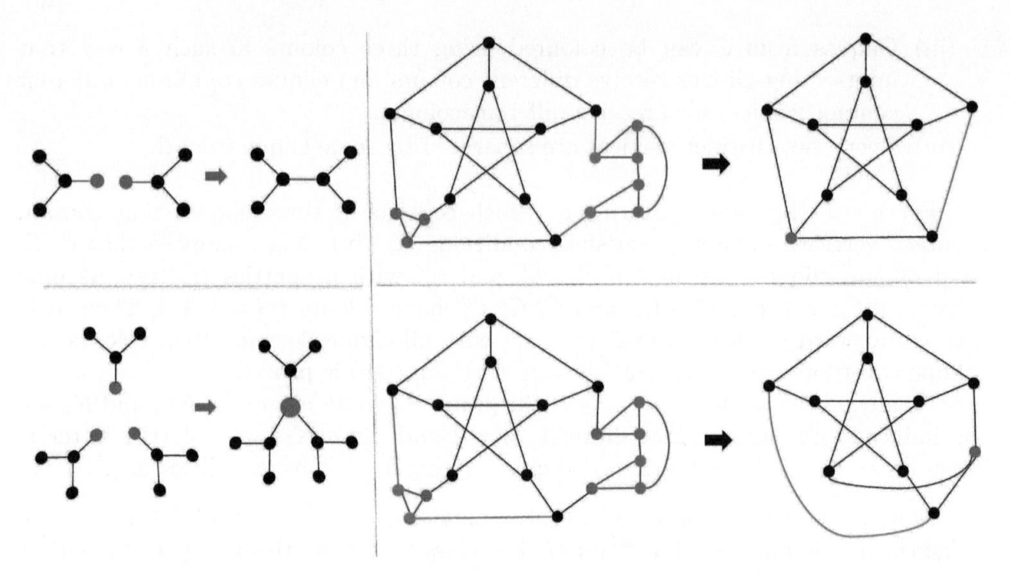

Figure 6.3 *Left*: Pendant edge identification (top) and pendant triple contraction (bottom) operations. Identified vertices and edges are highlighted in red. *Right*: Transformations of a cubic graph G. For each transformation, removed edges are highlighted in blue, vertices involved in the pendant triple contraction are highlighted in red, and vertices and edges involved in the pendant edge identification are highlighted in green. The top transformation converts G into the Petersen snark, which is a fractal. However, G is not a fractal because the bottom transformation converts it into a 3-edge-colourable cubic graph.

[11]. Snarks constitute an important class of graphs, which has been studied for more than a century and whose structural properties continue to puzzle researchers to this day [11]. The discovery of new non-trivial snarks is a valuable scientific result, as suggested by the name of this graph class introduced by Gardner [19]. This is similar to the discovery of new fractals, with many known snarks also possessing the high degree of symmetry and being constructed by certain recursive procedures. According to Proposition 6.4, this analogy is well justified because non-trivial snarks are indeed fractals according to our definition. As shown in Theorem 6.9, in some cases the inverse relation also holds as cubic fractals could be reduced to snarks.

The connection between fractality and class 2 graphs continues to hold for a wider class of subcubic graphs; that is, for graphs with $\Delta(G) \leq 3$. Indeed, consider the graph G_{-3} obtained from G by removal of edges of all its triangles. Given that $\dim_L(G) \leq 2$, the following theorem describes subcubic fractals:

Theorem 6.8 [34] *Let G be a subcubic connected graph with $n \geq 5$ vertices.*

(1) G is a 1-fractal if and only if it is claw-free but contains the diamond or an odd hole as an induced subgraph.

(2) G is a 2-fractal if and only if it contains the claw as an induced subgraph and G_{-3} is of class 2.

Proof: First, we will prove (1). By Theorem 6.1,

$$\dim_L(G) = \dim_R(G) - 1 = 1$$

if and only if G is the line graph of a multigraph. Such graphs are characterized by the list of seven forbidden induced subgraphs, only one of which ($K_{1,3}$) has the maximal degree that does not exceed 3 [5]. Therefore, $\dim_R(G) = 2$ if and only if G is claw-free. By Theorem 6.2,

$$\dim_H(G) = \dim_P(\overline{G}) - 1 = 1$$

if and only if G is the line graph of a bipartite graph. These graphs are exactly (claw, diamond, odd-hole)-free graphs [9]. By combining the above facts, we obtain that $\dim_R(G) = 2$, $\dim_P(\overline{G}) = 3$ if and only if G is claw-free but it contains the diamond or an odd hole.

Now, let us prove (2). Suppose that G contains the claw or, synonymously, some vertex of G does not belong to a triangle. In this case, $\dim_L(G) = 2$, $\Delta(G_{-3}) = 3$ and, therefore, G_{-3} is either 3- or 4-edge colourable. We will demonstrate that

$$\dim_H(G) = \dim_P(\overline{G}) - 1 = 2$$

if and only if G_{-3} is of class 1. The necessity is obvious, so it remains to prove the sufficiency. Let $\lambda : E(G_{-3}) \to \{1, 2, 3\}$ be a 3-edge colouring of G_{-3}, and $\mathcal{C} = E(G_{-3})$ be the clique covering of G_{-3} with all clusters being single edges, whose colours are set by λ. The covering $\mathcal{C}$ could be extended to an equivalent 3-covering of G as follows. Given that $\Delta(G) = 3$, there are two possible arrangements between pairs of triangles in G.

(A) There are two triangles $T_1 = \{a, b, c\}$ and $T_2 = \{b, c, d\}$ that share the edge bc. Suppose also that $a \sim e$, $b \sim f$ and $e, f \notin \{a, b, c, d\}$ (it is possible that $e = f$). If, without loss of generality, $\lambda(\{a, e\}) = \lambda(\{d, f\}) = 1$, then the edges of T_1 and T_2 could be covered by single-edge cliques, whose colours could be set as

$$\lambda(\{b, c\}) = \lambda(\{a, b\}) = \lambda(\{c, d\}) = 1,$$

$$\lambda(\{a, b\}) = \lambda(\{c, d\}) = 2,$$

$$\lambda(\{a, c\}) = \lambda(\{b, d\}) = 3.$$

If, say, $\lambda(\{a, e\}) = 1$ and $\lambda(\{d, f\}) = 2$, then we may cover T_1 and T_2 by the cliques $\{a, b\}$, $\{a, c\}$, $\{b, c, d\}$ with colours $\lambda(\{a, b\}) = 2$, $\lambda(\{a, c\}) = 3$, $\lambda(\{b, c, d\}) = 1$.

(B) The triangle $T = \{a, b, c\}$ does not share edges with other triangles. Suppose that $a \sim d$, $b \sim e$, $c \sim f$ and $d, e, f \notin \{a, b, c\}$. All these vertices are distinct, and ad, be and df do not belong to any triangles and, therefore, are present in G_{-3}. If the colours $\lambda(\{a, d\}), \lambda(\{b, e\})$ and $\lambda(\{d, f\})$ are distinct, then we cover T by single-edge cliques with colours

$$\lambda(\{a,b\}) = \lambda(\{c,f\}), \quad \lambda(\{b,c\}) = \lambda(\{a,d\}), \quad \lambda(\{a,c\}) = \lambda(\{b,e\}).$$

If, alternatively, some of these colours are identical, then there is a colour $i \in \{1,2,3\}$ not present among them. In this case, we cover T with the single clique $\{a,b,c\}$ of the colour $\lambda(\{a,b,c\}) = i$.

If in the resulting covering some vertex v is covered by a single triangle T, add the single-vertex clique $\{v\}$ with an appropriate colour $\lambda(\{v\}) \neq \lambda(T)$. Thus, the constructed covering is a separating equivalent k-covering of G. This concludes the proof. □

Theorem 6.8 states that subcubic 1-fractals are reducible to the diamond and odd cycles, whereas subcubic 2-fractals could be reduced to class 2 graphs. The next theorem will demonstrate that cubic 2-fractals could be reduced to snarks.

Let G be a cubic graph with $\dim_L(G) = 2$. In this case, every vertex of G_{-3} has degree 0, 1 or 3. Vertices of degree 1 are further referred to as *pendant vertices*, and edges incident to pendant vertices as *pendant edges*. We will establish a deeper relation between the topology of general cubic fractals and snarks. By Theorem 6.8, the case of 1-fractals is rather simple, so we will concentrate on 2-fractals. Thus, we will assume that G contains a claw. Consider the following graph operations:

(O1) *Pendant triple contraction* consists in replacement of pendant vertices u, v and w by a single vertex x, which is adjacent to all neighbours of u, v and w (see Figure 6.3).

(O2) *Pendant edge identification* of two edges u_1v_1 and u_2v_2 with $\deg(v_1) = \deg(v_2) = 1$ and $u_1 \neq u_2$ consists in removal of v_1 and v_2 and replacement of u_1v_1 and u_2v_2 by the edge u_1u_2 (see Figure 6.3).

Let G'_{-3} be the graph obtained from G_{-3} by removal of isolated vertices and edges. The graph G_{-3} is of class 1 if and only if G'_{-3} is as well.

Lemma 6.4 [34] *Suppose that G'_{-3} is of class 1. Let λ be its 3-edge colouring and p_1, p_2 and p_3 be the numbers of pendant edges with colours 1, 2 and 3, respectively. Then p_1, p_2, p_3 are either all odd or all even.*

Proof: Let $p_i = 2\alpha_i + \beta_i$, where $\beta_i = p_i \bmod 2$. For each colour i, let us consider α_i pairs of i-coloured pendant edges, identify the edges from each pair and assign the colour i to each newly added edge. So, the resulting graph G''_{-3} is of class 1 too, has all vertex degrees equal to 1 or 3 and contains β_i pendant edges of colour i.

Now, let us consider the subgraph $H_{i,j}$ of G''_{-3} formed by edges of colours i and j. Obviously, $H_{i,j}$ is a disjoint union of even cycles and, possibly, a single path with distinctly coloured end-edges. If the path is not present, then $\beta_i = \beta_j = 0$, otherwise $\beta_i = \beta_j = 1$. □

Theorem 6.9 [34] *A cubic graph G is a 2-fractal if and only if it contains a claw and any cubic graph obtained from G'_{-3} by pendant edge identifications and pendant triple contractions either has a bridge or is a snark.*

Proof: It can be easily shown that if a cubic graph has a bridge, then it is of class 2 [11]. Thus, the statement of the theorem is equivalent to the following statement: the cubic graph G that contains a claw is not a 2-fractal if and only if it is possible to construct a cubic graph of class 1 from G'_{-3} by pendant edge identifications and pendant triple contractions.

To prove the necessity, suppose that G is not a 2-fractal, for example the graph G'_{-3} is of class 1. Let us consider any 3-edge colouring of G'_{-3} and identify the pendant edges of the same colour as described in Lemma 6.4. If, after this operation, all pendant edges are eliminated, then the desired graph H of class 1 is constructed. Otherwise, by Lemma 6.4, H contains three pendant edges of pairwise distinct colours. Then, the desired graph can be constructed by contracting the pendant end-vertices of these edges.

Conversely, suppose that the graph H of class 1 is obtained from G'_{-3} by pendant edge identifications and pendant triple contractions. Consider any 3-edge colouring λ of H. Obviously, λ could be transformed into a 3-edge colouring of G'_{-3} by assigning the colour $\lambda(u_2v_2)$ to the identified edges u_1v_1 and u_2v_2. $\square$

Thus, Theorem 6.9 states that the biconnected cubic graph is a 2-fractal whenever any feasible sequence of operations transforms it into a snark. The feasible sequence of operations includes: removal of triangle edges, isolated edges and vertices; and pendant triple contractions and pendant edge identifications, which preserve the graph connectivity. Figure 6.3 provides an example of such transformations.

6.3.2 Graph Fractality and Graph Complexity

Theorems 6.1 and 6.2 allow to interpret graph Lebesgue and Hausdorff dimensions and fractality from the information-theoretic point of view. Indeed, graphs G with $\dim_L(G) = k$ could be described by assigning to every vertex a set of $k-1$ integer 'coordinates' represented by hyperedges of a k-uniform hypergraph H such that $G = L(H)$. Importantly, these coordinates are non-ordered, and the edges of G are determined by a presence of a shared coordinate for their end-vertices. In contrast, graphs with $\dim_H(G) = k$ are determined by ordered vectors of coordinates (Theorem 6.2 (4)), and adjacency of a pair of vertices is determined by a presence of a shared coordinate on the same position. Thus, non-fractal graphs are the graphs for which the set and vector representations are equivalent, while fractal graphs have additional structural properties that manifest themselves in extra dimensions needed to describe them using a vector representation. The whole concept is illustrated in Figure 6.4.

Formal relations between graph dimension and information complexity could be analysed using Kolmogorov complexity. Informally, Kolmogorov complexity of a string s could be described as the length of its shortest lossless encoding. Formally, let $\mathbb{B}^*$ be the set of all finite binary strings and $\Phi : \mathbb{B}^* \to \mathbb{B}^*$ be a computable function. *Kolmogorov complexity*, denoted $K(s) = K_\Phi(s)$, of a binary string s with respect to Φ is the minimal length of a string s' such that $\Phi(s') = s$. Since Kolmogorov complexities with respect to any two functions differ by an additive

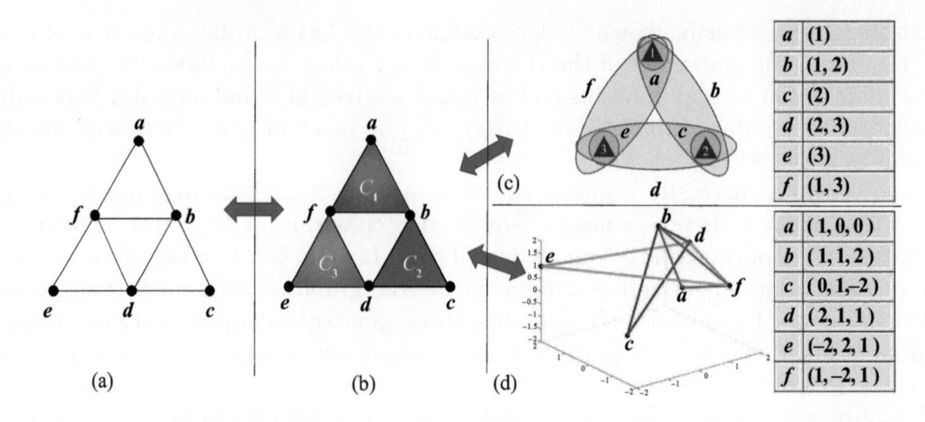

Figure 6.4 (a) Sierpiński gasket graph S_2. (b) Its optimal equivalent separating 3-covering. Clusters of the same colour are highlighted in brown, green and blue. (c) Left: hypergraph H such that $S_2 = L(H)$. The edges of H correspond to the vertices of S_2, with two vertices being adjacent if and only if the corresponding edges intersect. *Right*: table with the corresponding unordered set coordinates encoding the graph S_2. (d) Left: embedding of the graph S_2 into 3-dimensional space $\mathbb{Z}^3$ such that two vertices are adjacent whenever they share a coordinate. Different colours highlight different clusters. *Right*: table with the corresponding ordered vector coordinates encoding the graph S_2. Because S_2 is a fractal, the dimensionalities of encodings (c) and (d) differ.

constant [27], it can be assumed that some canonical function Φ is fixed. For two strings $s, t \in \mathbb{B}^*$, *conditional Kolmogorov complexity* $K(s|t)$ is the length of the shortest encoding of s if t is known in advance as an auxiliary input.

Every connected graph G can be encoded using the string representation of an upper triangle of its adjacency matrix. Kolmogorov complexity $K(G)$ of a graph G could be defined as Kolmogorov complexity of that string [30]. In addition, the conditional graph Kolmogorov complexity $K(G|n)$ is often considered, which is the complexity given that the number of vertices is known. Obviously, $K(G) = O(n^2)$ and $K(G|n) = O(n^2)$. Alternatively, an n-vertex connected labelled graph can be represented as a list of edges with ends of each edge encoded using their binary representations concatenated with a binary representation of n. It gives the following estimations [27, 30]:

$$K(G) \leq 2m\log(n) + \log(n) = O(m\log(n)),$$

$$K(G|n) \leq 2m\log(n) = O(m\log(n)).$$

Suppose that

$$\dim_H(G) = \dim_P(\overline{G}) - 1 = d - 1 \quad \text{and} \quad \mathcal{H}^d(G) = h.$$

Then, $\overline{G}$ is an induced subgraph of the product

$$K_{p_1} \times K_{p_2} \times \ldots \times K_{p_d}, \tag{6.10}$$

where $h = p_1 \times p_2 \times \ldots \times p_d$. Thus, by Theorem 6.2, G and $\overline{G}$ could be encoded using a collection of the vectors

$$\phi(v) = (\phi_1(v), \phi_2(v), \ldots, \phi_d(v)), \quad v \in V(G), \quad \phi_j(v) \in [p_j].$$

Such an encoding could be stored as a string containing binary representations of the coordinates $\phi_j(v)$ using $\log(p_j)$ bits concatenated with binary representations of n and p_j, $j = 1, 2, \ldots, n$. The length of this string is

$$(n+1)\sum_{j=1}^{d} \log(p_j) + \log(n).$$

Analogously, if n and p_j are given, then the length of the encoding is $n\sum_{j=1}^{d} \log(p_j)$.

Proposition 6.5 [34] *The following estimations hold:*

$$K(G) \leq (n+1)\log(\mathcal{H}^{\mathrm{d}}(G)) + \log(n), \tag{6.11}$$

$$K(G|n, p_1, p_2, \ldots, p_d) \leq n\log(\mathcal{H}^{\mathrm{d}}(G)). \tag{6.12}$$

Let $p^* = \max_j p_j$. Then,

$$K(G) \leq (n+1)d\log(p^*) + \log(n)$$

and

$$K(G|n, p_1, p_2, \ldots, p_d) \leq nd\log(p^*).$$

By minimality of the representation (6.10), we have $p^* \leq n$. Thus,

$$K(G) = O(dn\log(n))$$

and

$$K(G|n, p_1, p_2, \ldots, p_d) = O(dn\log(n)).$$

So, Hausdorff (Prague) dimension could be considered as a measure of descriptive complexity of a graph.

The relations between Hausdorff (Prague) dimension and Kolmogorov complexity could be used to derive a lower bound for Hausdorff dimension of a dense Erdős–Rényi random graph. Formally, let X be a graph property and $\mathcal{P}_n(X)$ be the set of labelled n-vertex graphs having this property. The property X holds for *almost all graphs* [15] if

$$|\mathcal{P}_n(X)|/2^{\binom{n}{2}} \to 1 \quad \text{as} \quad n \to \infty.$$

This means that the probability that the sparse Erdős–Rényi random graph $G(n, \frac{1}{2})$ has the property X converges to 1 when $n \to \infty$. We will use the following lemma:

Lemma 6.5 [10] *For every $n>0$ and $\delta : \mathbb{N} \to \mathbb{N}$, there are at least*

$$2^{\binom{n}{2}}(1 - 2^{-\delta(n)})$$

n-vertex labelled graphs G such that

$$K(G|n) \geq \frac{n(n-1)}{2} - \delta(n).$$

The next result states that almost all sparse Erdős–Rényi graphs have large Hausdorff dimension:

Theorem 6.10 [34] *For every $\epsilon > 0$, almost all sparse Erdős–Rényi graphs have Hausdorff dimension such that*

$$\frac{1}{1+\epsilon}\left(\frac{n-1}{2\log(n)} - \frac{1}{n}\right) - 1 \leq \dim_H(G) \leq C\frac{n}{\log(n)}, \tag{6.13}$$

where C is a constant.

Proof: The upper bound was proved in [20]; the corresponding proof is rather complicated. Hence, we will only provide the proof of the lower bound. Let $n_\epsilon = \left\lceil \frac{2}{\epsilon} \right\rceil$. Consider a graph G with $n \geq n_\epsilon$. From (6.12), we obtain

$$K(G) \leq (n+1)d\log(n) + \log(n).$$

Using the fact that $\frac{1}{n} + \frac{1}{nd} \leq \frac{2}{n} \leq \epsilon$, it is straightforward to check that

$$(n+1)d\log(n) + \log(n) \leq (1+\epsilon)nd\log(n).$$

Therefore, we have

$$K(G) \leq (1+\epsilon)nd\log(n). \tag{6.14}$$

Let X be the set of all graphs G such that

$$K(G|n) \geq \frac{n(n-1)}{2} - \log(n). \tag{6.15}$$

Using Lemma 6.5 with $\delta(n) = \log(n)$, we conclude that

$$|\mathcal{P}_n(X)|/2^{\binom{n}{2}} \geq 1 - \frac{1}{n},$$

and hence almost all graphs have the property X. Next, it is easy to see that for graphs with the property X and $n \geq n_\epsilon$, inequality (6.13) holds. Now, the result follows by combining inequalities (6.14) and (6.15) and using the fact that $K(G|n) \leq K(G)$. □

Acknowledgements

This chapter is partially based on the article in the Journal of Complex Networks, 8 (4), P. Skums and L. Bunimovich, Graph fractal dimension and the structure of fractal networks, cnaa037, ©2020, with permission from Oxford University Press.

6.4 References

[1] Y.-Y. Ahn, J. P. Bagrow and S. Lehmann, Link communities reveal multiscale complexity in networks, *Nature*, **466**(7307) (2010), 761–764.

[2] N. Alon, Covering graphs by the minimum number of equivalence relations, *Combinatorica*, **6** (3)(1986), 201–206.

[3] L. Babai and P. Frankl, *Linear Algebra Methods in Combinatorics with Applications to Geometry and Computer Science*, Chicago, IL: Department of Computer Science, The University of Chicago, 1992.

[4] A. Babaitsev and R. I. Tyshkevich, k-Dimensional graphs, *Izvestiya Akademii Nauk Belarusi*, Ser. fiz.-mat. nauk, **3** (1996), 75–82.

[5] C. Berge, *Hypergraphs: Combinatorics of Finite Sets*, Amsterdam, The Netherlands: Elsevier Science Publishers B.V., 1989.

[6] B. Bollobás, *Random Graphs*, vol. 73, Cambridge, UK: Cambridge University Press, 2001.

[7] B. Bollobás and O. M. Riordan, Mathematical results on scale-free random graphs. *Handbook of Graphs and Networks: From the Genome to the Internet* (Stefan Bornholdt and Hans Georg Schuster eds), Weinheim, Germany: Wiley-VCH, 2003, pp. 1–34.

[8] G. Bounova and O. De Weck, Overview of metrics and their correlation patterns for multiple-metric topology analysis on heterogeneous graph ensembles, *Physical Review E*, **85** (1)(2012), 016117.

[9] A. Brandstädt, V. B. Le and J. P. Spinrad, *Graph Classes: A Survey*, Philadelphia, PA: SIAM, 1999.

[10] H. Buhrman, M. Li, J. Tromp and P. Vitányi, Kolmogorov random graphs and the incompressibility method, *SIAM Journal on Computing*, **29** (2)(1999), 590–599.

[11] M. Chladný and M. Škoviera, Factorisation of snarks, *The Electronic Journal of Combinatorics*, **17** (1)(2010), R32.

[12] F. Chung and L. Lu, Connected components in random graphs with given expected degree sequences, *Annals of Combinatorics*, **6** (2) (2002), 125–145.

[13] J. R. Cooper, *Product Dimension of a Random Graph*, PhD thesis, Miami University, 2010.

[14] G. Edgar, *Measure, Topology, and Fractal Geometry*, New York, NY: Springer Science & Business Media, 2007.

[15] P. Erdős and R. J. Wilson, On the chromatic index of almost all graphs, *Journal of Combinatorial Theory, Ser. B*, **23** (2–3)(1977), 255–257.

[16] A. V. Evako, Dimension on discrete spaces, *International Journal of Theoretical Physics*, **33** (7)(1994), 1553–1568.

[17] K. Falconer, *Fractal Geometry: Mathematical Foundations and Applications*, Chichester, England: John Wiley & Sons, 2004.

[18] A. Flaxman, A. Frieze and T. Fenner, High degree vertices and eigenvalues in the preferential attachment graph, *Internet Mathematics*, **2** (1)(2005), 1–19.

[19] M. Gardner, Mathematical games, *Scientific American*, **236** (2)(1977), 121–127.

[20] H. Guo, K. Patton and L. Warnke, Prague dimension of random graphs, arXiv preprint arXiv:2011.09459 (2020).

[21] P. Hell and J. Nešetril, *Graphs and Homomorphisms*, vol. 28, Oxford Lecture Series in Mathematics and Its Applications, Oxford, UK: Oxford University Press, 2004.

[22] I. Holyer, The NP-completeness of edge-colouring, *SIAM Journal on Computing*, **10** (4)(1981), 718–720.

[23] S. Janson, T. Łuczak and I. Norros, Large cliques in a power-law random graph, *Journal of Applied Probability*, **47** (4)(2010), 1124–1135.

[24] S. Klavžar, I. Peterin and S. S. Zemljič, Hamming dimension of a graph—The case of Sierpiński graphs, *European Journal of Combinatorics*, **34** (2)(2013), 460–473.

[25] D. König, Über graphen und ihre anwendung auf determinantentheorie und mengenlehre, *Mathematische Annalen*, **77** (4)(1916), 453–465.

[26] L. Li, D. Alderson, J. C. Doyle and W. Willinger, Towards a theory of scale-free graphs: Definition, properties, and implications, *Internet Mathematics*, **2** (4)(2005), 431–523.

[27] M. Li and P. Vitányi, *An Introduction to Kolmogorov Complexity and Its Applications*, New York, NY: Springer Science & Business Media, 2009.

[28] C. McDiarmid and F. Skerman, Modularity of regular and treelike graphs, *Journal of Complex Networks*, **6** (4)(2017), 596–619.

[29] Y. Metelsky and R. I. Tyshkevich, Line graphs of Helly hypergraphs, *SIAM Journal on Discrete Mathematics*, **16** (3)(2003), 438–448.

[30] A. Mowshowitz and M. Dehmer, Entropy and the complexity of graphs revisited, *Entropy*, **14** (3)(2012), 559–570.

[31] M. E. Newman, The structure and function of complex networks, *SIAM Review*, **45** (2)(2003), 167–256.

[32] S. Poljak, V. Rödl and D. Turzik, Complexity of representation of graphs by set systems, *Discrete Applied Mathematics*, **3** (4)(1981), 301–312.

[33] O. Shanker, Defining dimension of a complex network, *Modern Physics Letters B*, **21** (06)(2007), 321–326.

[34] P. Skums and L. Bunimovich, Graph fractal dimension and the structure of fractal networks, *Journal of Complex Networks*, **8** (4)(2020), cnaa037.

[35] M. B. Smyth, R. Tsaur and I. Stewart, Topological graph dimension, *Discrete Mathematics*, **310** (2)(2010), 325–329.

[36] C. Song, L. K. Gallos, S. Havlin and H. A. Makse, How to calculate the fractal dimension of a complex network: the box covering algorithm, *Journal of Statistical Mechanics: Theory and Experiment*, (03)(2007), P03006.

[37] C. Song, S. Havlin and H. A. Makse, Self-similarity of complex networks, *Nature*, **433**(7024) (2005), 392–395.
[38] R. I. Tyshkevich, Matroid decompositions of graphs, *Discretnaya Matematika*, **1** (3)(1989), 129–138.
[39] V. G. Vizing, On an estimate of the chromatic class of a p-graph, *Discretnyi Analiz*, **3** (1964), 25–30.
[40] W. Willinger, D. Alderson and J. C. Doyle, Mathematics and the internet: A source of enormous confusion and great potential, *Notices of the American Mathematical Society*, **56** (5)(2009), 586–599.
[41] F. G. Woodhouse, A. Forrow, J. B. Fawcett and J. Dunkel, Stochastic cycle selection in active flow networks, *Proceedings of the National Academy of Sciences*, **113** (29)(2016), 8200–8205.

Index